Quantum Technology and its Applications

Quantum Technology and its Applications

Edited by **Glenn Rogers**

Willford Press
New York

Published by Willford Press,
118-35 Queens Blvd., Suite 400,
Forest Hills, NY 11375, USA
www.willfordpress.com

Quantum Technology and its Applications
Edited by Glenn Rogers

International Standard Book Number: 978-1-68285-239-2 (Hardback)

Printed in China.

Contents

Preface

This book has been an outcome of determined endeavour from a group of educationists in the field. The primary objective was to involve a broad spectrum of professionals from diverse cultural background involved in the field for developing new researches. The book not only targets students but also scholars pursuing higher research for further enhancement of the theoretical and practical applications of the subject.

Quantum technology may have emerged as a specialized field only in the last two decades but it is widely applied across various industries presently. Quantum computing, quantum measurement, quantum optics, etc. are being used in various industries to develop and engineer complex systems. Quantum technology has crucial implications in the field of secure communication technologies, metrology, sensor instrumentation, optics and simulation. This book aims at providing an in-depth analysis of the quantum technology and its applications in diverse fields at present. Some of the important topics covered in this text are quantum thermodynamics, quantum sensing, quantum measurement, hybrid quantum systems, etc. It is a highly elucidative resource book that assesses both primary and contemporary issues in quantum technology as well as brings forth advanced topics for further research and discussion for all aspiring researchers and students involved in this field.

It was an honour to edit such a profound book and also a challenging task to compile and examine all the relevant data for accuracy and originality. I wish to acknowledge the efforts of the contributors for submitting such brilliant and diverse chapters in the field and for endlessly working for the completion of the book. Last, but not the least; I thank my family for being a constant source of support in all my research endeavours.

Editor

1

Controlling the dynamic range of a Josephson parametric amplifier

Christopher Eichler* and Andreas Wallraff

*Correspondence:
eichlerc@phys.ethz.ch
Department of Physics, ETH Zürich, Zürich, 8093, Switzerland

Abstract

One of the central challenges in the development of parametric amplifiers is the control of the dynamic range relative to its gain and bandwidth, which typically limits quantum limited amplification to signals which contain only a few photons per inverse bandwidth. Here, we discuss the control of the dynamic range of Josephson parametric amplifiers by using Josephson junction arrays. We discuss gain, bandwidth, noise, and dynamic range properties of both a transmission line and a lumped element based parametric amplifier. Based on these investigations we derive useful design criteria, which may find broad application in the development of practical parametric amplifiers.

1 Introduction

Due to the rapidly evolving field of quantum optics and information processing with superconducting circuits the interest in low-noise amplifiers has dramatically increased in the past five years and has lead to a body of dedicated research on Josephson junction based amplifiers [1–15]. The most successful quantum limited detectors which have so far been realized in the microwave frequency range are based on the principle of parametric amplification [16–19]. Josephson parametric amplifiers (JPAs) have not only been used to generate squeezed radiation [6, 13, 20–22], but moreover enabled the realization of quantum feedback and post-selection based experiments [23–26], the efficient displacement measurement of nanomechanical oscillators [27] and the exploration of higher order photon field correlations [28, 29].

While JPAs have been demonstrated to operate close to the quantum limit, their performance is to date mostly limited by their relatively small dynamic range, *i.e.* the saturation of the gain for large input signals. Here, we discuss the control of the dynamic range by making use of Josephson junctions arrays in the parametric amplifier circuit, which we have already employed in recent experiments [29, 30]. After reviewing the principles of parametric amplification we discuss bandwidth and noise constraints in dependence on the circuit design, based on which we derive simple strategies for optimized circuit design.

2 Principles of parametric amplification

2.1 Parametric processes at microwave frequencies

In quantum optics the word *parametric* is used for processes in which a nonlinear refractive medium is employed for mixing different frequency components of light. Such processes are parametric in the sense that a coherent pump field, applied to a nonlinear medium, modulates its refractive index, which appears as a *parameter* in a semi-classical treatement. This time-varying parameter is affecting modes with frequencies detuned from the frequency of the pump field and can stimulate their population with photons. The energy for creating these photons is provided by the pump field.

The refractive index in optics is equivalent to the impedance of electrical circuits. In order to realize parametric processes at microwave frequencies we therefore modulate an effective impedance. This is achieved by varying the parameters of either a capacitive or an inductive element in time. Although there have been early proposals for fast time-varying capacitances [31], it now is considered to be more convenient to make use of dissipationless Josephson junctions for this purpose. In a regime in which the current I flowing through a Josephson junction is much smaller than its critical current $I_C \equiv 2eE_J/\hbar$ its associated inductance is approximately $L \approx L_J(1 + \frac{1}{6}(I(t)/I_C)^2)$. Applying an AC current through the junction using appropriate microwave drive fields therefore leads to the desired time-varying impedance. Because of the proportionality of the inductance L to the square of the current $I^2(t)$, such a drive results in a four-wave mixing process [32].

The effective impedance can alternatively be modulated by varying the magnetic flux threading a superconducting quantum interference device (SQUID) loop [5] such that the effective inductance is approximately modulated proportionally to the AC current $I(t)$ flowing in the loop, $L \approx L_J(1 + I(t)/I_0)$. The quantity I_0 in this expression depends on the DC flux bias point of the SQUID loop. Since the relation between current and inductance is in this case linear, the magnetic flux drive results in a three-wave mixing process [33].

In order to enhance parametric amplification in a well-controlled frequency band while suppressing it for frequencies out of this band, the modulated Josephson inductance is frequently integrated into a microwave frequency resonator. This is the simplest way to control the band in which parametric amplification occurs. A number of variations of this basic idea are now explored. The circuit design has recently been modified to achieve a spatial separation of signal and idler modes [9, 21, 34–36] and to build traveling wave amplifiers, in which a field is amplified while propagating in forward direction coaxially with a pump field [37, 38]. Various drive mechanisms ranging from single and double pumps [8] to magnetic flux drives [5, 39, 40] have been explored. Being aware of this variety of possible approaches, we focus here on a single mode (degenerate) parametric amplifier driven with one pump tone close to its resonance frequency.

2.2 Circuit QED implementation of a parametric amplifier

The JPA essentially is a weakly nonlinear oscillator, in which the nonlinearity is provided by Josephon tunnel junctions. In practice, this is typically realized either as a transmission line resonator shunted by a SQUID [7, 12, 13], see Figure 1(a), or as a lumped element nonlinear oscillator [11]. The use of a SQUID instead of single tunnel junction guarantees tunability of the resonance frequency. Since resonator-based parametric amplifiers provide amplification in a narrow band only, tunability is highly desirable to match the band of amplification with the frequency of the signal to be amplified.

The relevant part of the Hamiltonian which describes the parametric amplifier considered here can be written as

$$H_{\text{JPA}} = \hbar\tilde{\omega}_0 A^\dagger A + \hbar\frac{K}{2}(A^\dagger)^2 A^2, \tag{1}$$

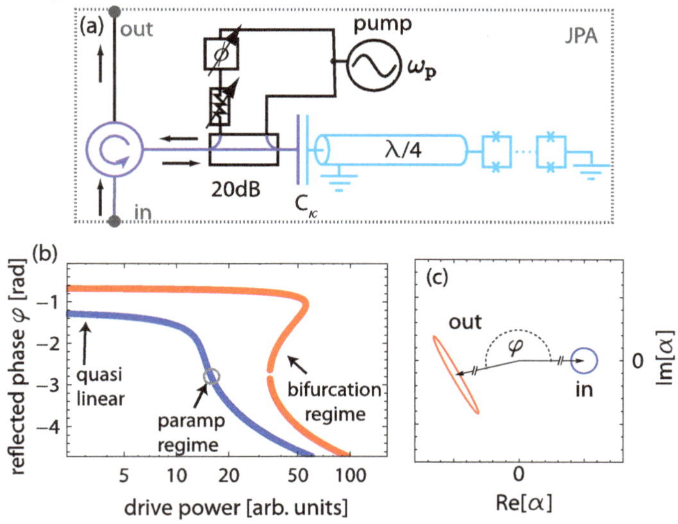

Figure 1 Schematic and operation of the parametric amplifier. (**a**) Circuit diagram of a transmission line resonator based parametric amplifier. The resonator is coupled with capacitance C_κ to a transmission line where input and output modes are spatially separated using a circulator. A 20 dB directional coupler between the λ/4-resonator and the circulator is used to apply the pump field required for modulating the SQUID inductance. The second port of the directional coupler can be used to interferometrically cancel out the pump tone reflected from the sample. (**b**) Phase of the reflected probe signal *vs.* drive power for two characteristic drive frequencies below (blue) and above (red) the bifurcation threshold. (**c**) Illustration of the nonlinear oscillator response in the quadrature plane. The blue circle represents various input fields α_{in} close to the one indicated by the gray circle in (b). Due to the nonlinear response of the resonator they are transformed into output fields α_{out} indicated by the red ellipse.

where A labels the annihilation operator of the intra-resonator field. Expressions for the resonance frequency $\tilde{\omega}_0/2\pi$ and the effective Kerr nonlinearity K are derived in Section 4 based on the full circuit model. In the following section we analytically study the dynamics of this system using the input-output formalism. Before presenting the mathematical derivations, we qualitatively describe different dynamical regimes of this nonlinear oscillator and explain the mechanism which leads to amplification.

If we assume for the moment that the JPA has no internal losses, all the incident power is reflected from the resonator and the classical response (*i.e.* reflection coefficient) is completely specified by the phase φ of the reflected field. In contrast to a linear system, where φ only depends on the frequency $\omega/2\pi$, it also depends on the power of the probe field in the case of a nonlinear oscillator. In Figure 1(b), the theoretically expected value of φ is plotted as a function of the probe amplitude for two characteristic drive frequencies. While the phase is constant for low drive powers (quasi-linear response), the phase changes significantly for increased drive power. Depending on the probe frequency we either find a bistable regime where two stable solutions exist [41, 42] or a regime where the phase has a unique solution (red and blue data sets in Figure 1(b)). In both cases the phase significantly depends on the input power. The bistable response can for example be used to realize a bifurcation amplifier [43, 44] and for nonlinear dispersive readout [45], which has been intensely studied in the context of circuit QED.

Since we are particularly interested in *linear* amplification the following discussion is focused on the regime, in which the response has a unique solution (blue data set). The mechanism of amplification can be understood qualitatively in the following way. If we

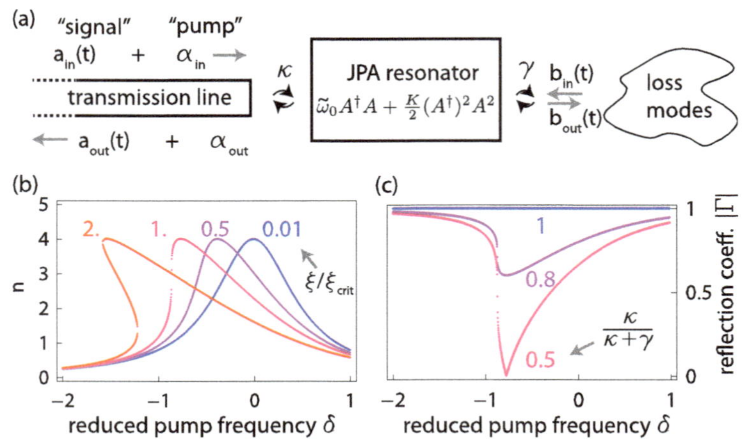

Figure 2 Input-output model. (**a**) Schematic of the input-output model used for calculating the response of the parametric amplifier in the presence of additional loss modes. (**b**) Normalized pump field photon number n in the resonator as a function of reduced pump frequency δ for effective drive strengths $\xi/\xi_{\text{crit}} = 0.01, 0.5, 1, 2$, where $\xi_{\text{crit}} = -1/\sqrt{27}$. (**c**) Absolute value of the reflection coefficient $|\Gamma|$ for different coupling ratios $\kappa/(\kappa+\gamma) = 1, 0.8, 0.5$.

imagine that the device is constantly driven at a frequency and power at which the reflected phase φ depends sensitively on power (see gray circle in Figure 1(b)), the system will strongly react to small perturbations. Such perturbations, which could be caused by an additional small signal field for example, are therefore translated into a large change of the output field.

We illustrate this process leading to amplification by plotting the resonator response for input fields α_{in} with slightly varying amplitude and phase. In Figure 1(c) we indicate the input fields by a blue circle around the mean value (arrow). The small differences in amplitude of the input field translate into large changes in φ of the output field α_{out} (red ellipse). If we interpret the arrow in Figure 1(c) as a constant pump field and its difference to the points on the blue circle as an additional signal, the signal is either amplified or deamplified depending on its phase relative to the pump.

The mechanism of amplification can thus be understood intuitively by considering the nonlinear response to a monochromatic drive field. In order to characterize the exact behavior of input fields with finite bandwidth we analyze the response in more detail below.

3 Input-output relations for the parametric amplifier

3.1 Classical nonlinear response

Here, we employ the input-output formalism [46, 47] to calculate the nonlinear resonator response discussed qualitatively in the previous section. The derivation presented here is inspired by Ref. [1]. A schematic of the input-output model is shown in Figure 2. The nonlinear resonator is coupled with rate κ to a transmission line, through which the pump and signal fields propagate. Based on this model and the Hamiltonian in Eq. (1) we obtain the following equation of motion for the intra-resonator field

$$\dot{A} = -i\tilde{\omega}_0 A - iKA^\dagger AA - \frac{\kappa+\gamma}{2}A + \sqrt{\kappa}A_{\text{in}}(t) + \sqrt{\gamma}b_{\text{in}}(t). \tag{2}$$

In addition to the coupling to transmission line modes $A_{\rm in}(t)$ with rate κ we account for potential radiation loss mechanisms by introducing the coupling to modes $b_{\rm in}(t)$ with loss rate γ, compare Figure 2(a). A boundary condition equivalent to

$$A_{\rm out}(t) = \sqrt{\kappa}A(t) - A_{\rm in}(t), \tag{3}$$

also holds for the loss modes. When operating the device as a parametric amplifier, the input field $A_{\rm in}$ is typically a sum of a strong coherent pump field and an additional weak signal field. Since this signal carries at least the vacuum noise, it is treated as a quantum field. In this formalism this particular situation is accounted for by decomposing each field mode into a sum of a classical part and a quantum part

$$\begin{aligned} A_{\rm in}(t) &= \big(a_{\rm in}(t) + \alpha_{\rm in}\big)e^{-i\omega_p t}, \\ A_{\rm out}(t) &= \big(a_{\rm out}(t) + \alpha_{\rm out}\big)e^{-i\omega_p t}, \\ A(t) &= \big(a(t) + \alpha\big)e^{-i\omega_p t}, \end{aligned} \tag{4}$$

where α, $\alpha_{\rm in}$, $\alpha_{\rm out}$ represent the classical parts of the field which are associated with the pump, while a, $a_{\rm in}$, $a_{\rm out}$ account for the quantum signal fields. Since all α's are complex numbers the modes a satisfy the same bosonic commutation relations as modes A do. By multiplying the field modes defined in Eq. (4) with the additional exponential factor $e^{-i\omega_p t}$, one works in a frame rotating at the pump frequency ω_p. The strategy is to first solve the classical response for the pump field α exactly and then linearize the equation of motion for the weak quantum field a in the presence of the pump. Finally, we derive a scattering relation between input modes $a_{\rm in}$ and reflected modes $a_{\rm out}$.

The steady state solution for the coherent pump field is determined by

$$\left(i(\tilde{\omega}_0 - \omega_p) + \frac{\kappa + \gamma}{2}\right)\alpha + iK\alpha^2\alpha^* = \sqrt{\kappa}\alpha_{\rm in}, \tag{5}$$

which follows immediately by substituting Eq. (4) into Eq. (2) and collecting only the c-number terms. By multiplying both sides with their complex conjugate we get to the equation

$$\frac{\kappa}{(\kappa+\gamma)^2}|\alpha_{\rm in}|^2 = \left(\left(\frac{\omega_p - \tilde{\omega}_0}{\kappa+\gamma}\right)^2 + \frac{1}{4}\right)|\alpha|^2 - \frac{2(\omega_p - \tilde{\omega}_0)K}{(\kappa+\gamma)^2}|\alpha|^4 + \left(\frac{K}{\kappa+\gamma}\right)^2|\alpha|^6, \tag{6}$$

which determines the average number of pump photons $|\alpha|^2$ in the resonator. Eq. (6) reduces to

$$1 = \left(\delta^2 + \frac{1}{4}\right)n - 2\delta\xi n^2 + \xi^2 n^3, \tag{7}$$

by defining the scale invariant quantities

$$\delta \equiv \frac{\omega_p - \tilde{\omega}_0}{\kappa+\gamma}, \qquad \tilde{\alpha}_{\rm in} \equiv \frac{\sqrt{\kappa}\alpha_{\rm in}}{\kappa+\gamma}, \qquad \xi \equiv \frac{|\tilde{\alpha}_{\rm in}|^2 K}{\kappa+\gamma}, \qquad n \equiv \frac{|\alpha^2|}{|\tilde{\alpha}_{\rm in}|^2}. \tag{8}$$

δ is the detuning between pump and resonator frequency in units of the total resonator linewidth, $\tilde{\alpha}_{\rm in}$ is the dimensionless drive amplitude, and ξ is the product of drive power and nonlinearity, also expressed in dimensionless units. Finally, n is the mean number of pump photons in the resonator relative to the incident pump power. As an important consequence, we notice from Eq. (8) that only the product of drive power and nonlinearity determines the dynamics but not each quantity itself. Therefore, a small nonlinearity can at least in principle be compensated by increasing the drive power. Properties such as the gain-bandwidth product are therefore independent of the strength of the nonlinearity as long as the pump power is much larger than the power of amplified fluctuations. Furthermore, the solutions of Eq. (7) for negative ξ values are identical to those for positive ξ up to a sign change in δ. Since ξ is negative for the Josephson parametric amplifier, we focus on this particular case.

Equation (7) is a cubic equation in n and can therefore be solved analytically. We do not present the lengthy solutions here explicitly, but assume in the following that we have an explicit analytical expression for n in terms of δ and ξ. In Figure 2(b) we plot n for various parameters ξ as a function of δ. At the critical value $\xi_{\text{crit}} = -1/\sqrt{27}$ the derivative $\partial n/\partial\delta$ diverges and thus the response of the parametric amplifier becomes extremely sensitive to small changes. For even stronger effective drive powers $\xi/\xi_{\text{crit}} > 1$ the cubic Eq. (7) has three real solutions. The solutions for the high and low photon numbers are stable, while the intermediate one is unstable. The system bifurcates in this regime as mentioned earlier. The critical detuning below which the system becomes bistable is $\delta_{\text{crit}} = -\sqrt{3}/2$. The critical point $(\xi_{\text{crit}}, \delta_{\text{crit}})$ is the one at which both $\partial\delta/\partial n$ and $\partial^2\delta/\partial^2 n$ vanish. In scale invariant units the maximal value of n is 4, which is reached at the detuning $\delta = 4\xi$.

Experimentally, the system parameters are characterized by measuring the complex reflection coefficient $\Gamma \equiv \alpha_{\text{out}}/\alpha_{\text{in}}$. Based on the input-output relation $\alpha_{\text{out}} = \sqrt{\kappa}\alpha - \alpha_{\text{in}}$ and Eq. (5) we evaluate this reflection coefficient as

$$\Gamma = \frac{\kappa}{\kappa + \gamma} \frac{1}{\frac{1}{2} - i\delta + i\xi n} - 1. \tag{9}$$

In Figure 2(c) we plot the absolute value of the reflection coefficient at $\xi = \xi_{\text{crit}}$ for various loss rates γ. For vanishing losses $\gamma = 0$ all the incident drive power is reflected from the device and $|\Gamma| = 1$. Note that also in this case the resonance is clearly visible in the phase of the reflected signal (not shown here). When the loss rate γ becomes similar to the external coupling rate κ part of the radiation is dissipated into the loss modes. In the case of critical coupling $\gamma = \kappa$ all the coherent power is transmitted into the loss modes at resonance. This is equivalent to the case of a symmetrically coupled $\lambda/2$ resonator, for which the transmission coefficient is one at resonance [48].

3.2 Linearized response for weak (quantum) signal fields

Under the assumption that the photon flux associated with the signal $\langle a_{\text{in}}^\dagger a_{\text{in}} \rangle$ is much smaller than the photon flux of the pump field $|\alpha_{\text{in}}|^2$, we can drop terms such as $Ka^\dagger a\alpha$, because they are small compared to the leading terms $Ka^\dagger\alpha^2$ and $Ka|\alpha|^2$. By neglecting these terms we obtain a linearized equation of motion for a in the presence of the pump field. In order to preserve the validity of this approximation even for larger input signals, the amplitude α of the pump field needs to be increased. Experimentally, this can be achieved by reducing the strength of the nonlinearity K as discussed in Section 3.3 in more detail.

Substituting Eq. (4) into Eq. (2) and keeping only terms which are linear in a one finds

$$\dot{a}(t) = i\left(\omega_p - \tilde{\omega}_0 - 2K|\alpha|^2 + i\frac{\kappa+\gamma}{2}\right)a(t) - iK\alpha^2 a^\dagger(t) + \sqrt{\kappa}a_{\text{in}}(t) + \sqrt{\gamma}b_{\text{in}}(t). \quad (10)$$

Since Eq. (10) is linear, we can solve it by decomposing all modes into their Fourier components

$$a(t) \equiv \frac{\kappa+\gamma}{\sqrt{2\pi}}\int_{-\infty}^{\infty} \mathrm{d}\Delta e^{-i\Delta(\kappa+\gamma)t} a_\Delta \quad (11)$$

and equivalently for $a_{\text{in},\Delta}$ and $b_{\text{in},\Delta}$. Note that the detuning Δ between signal frequencies and the pump frequency, is expressed here in units of the linewidth $\kappa+\gamma$. Substituting the Fourier decompositions into Eq. (10) and comparing the coefficients of different harmonics, results in

$$0 = \left(i(\delta - 2\xi n + \Delta) - \frac{1}{2}\right)a_\Delta - i\xi n e^{2i\phi}a^\dagger_{-\Delta} + \tilde{c}_{\text{in},\Delta}, \quad (12)$$

where $\tilde{c}_{\text{in},\Delta} \equiv (\sqrt{\kappa}a_{\text{in},\Delta} + \sqrt{\gamma}b_{\text{in},\Delta})/(\kappa+\gamma)$ is the sum of all field modes incident on the resonator. Furthermore, in Eq. (12) ϕ is the phase of the intra-resonator pump field, defined by $\alpha = |\alpha|e^{i\phi}$. The fact that Eq. (12) couples modes a_Δ and $a^\dagger_{-\Delta}$ can be interpreted as a wave mixing process. In order to express a_Δ in terms of the input fields $c_{\text{in},\Delta}$, Eq. (12) is rewritten as a matrix equation

$$\begin{pmatrix} \tilde{c}_{\text{in},\Delta} \\ \tilde{c}^\dagger_{\text{in},-\Delta} \end{pmatrix} = \begin{pmatrix} i(-\delta + 2\xi n - \Delta) + \frac{1}{2} & i\xi n e^{i2\phi} \\ -i\xi n e^{-i2\phi} & i(\delta - 2\xi n - \Delta) + \frac{1}{2} \end{pmatrix} \begin{pmatrix} a_\Delta \\ a^\dagger_{-\Delta} \end{pmatrix}. \quad (13)$$

By inverting the matrix on the right hand side, the quantum part of the intra-resonator field a_Δ is expressed in terms of the incoming field $\tilde{c}_{\text{in},\Delta}$

$$a_\Delta = \frac{i(\delta - 2\xi n - \Delta) + \frac{1}{2}}{(i\Delta - \lambda_-)(i\Delta - \lambda_+)}\tilde{c}_{\text{in},\Delta} + \frac{-i\xi n e^{2i\phi}}{(i\Delta - \lambda_-)(i\Delta - \lambda_+)}\tilde{c}^\dagger_{\text{in},-\Delta} \quad (14)$$

with $\lambda_\pm = \frac{1}{2} \pm \sqrt{(\xi n)^2 - (\delta - 2\xi n)^2}$. Using Eq. (3), the final transformation between input and output modes is

$$a_{\text{out},\Delta} = g_{S,\Delta}a_{\text{in},\Delta} + g_{I,\Delta}a^\dagger_{\text{in},-\Delta} + \sqrt{\frac{\gamma}{\kappa}}(g_{S,\Delta}+1)b_{\text{in},\Delta} + \sqrt{\frac{\gamma}{\kappa}}g_{I,\Delta}b^\dagger_{\text{in},-\Delta} \quad (15a)$$

$$\overset{\gamma/\kappa\to 0}{=} g_{S,\Delta}a_{\text{in},\Delta} + g_{I,\Delta}a^\dagger_{\text{in},-\Delta}, \quad (15b)$$

with

$$g_{S,\Delta} = -1 + \frac{\kappa}{\kappa+\gamma}\frac{i(\delta - 2\xi n - \Delta) + \frac{1}{2}}{(i\Delta - \lambda_-)(i\Delta - \lambda_+)} \quad (16)$$

and

$$g_{I,\Delta} = \frac{\kappa}{\kappa+\gamma}\frac{-i\xi n e^{2i\phi}}{(i\Delta - \lambda_-)(i\Delta - \lambda_+)}. \quad (17)$$

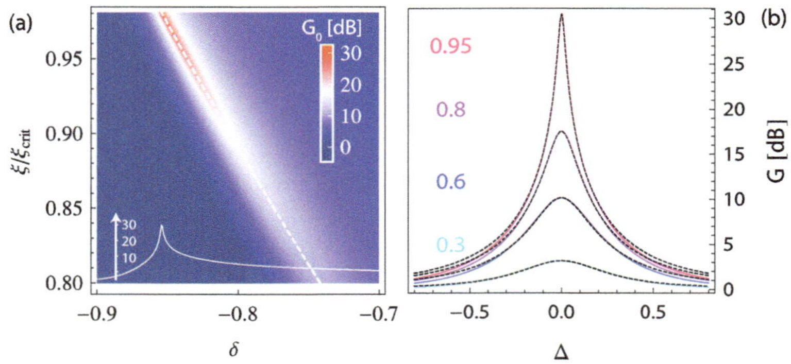

Figure 3 Parametric amplifier gain. (**a**) $G_\Delta = |g_{S,\Delta}|^2$ *vs.* pump tone detuning δ and drive strength ξ at zero signal detuning Δ and for $\kappa = \gamma$. For increasing drive strength ξ the detuning for maximum gain is indicated by the dashed white line. A cut through the data for the highest value $\xi = 0.98\xi_{crit}$ is shown as the solid white line in the bottom part. (**b**) Gain as a function of signal detuning Δ for the indicated drives strengths ξ/ξ_{crit} and optimal pump detuning. The exact gain curves (solid lines) are well approximated by Lorentzian lines (black dashed lines).

Eq. (15b) is the central result of this calculation. The output field at detuning Δ from the pump frequency is a sum of the input fields at frequencies Δ and $-\Delta$ multiplied with the signal gain factor $g_{S,\Delta}$ and the idler gain factor $g_{I,\Delta}$, respectively. The additional noise contributions introduced via the loss modes $b_{\mathrm{in},\Delta}$ vanish in the limit $\gamma/\kappa \to 0$. In the ideal case $\gamma = 0$, the coefficients $g_{S,\Delta}$ and $g_{I,\Delta}$ satisfy the relation

$$G_\Delta \equiv |g_{S,\Delta}|^2 = |g_{I,\Delta}|^2 + 1 \tag{18}$$

and Eq. (15b) is identical to a two-mode squeezing transformation [19, 49] with gain G_Δ. The two-mode squeezing transformation describes a linear amplifier in its minimal form (compare Ref. [50]), of which we discuss characteristic properties in the following section.

3.3 Gain, bandwidth, noise and dynamic range

For simplicity we consider the case of no losses $\gamma = 0$, for which the parametric amplifier response is described by Eq. (15b). An incoming signal at detuning Δ is thus amplified by the power gain $G_\Delta = |g_{S,\Delta}|^2$ and mixed with the frequency components at the opposite detuning from the pump. Characteristic properties of the parametric amplifier, such as the maximal gain and the bandwidth, are thus encoded in the quantity $g_{S,\Delta}$ as a function of pump-resonator detuning δ, effective drive strength ξ and detuning between signal and pump Δ.

In Figure 3(a) we plot the gain G_0 for zero signal detuning $\Delta = 0$ as a function of δ and ξ. We find that the maximal gain increases with increasing drive strength ξ while the optimal value for δ at which this gain is reached, shifts approximately linearly with increasing ξ. The optimal values for δ are indicated as a dashed white line in Figure 3(a). Mathematically, the gain diverges when ξ approaches the critical value ξ_{crit}. In practice, the gain is limited to finite values due to the breakdown of the stiff pump approximation (see discussion below).

By changing the pump parameters ξ and δ we can adjust the gain G_0 to a desirable value, which is typically about 20 dB. Note that the gain can take values smaller than one, in the presence of finite internal losses $\gamma > 0$. Once the pump parameters are fixed we characterize the bandwidth of the amplifier by analyzing the gain as a function of the signal detuning Δ. In Figure 3(b) we plot the gain as a function of Δ for the indicated values of ξ/ξ_{crit} and the corresponding optimal pump detunings δ (compare dashed white line in (a)). The gain curves are well approximated by Lorentzian lines as indicated by the dashed black lines in Figure 3(b). When the gain is increased, the band of amplification becomes narrower. This is quantitatively expressed by the gain-bandwidth relation $\sqrt{G_0}B \approx 1$, where B is the detuning Δ for which the gain reaches half of its maximal value. This gain-bandwidth relation follows from the Lorentzian approximation of the gain curves shown as black dashed lines in Figure 3(b)) and holds for gain values above a few dB. Remember that Δ is defined in units of the resonator linewidth $\kappa + \gamma$, which means that the amplifier bandwidth equals approximately the resonator linewidth divided by the square root of the gain.

When operating the JPA, we also have to understand its behavior in terms of added noise. In the ideal case with zero loss rate ($\gamma = 0$), the input-output relation of the parametric amplifier in Eq. (15b) has the minimal form of a scattering mode amplifier [19]. The amplification process reaches the vacuum limit as long as the input modes are cooled into the vacuum. In practice, however, the device may have finite loss γ which increases the effectively added noise by a factor of $(\kappa + \gamma)/\kappa$. This is due to the additional amplified noise, which originates from the modes $b_{\text{in},\Delta}$ and contributes to the output field $a_{\text{out},\Delta}$ (compare Eq. (15a)). Another potential source of noise is related to the stability of the resonance frequency of the parametric amplifier. Magnetic flux noise in the SQUID loop may lead to a fluctuating resonance frequency and thus a fluctuating effective gain.

In the derivation made in the previous sections we have assumed that the solution of the classical drive field is unaffected by the presence of additional signal and quantum fluctuations at the input. This is known as the stiff pump approximation [8], which assumes that the pump power at the output is equal to the pump power at the input of the JPA. This is of course an approximation, since the pump field provides the energy which is necessary for amplifying the input signal. The stiff pump approximation is valid as long as the pump power is significantly larger than the total output power of all amplified (quantum) signal and vacuum fields. In order to quantitatively analyze the pump depletion due to the presence of amplified fields we add the terms $2iK\langle a^\dagger a\rangle\alpha$ and $iK\langle a^2\rangle\alpha^*$ to the left hand side of Eq. (5) and solve Eq. (5) and Eq. (10) self-consistently. This mean-field approach is similar to the one used in Ref. [8]. For our calculation we model the incoming signal field, which is to be amplified, as white noise with average photon number n_{th} per unit time and bandwidth. Based on this model we find that the gain decreases when the signal strength exceeds a certain value, see Figure 4(a). The number of input photons n_{th} at which this happens becomes smaller with decreasing ratio $\kappa/|K|$. This is expected because the pump power close to the bifurcation point is proportional to $\kappa/|K|$ and provides the energy required for amplification. For small values $\kappa/|K|$ the gain is reduced even for $n_{\text{th}} = 0$ due to the amplification of vacuum fluctuations (see blue data points in Figure 4(a)). As a measure of the dynamic range we specify the 1 dB compression point of the JPA, *i.e.* the value $n_{1\,\text{dB}}$ of input photons n_{th} at which the JPA gain G_0 decreases by 1 dB compared to the stiff pump approximated gain value. As shown in Figure 4(b), the 1 dB compression point increases proportionally to $\kappa/|K|$ in the limit of $n_{1\,\text{dB}} \gg 1$. The presence of constant vacuum fluctuations leads to a gain compression by more than 1 dB even for $n_{\text{th}} = 0$ when $\kappa/|K|$ becomes smaller. We can qualitatively explain this behavior by comparing the pump

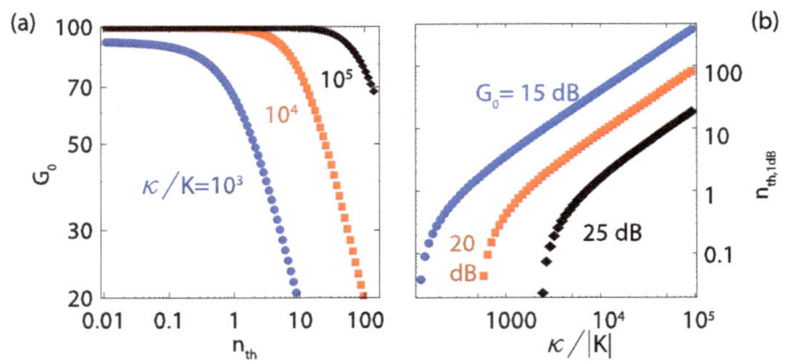

Figure 4 Dynamic range of the JPA. (**a**) Gain G_0 as a function of the photon number n_{th} of the input signal field for the three indicated ratios $\kappa/|K|$. (**b**) 1 dB compression point as a function of $\kappa/|K|$ for various gain values G_0.

power $P_p = \hbar\omega_p|\alpha_{\text{in}}|^2$ with the power of amplified signal and vacuum fields

$$P_{\text{out}} \overset{\gamma=0}{=} \hbar\omega_p\kappa(2n_{\text{th}}+1)\int\frac{\mathrm{d}\Delta}{2\pi}(G_\Delta - 1). \tag{19}$$

Making use of the gain-bandwidth relation we find the following scaling of the ratio between the two powers

$$\frac{P_{\text{out}}}{P_p} \propto (2n_{\text{th}}+1)\frac{|K|}{\kappa}\sqrt{G_0}. \tag{20}$$

The results shown in Figure 4(b) together with Eq. (20) indicate that the validity of the stiff pump approximation is essentially determined by the ratio P_{out}/P_p. The calculations furthermore show that the dynamic range can be increased by reducing the ratio $|K|/\kappa$ of the JPA, which seems to be the case also for flux driven parametric amplifiers [40]. In Section 5 we discuss how to achieve small nonlinearities by making use of multiple SQUIDs connected in series.

4 Effective system parameters from distributed circuit model

In the previous section we have analyzed the model of a nonlinear resonator with resonance frequency $\tilde{\omega}_0$, Kerr nonlinearity K and decay rate κ. Here, we explicitly derive this effective Hamiltonian from the full circuit model of a $\lambda/4$ - transmission line resonator, which is terminated by a SQUID loop at the short-circuited end and coupled capacitively to a transmission line, see Figure 1(a). These calculations allow us to determine $\tilde{\omega}_0$, K, κ from the distributed circuit parameters and give insight into potential limitations of the effective model. We also compare the obtained parameter relations with those of a lumped element parametric amplifier.

4.1 Resonator mode structure in the linear regime

In order to find the normal mode structure of the system, we first neglect its capacitive coupling to the transmission line as indicated in Figure 5. The derivation is similar to the ones in Refs. [40, 51, 52]. Dissipation effects due the environment are discussed in Section 4.3.

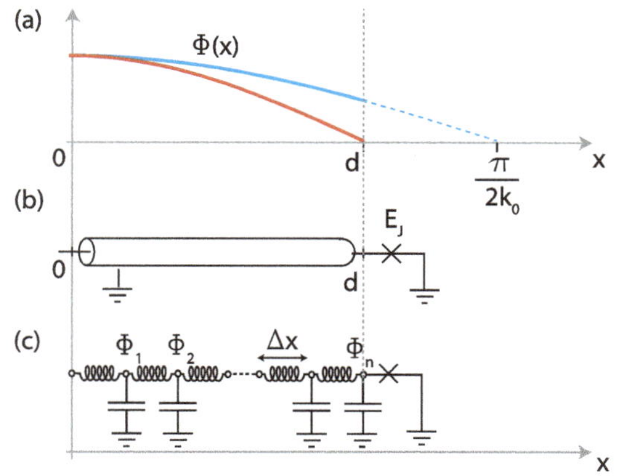

Figure 5 Circuit model. (**a**) Distribution of the magnetic flux field $\Phi(x)$ along the $\lambda/4$-resonator for the fundamental resonator mode $j = 0$, without (red) and with (blue) a Jospehson junction. The additional Josephson inductance changes the boundary condition such that neither the current nor the voltage is zero at position $x = d$. The resulting increase in the effective wavelength $\pi/(2k_0)$ is indicated by the dashed blue line. (**b**) Transmission line resonator of length d with a Josephson junction at the grounded end. (**c**) Lumped element representation with indicated discretized magnetic flux field Φ_j as used in Eq. (21).

The total Lagrangian of the system in the magnetic flux field $\Phi(x)$ has a transmission line part and a term which describes the SQUID at position $x = d$ (Figure 5).

$$\mathcal{L} = \int_0^d \mathrm{d}x \left\{ \frac{c}{2} \left(\partial_t \Phi(x) \right)^2 - \frac{1}{2l} \left(\partial_x \Phi(x) \right)^2 \right\} + E_J \cos\left(\frac{\Phi(d)}{\varphi_0} \right), \tag{21}$$

with the reduced flux quantum $\varphi_0 = \hbar/2e$. Since we work in a limit in which the plasma frequency of the SQUID is much larger than the resonance frequencies of interest, we neglect the self-capacitance of the SQUID. Note that Ref. [52] provides a detailed study of the effect of the self-capacitance in various parameter regimes. We furthermore describe the SQUID as a single junction with tunable effective Josephson energy E_J.

We first investigate the linear regime of the system, in which the cosine potential of the SQUID is approximated as a quadratic potential.

$$E_J \cos\left(\frac{\Phi(d)}{\varphi_0} \right) \approx \text{const} - \frac{1}{2} \left(\frac{\Phi(d)}{\varphi_0} \right)^2. \tag{22}$$

Due to the spatial derivative in the Lagrangian in Eq. (21) all local fields in the chain are coupled to their nearest neighbors and the normal mode structure is found by solving the Euler-Lagrange equation $\partial_t(\delta\mathcal{L}/\delta\dot{\Phi}) - \delta\mathcal{L}/\delta\Phi = 0$ of the transmission line resonator. This results in the wave equation

$$v^2 \partial_x^2 \Phi(x) - \partial_t^2 \Phi(x) = 0, \tag{23}$$

with the phase velocity $v = 1/\sqrt{cl}$, of which the general solution can be written as a sum of normal modes

$$\Phi(x) = \sum_{j=0}^{\infty} \phi_j \cos(k_j x). \tag{24}$$

The valid wavevectors k_j are determined by the boundary conditions at the two ends of the $\lambda/4$ transmission line. The open end at $x = 0$ requires that the current $\partial_x \Phi(x)/l$ vanishes, which is implicitly satisfied by choosing the cosine ansatz in Eq. (24). On the shorted end the boundary condition is modified by the presence of the Josephson junction. In order to determine this boundary condition, we evaluate the Euler-Lagrange equation at position $x = d$. For this purpose it is convenient to write the Lagrangian in a discretized form, see Figure 5(c) and compare Ref. [51]:

$$\mathcal{L} = \lim_{n\to\infty} \sum_{j=1}^{n} \Delta x \left\{ \frac{c}{2} (\partial_t \Phi_j)^2 - \frac{1}{l} \frac{(\Phi_j - \Phi_{j-1})^2}{\Delta x^2} \right\} - \frac{1}{2} E_J \left(\frac{\Phi_n}{\varphi_0} \right)^2, \tag{25}$$

where $\Phi_n = \Phi(x = d)$ and $\Delta x = d/n$. Evaluating $\partial_t(\partial\mathcal{L}/\partial\dot{\Phi}_n) - \partial\mathcal{L}/\partial\Phi_n = 0$ leads to the equation

$$\frac{1}{l}\partial_x \Phi(d) + E_J \frac{\Phi(d)}{\varphi_0^2} = 0. \tag{26}$$

Substituting the ansatz (24) into Eq. (26) and comparing the resulting coefficients of the independent variables ϕ_j, results in the transcendental equation

$$k_j d \tan(k_j d) = ld \frac{E_J}{\varphi_0^2} \equiv \frac{ld}{L_J}. \tag{27}$$

Here, we have defined the Josephson inductance $L_J = \varphi_0^2/E_J$. The infinite set of solutions k_j of this equation determines the normal modes structure of the system in the linear regime. In the limit in which the SQUID inductance L_J vanishes, Eq. (27) is solved by the poles of $\tan(k_j d)$, and we recover the normal modes of the $\lambda/4$ resonator

$$k_j^{(0)} d = \frac{\pi}{2}(1 + 2j) \quad \text{with } j \in \{0, 1, 2, 3, \ldots\}. \tag{28}$$

As a first order correction to this result in the limit of $L_J/ld \ll 1$, we expand Eq. (27) to first order in $(k_j^{(0)} - k_j)d$ and find $k_j L_J/ld = (k^{(0)} - k_j)$ or equivalently

$$k_j \approx \frac{k_j^{(0)}}{1 + L_J/ld}. \tag{29}$$

For the fundamental mode with $j = 0$ this linearized approximation is typically accurate even for inductance ratios up to $L_J/ld \approx 0.5$, whereas for the higher harmonic modes the linearized equation breaks down for much smaller values of L_J/ld. A comparison between the exact solution based on Eq. (27) and the approximate solution in Eq. (29) is shown in Figure 6 for the first three resonant modes. When higher harmonics are expected to be relevant one should solve Eq. (27) numerically in order to determine the exact wave numbers k_j.

4.2 Kerr nonlinear terms and effective Hamiltonian

Using the normal mode decomposition in Eq. (24) we reexpress the Lagrangian in Eq. (21) as a sum of oscillators which are only coupled via the boundary condition imposed by the

Figure 6 Resonance frequencies of the first three modes as a function of Josephson energy. The solid line results from the exact numerical solution of Eq. (27) while the dashed line shows the linearized solution in Eq. (29). The bare resonance frequency is chosen to be 7 GHz and the impedance of the transmission line resonator 50 Ω.

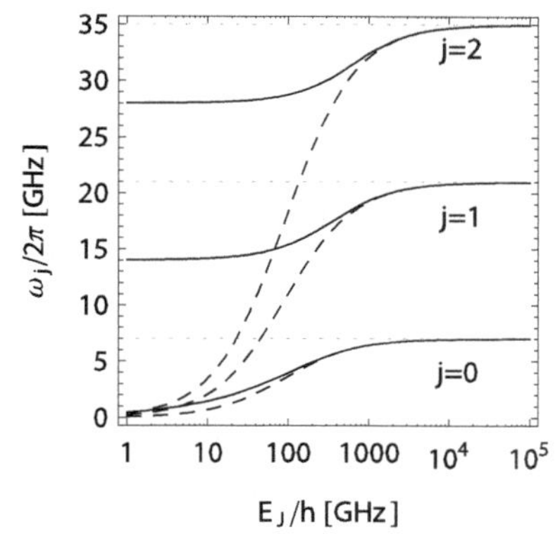

SQUID. For the purposes of parametric amplification the phase drop across the junction is desired to be small, $\Phi_n/\varphi_0 < 1$, *i.e.* the current flowing through the Josephson junction is small compared to its critical current. We can therefore expand the SQUID cosine potential and take into account only the first non-quadratic correction

$$E_\mathrm{J}\cos\left(\frac{\Phi_n}{\varphi_0}\right) = \mathrm{const} - \frac{1}{2}E_\mathrm{J}\left(\frac{\Phi_n}{\varphi_0}\right)^2 + \frac{1}{24}E_\mathrm{J}\left(\frac{\Phi_n}{\varphi_0}\right)^4 + \cdots. \tag{30}$$

In Section 5 we discuss under which circumstances such an approximation may break down. Substituting the normal mode decomposition Eq. (24) into the Taylor expansion of the Lagrangian results in

$$\mathcal{L} = \frac{1}{2}\sum_{i=1}^{\infty}\left\{\dot{\phi}_i C_i \dot{\phi}_i - \phi_i L_i^{-1}\phi_i\right\} + \sum_{j,i,k,l=1}^{\infty} N_{ijkl}\phi_i\phi_j\phi_k\phi_l \tag{31}$$

with the effective capacitances and inductances [51]

$$\begin{aligned} C_i &= c\int_0^d \mathrm{d}x\,\cos^2(k_i x) = \frac{cd}{2}\left(1+\frac{\sin(2k_i d)}{2k_i d}\right),\\ L_i^{-1} &= L_\mathrm{J}^{-1}\cos^2(k_i d) + \frac{k_i^2}{l}\int_0^d \mathrm{d}x\,\sin^2(k_i x)\\ &\overset{\text{Eq. (27)}}{=} \frac{(k_i d)^2}{2ld}\left(1+\frac{\sin(2k_i d)}{2k_i d}\right), \end{aligned} \tag{32}$$

and the nonlinearity coefficients

$$N_{ijkl} = \frac{1}{24}E_\mathrm{J}\varphi_0^{-4}\prod_{m\in\{i,j,k,l\}}\cos(k_m d). \tag{33}$$

As expected the linear part of the Lagrangian is diagonal in the normal mode basis. It describes a set of uncoupled *LC* oscillators for which the effective resonance frequencies coincide with the product of phase velocity and wave vector $\omega_j = k_j v = 1/\sqrt{L_j C_j}$.

Based on the Lagrange function (31) we derive the Hamiltonian by introducing the conjugate charge variables $q_i = \delta\mathcal{L}/\delta\dot{\phi}_i = C_i\dot{\phi}_i$. Performing a Legendre transformation and taking only self-interactions and two-mode interactions into account, results in the Hamiltonian

$$\mathcal{H} = \frac{1}{2}\sum_{i=1}^{\infty}\left\{q_i C_i^{-1} q_i + \phi_i L_i^{-1}\phi_i\right\} - 3\sum_{j\neq i}^{\infty} N_{iijj}\phi_i^2\phi_j^2 - \sum_{i}^{\infty} N_{iiii}\phi_i^4. \tag{34}$$

In a quantum regime q_i and ϕ_i are operators which satisfy the commutation relation $[\phi_j, q_k] = \delta_{kj}\hbar/i$ and it is convenient to write the Hamiltonian in terms of normal mode annihilation and creation operators [53]

$$\phi_j = i\phi_{\mathrm{zpf},j}\left(a_j^\dagger - a_j\right), \qquad q_j = q_{\mathrm{zpf},j}\left(a_j + a_j^\dagger\right) \tag{35}$$

with $q_{\mathrm{zpf},i} = \sqrt{\hbar\omega_i C_i/2}$ and $\phi_{\mathrm{zpf},i} = \sqrt{\hbar/2\omega_i C_i}$. The abbreviation zpf stands for zero point fluctuations. Performing a rotating wave approximation (*i.e.* removing all terms with an unequal number of creation and annihilation operators), and neglecting the small photon number independent frequency shifts due to the nonlinear terms (*i.e.* Lamb shifts) we arrive at

$$\mathcal{H} = \sum_{i=1}^{\infty}\hbar\omega_i a_i^\dagger a_i + \hbar\frac{K_{ii}}{2}a_i^\dagger a_i^\dagger a_i a_i + \sum_{j\neq i}^{\infty}\hbar K_{ij} a_i^\dagger a_i a_j^\dagger a_j \tag{36}$$

with

$$K_{ij} = -\frac{E_\mathrm{J}}{2\hbar}\left(\frac{\phi_{\mathrm{zpf}}}{\varphi_0}\right)^4 \cos^2(k_i d)\cos^2(k_j d). \tag{37}$$

The quantity $K = K_{00}$ is the Kerr nonlinearity of the fundamental mode, which is used for the parametric amplification process. The terms proportional to K_{ij} with unequal $i \neq j$ are cross Kerr interaction terms which couple different modes to each other. Such an interaction can for example be used for counting the number of photons in one mode by probing another one with a coherent field [54–61], similarly to a dispersive qubit measurement. Note that the values resulting from Eq. (37) are divided by the square of the number of SQUIDs, if an array is used instead of a single SQUID, as discussed in the following section.

4.3 Decay rate and resonance frequency correction for low Q resonators

Since the parametric amplifier bandwidth is proportional to the decay rate κ, typical devices are designed to have a low external quality factor, which is achieved by increasing the coupling capacitance C_κ between transmission line and resonator (Figure 1). The coupling of an oscillator to the environment shifts its resonance frequency $\omega_j \to \tilde{\omega}_j$ [48], which can be significant if the coupling rate is large. When designing parametric amplifier devices, it is therefore necessary to take these shifts into account. Based on the effective inductance and capacitances calculated in Eq. (32) we find

$$\tilde{\omega}_j^2 \approx \frac{\omega_j^2}{1 + C_\kappa/C_j} = \frac{1}{(C_j + C_\kappa)L_j} \quad \text{and} \quad \kappa_j \approx \frac{\tilde{\omega}_j^2 C_\kappa^2 R}{C_\kappa + C_j} \tag{38}$$

for resonance frequency and decay rate of the jth mode of the parametric amplifier device. The external quality factor is given by $Q_j \equiv \tilde{\omega}_j/\kappa_j$.

4.4 Lumped element JPA

As already mentioned in the introduction, a JPA can also be realized as a lumped element resonator by shunting a SQUID with a large capacitance C_J [10, 11]. In this case the resonator is described by the transmon Hamiltonian [62], which in the deep transmon limit $E_J \gg E_C \equiv e^2/2C_J$ takes the form of Eq. (1) with anharmonicity $K \approx E_C/\hbar$ and resonance frequency $\tilde{\omega}_0 \approx 1/\sqrt{L_J C_J}$. Also for this type of resonators the coupling rate κ to the transmission line can be designed independently of E_J and E_C by designing an appropriate capacitive network. Similarly as for the transmission line JPA, the description in terms of the effective Hamiltonian Eq. (1) is based on the assumption that for relevant resonator fields the phase drop across the Josephson junctions is small (compare Eq. (30)). In the following section we study the validity of this approximation when the resonator is driven close to the bifurcation point where we expect parametric amplification to occur and analyze its implications for realizing a parametric amplifier with large bandwidth and dynamic range.

5 Bandwidth and dynamic range constraints

5.1 Validity of the quartic approximation

In Section 3.3 we have shown that the dynamic range of the JPA scales with $\kappa/|K|$. Also the bandwidth becomes larger with increasing κ, which indicates that a large κ is desirable for JPAs. However, there are limitations on the maximal possible value for κ as discussed in the following.

For deriving the Hamiltonian in Eq. (1), or more generally Eq. (36), we have expanded the SQUID cosine potential to quartic order in the dimensionless flux variable Φ_n/φ_0, where $\Phi_n \equiv \Phi(x = d)$ is the phase drop across the SQUID. To guarantee that this approximation holds when we operate the device in the parametric amplification regime, we have to make sure that Φ_n/φ_0 is small even when it is driven close to the bifurcation point. This is equivalent to keeping the current flowing through the SQUID small compared to the critical current.

To characterize the validity of the low order expansion of the cosine potential, we define the maximal coherent field inside the resonator $\alpha_{\max}$ as the one for which $\Phi_n = \varphi_0$. This is the coherent amplitude, at which the current flowing through the SQUID equals its critical current. According to Eq. (35) and Eq. (24) a coherent field α in mode j leads to a maximal amplitude of $\Phi_n = \phi_{\mathrm{zpf},j} 2\alpha \cos(k_j d)$ across the tunnel junctions, based on which we define the critical amplitude as

$$\alpha_{\max,j} = \frac{\varphi_0}{\phi_{\mathrm{zpf},j}} \frac{1}{2\cos(k_j d)} = \frac{\varphi_0}{\phi_{\mathrm{zpf},j}} \frac{ld}{2L_J k_j d \sin(k_j d)}. \tag{39}$$

The low order expansion of the cosine potential is only valid if the field inside the resonator α is much smaller than this maximal amplitude $\alpha < \alpha_{\max,j}$. In Section 3.1 we have found that the photon number in a resonator mode at the bifurcation point is $N_{\mathrm{crit}} = (\kappa + \gamma)/\sqrt{3}K$. The ratio between N_{crit} and the maximal coherent photon number $N_{\max,j} \equiv |\alpha_{\max,j}|^2$ is therefore given by

$$\frac{N_{\mathrm{crit}}}{N_{\max,j}} = \frac{8\kappa}{\sqrt{3}} \frac{\hbar}{E_J \cos^2(k_j d)} \left(\frac{\varphi_0}{\phi_{\mathrm{zpf}}} \right)^2. \tag{40}$$

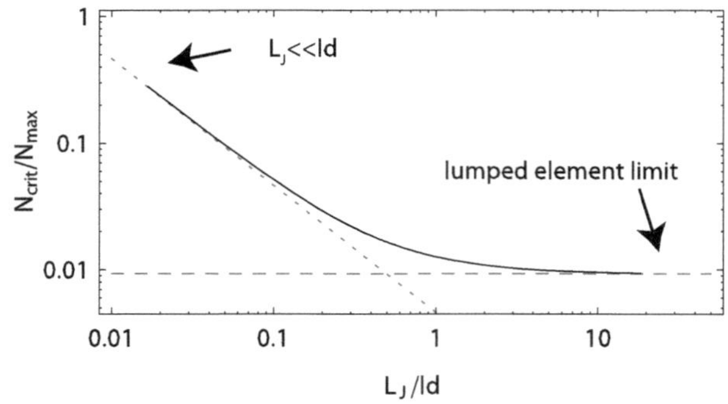

Figure 7 Validity of the quartic approximation. Ratio N_{crit}/N_{max} according to Eq. (40) (black), in the lumped element limit (dashed blue) and in the limit of small participation ratio $L_J \ll ld$ (dotted red) for the parameters $\{\tilde{\omega}_0/2\pi, Q\} = \{7\text{ GHz}, 1{,}000\}$.

In order to minimize the effect of higher order nonlinearities we want to keep this ratio small. It has the following two interesting limits

$$\frac{N_{crit}}{N_{max,j}} = \begin{cases} \frac{8}{\sqrt{3}} Q^{-1} \frac{ld}{L_J}, & \text{for } L_J \ll ld, \\ \frac{16}{\sqrt{3}} Q^{-1}, & \text{for lumped element JPA.} \end{cases} \tag{41}$$

This is an important result, which sets clear constraints on both the maximally achievable bandwidth and the dynamic range of the JPA. If we want $N_{crit}/N_{max,j}$ to be small, κ needs to be sufficiently small as well. One may also want to increase the dynamic range by reducing the Kerr nonlinearity $|K|$. While this can in principle be achieved by choosing a small ratio L_J/ld between Josephson and geometric inductance, care has to be taken when using this approach because additional geometric inductance leads to a larger ratio N_{crit}/N_{max} as illustrated in Figure 7.

Interestingly, we find that in the lumped element case the Josephson inductance L_J, and with it the Kerr nonlinearity K, can in principle be made smaller without affecting N_{crit}/N_{max}. However, in practice a small Josephson inductance has to be compensated by a large lumped element capacitor to retain the desired resonance frequency, which is challenging to realize without introducing additional parasitic geometric inductances. It therefore seems difficult to build a parametric amplifier with large bandwidth and high dynamic range at the same time using a single SQUID only. In the following we show how one can keep N_{crit}/N_{max} constant while decreasing the nonlinearity and thus increasing the dynamic range of the amplifier, by replacing the single SQUID with a serial array of M SQUIDs of M-times larger Josephson energy per SQUID (Figure 8).

5.2 Josephson junction arrays

For simplicity we assume that all SQUIDs in the array have the same effective Josephson energy ME_J. Since the spatial extent of the junction is still small compared to typical resonance wavelengths, we can treat the array as a lumped element. To derive the nonlinearity of the oscillator for this situation we investigate how the different terms in the Lagrangian scale with M.

Assuming that the phase drop from the flux node at the end of the transmission line resonator to the ground is homogeneously distributed over the array, we have the same

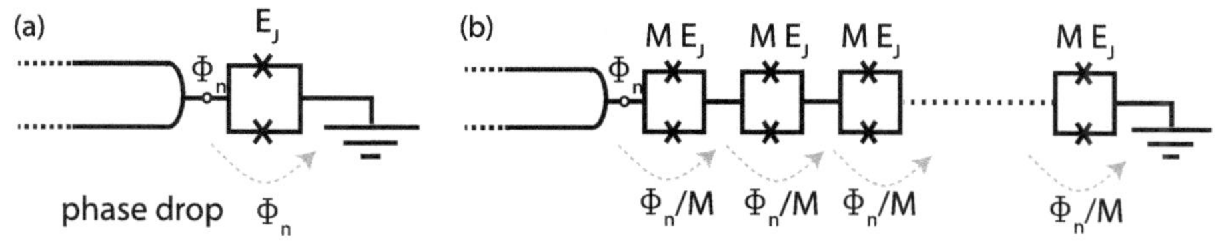

Figure 8 Schematic of the SQUID array. (**a**) The phase drop across a single SQUID junction is proportional to the node flux Φ_n (indicated by the circle) at the end of the transmission line. (**b**) If we replace the single junction by an serial array of M junctions with M times larger Josephson energy, the phase drop across each junction is by a factor of M smaller while the total effective Josephson inductance stays the same.

phase drop Φ_n/M across each SQUID, see Figure 8. As a result the quadratic term in the Lagrangian scales as

$$\frac{E_J}{2}\Phi_n^2 \overset{1\to M}{\longrightarrow} \sum_{i=1}^{M} \frac{ME_J}{2}\left(\frac{\Phi_n}{M}\right)^2 = \frac{E_J}{2}\Phi_n^2 \tag{42}$$

and thus remains constant. This agrees with our expectation, since the total linear Josephson inductance has not been changed. However, the quartic term scales like

$$\frac{E_J}{24}\Phi_n^4 \overset{1\to M}{\longrightarrow} \sum_{i=1}^{M} \frac{ME_J}{24}\left(\frac{\Phi_n}{M}\right)^4 = \frac{1}{M^2}\frac{E_J}{24}\Phi_n^4, \tag{43}$$

which leads to a quadratic decrease in the effective Kerr nonlinearity $K \to K/M^2$ and thus a quadratic increase in $N_{\text{crit}} \propto M^2$. Furthermore, the maximal photon number also scales as $N_{\max} \propto M^2$ since the critical current of each junction is larger by a factor of M. In other words, the ratio $N_{\text{crit}}/N_{\max}$ only depends on the total Josephson inductance whereas the bifurcation power increases quadratically in M. We thus conclude that the dynamic range of a JPA can be increased without affecting the amplifier bandwidth, by using an array of SQUIDs instead of a single SQUID. This conclusion is valid for both the transmission line JPA and the lumped element JPA.

In practice, the Josephson energies in the array are not all equal due to inhomogeneous coupling to the external magnetic flux and scatter in the critical current of Josephson junctions due to unavoidable variations in fabrication. A quantitative analysis of the influence of such variations of Josephson energies on the parametric amplifier characteristics could be an interesting task for future studies. This would help to quantify limitations in the accessible tuning range of the parametric amplifier and a realistic understanding of the breakdown of the low order expansion of the cosine potential. For such an approach the methods used in Ref. [63] could turn out to be useful.

6 Conclusion

In summary, we have presented a detailed analysis of Josephson junction based parametric amplifiers, including a discussion of bandwidth, noise and dynamic range. By establishing relations between basic JPA properties and designable circuit parameters we have been able to derive two simple design strategies to achieve optimized JPA performance. On the one hand the contribution of the Josephson inductance to the total effective inductance of the resonant circuit has to be chosen sufficiently large. On the other hand the use of

SQUID arrays instead of single SQUIDs provides the possibility to enhance the strength of the pump field close at the bifuraction point and with it the dynamic range of the JPA.

Competing interests

The authors declare they have no competing interests.

Authors' contributions

All authors contributed equally to the work in this article.

Acknowledgements

The authors would like to acknowledge helpful discussions with Alexandre Blais, Vitaly Shumeiko and Sebastian Schmidt. This work was supported by the European Research Council (ERC) through a Starting Grant and by ETHZ.

References

1. Yurke B, Buks E: *J Lightwave Technol* 2006, **24**:5054.
2. Castellanos-Beltran MA, Lehnert KW: *Appl Phys Lett* 2007, **91**:083509.
3. Tholén E, Ergül A, Doherty E, Weber F, Grégis F, Haviland D: *Appl Phys Lett* 2007, **90**:253509.
4. Kinion D, Clarke J: *Appl Phys Lett* 2008, **92**:172503.
5. Yamamoto T, Inomata K, Watanabe M, Matsuba K, Miyazaki T, Oliver WD, Nakamura Y, Tsai JS: *Appl Phys Lett* 2008, **93**:042510.
6. Castellanos-Beltran MA, Irwin KD, Hilton GC, Vale LR, Lehnert KW: *Nat Phys* 2008, **4**:929.
7. Palacios-Laloy A, Nguyen F, Mallet F, Bertet P, Vion D, Esteve D: *J Low Temp Phys* 2008, **151**:1034.
8. Kamal A, Marblestone A, Devoret M: *Phys Rev B* 2009, **79**:184301.
9. Bergeal N, Schackert F, Metcalfe M, Vijay R, Manucharyan VE, Frunzio L, Prober DE, Schoelkopf RJ, Girvin SM, Devoret MH: *Nature* 2010, **465**:64.
10. Hatridge M, Vijay R, Slichter DH, Clarke J, Siddiqi I: *Phys Rev B* 2011, **83**:134501.
11. Vijay R, Slichter DH, Siddiqi I: *Phys Rev Lett* 2011, **106**:110502.
12. Gao J, Vale LR, Mates JAB, Schmidt DR, Hilton GC, Irwin KD, Mallet F, Castellanos-Beltran MA, Lehnert KW, Zmuidzinas J, Leduc HG: *Appl Phys Lett* 2011, **98**:232508.
13. Eichler C, Bozyigit D, Lang C, Baur M, Steffen L, Fink JM, Filipp S, Wallraff A: *Phys Rev Lett* 2011, **107**:113601.
14. Abdo B, Kamal A, Devoret M: *Phys Rev B* 2013, **87**:014508.
15. Kamal A, Clarke J, Devoret MH: *Phys Rev B* 2012, **86**:144510.
16. Louisell WH, Yariv A, Siegman AE: *Phys Rev* 1961, **124**:1646.
17. Gordon JP, Louisell WH, Walker LR: *Phys Rev* 1963, **129**:481.
18. Mollow BR, Glauber RJ: *Phys Rev* 1967, **160**:1076.
19. Clerk AA, Devoret MH, Girvin SM, Marquardt F, Schoelkopf RJ: *Rev Mod Phys* 2010, **82**:1155.
20. Mallet F, Castellanos-Beltran MA, Ku HS, Glancy S, Knill E, Irwin KD, Hilton GC, Vale LR, Lehnert KW: *Phys Rev Lett* 2011, **106**:220502.
21. Flurin E, Roch N, Mallet F, Devoret MH, Huard B: *Phys Rev Lett* 2012, **109**:183901.
22. Menzel EP, Di Candia R, Deppe F, Eder P, Zhong L, Ihmig M, Haeberlein M, Baust A, Hoffmann E, Ballester D, Inomata K, Yamamoto T, Nakamura Y, Solano E, Marx A, Gross R: *Phys Rev Lett* 2012, **109**:250502.
23. Vijay R, Macklin C, Slichter DH, Weber SJ, Murch KW, Naik R, Korotkov AN, Siddiqi I: *Nature* 2012, **490**:77.
24. Johnson JE, Macklin C, Slichter DH, Vijay R, Weingarten EB, Clarke J, Siddiqi I: *Phys Rev Lett* 2012, **109**:050506.
25. Ristè D, van Leeuwen JG, Ku H-S, Lehnert KW, DiCarlo L: *Phys Rev Lett* 2012, **109**:050507.
26. Campagne-Ibarcq P, Flurin E, Roch N, Darson D, Morfin P, Mirrahimi M, Devoret MH, Mallet F, Huard B: *Phys Rev X* 2013, **3**:021008.
27. Teufel JD, Donner T, Li D, Harlow JW, Allman MS, Cicak K, Sirois AJ, Whittaker JD, Lehnert KW, Simmonds RW: *Nature* 2011, **475**:359.
28. Eichler C, Bozyigit D, Wallraff A: *Phys Rev A* 2012, **86**:032106.
29. Eichler C, Lang C, Fink JM, Govenius J, Filipp S, Wallraff A: *Phys Rev Lett* 2012, **109**:240501.
30. Steffen L, Salathe Y, Oppliger M, Kurpiers P, Baur M, Lang C, Eichler C, Puebla-Hellmann G, Fedorov A, Wallraff A: *Nature* 2013, **500**:319.
31. Louisell WH: *Coupled Mode and Parametric Electronics*. New York: John Wiley; 1960.
32. Slusher RE, Hollberg LW, Yurke B, Mertz JC, Valley JF: *Phys Rev Lett* 1985, **55**:2409.
33. Burnham DC, Weinberg DL: *Phys Rev Lett* 1970, **25**:84.
34. Bergeal N, Vijay R, Manucharyan VE, Siddiqi I, Schoelkopf RJ, Girvin SM, Devoret MH: *Nat Phys* 2010, **6**:296.
35. Bergeal N, Schackert F, Frunzio L, Devoret MH: *Phys Rev Lett* 2012, **108**:123902.
36. Roch N, Flurin E, Nguyen F, Morfin P, Campagne-Ibarcq P, Devoret MH, Huard B: *Phys Rev Lett* 2012, **108**:147701.
37. Ho Eom B, Day PK, LeDuc HG, Zmuidzinas J: *Nat Phys* 2012, **8**:623.
38. Yaakobi O, Friedland L, Macklin C, Siddiqi I: *Phys Rev B* 2013, **87**:144301.
39. Wilson CM, Johansson G, Pourkabirian A, Simoen M, Johansson JR, Duty T, Nori F, Delsing P: *Nature* 2011, **479**:376.
40. Wustmann W, Shumeiko V: *Phys Rev B* 2013, **87**:184501.
41. Dykman M, Krivoglaz M: *Phys A, Stat Mech Appl* 1980, **104**:480.
42. Marthaler M, Dykman MI: *Phys Rev A* 2006, **73**:042108.
43. Siddiqi I, Vijay R, Pierre F, Wilson CM, Metcalfe M, Rigetti C, Frunzio L, Devoret MH: *Phys Rev Lett* 2004, **93**:207002.
44. Vijay R, Devoret MH, Siddiqi I: *Rev Sci Instrum* 2009, **80**:111101.
45. Mallet F, Ong FR, Palacios-Laloy A, Nguyen F, Bertet P, Vion D, Esteve D: *Nat Phys* 2009, **5**:791.
46. Gardiner CW, Collett MJ: *Phys Rev A* 1985, **31**:3761.

47. Walls DF, Milburn GJ: *Quantum Optics*. Berlin: Springer; 1994.
48. Göppl M, Fragner A, Baur M, Bianchetti R, Filipp S, Fink JM, Leek PJ, Puebla G, Steffen L, Wallraff A: *J Appl Phys* 2008, **104**:113904.
49. Braunstein SL, van Loock P: *Rev Mod Phys* 2005, **77**:513.
50. Caves CM: *Phys Rev D* 1982, **26**:1817.
51. Wallquist M, Shumeiko VS, Wendin G: *Phys Rev B* 2006, **74**:224506.
52. Bourassa J, Beaudoin F, Gambetta JM, Blais A: *Phys Rev A* 2012, **86**:013814.
53. Girvin SM: **Lectures delivered at Ecole d'Eté Les Houches**. To be published by Oxford University Press, 2011.
54. Imoto N, Haus HA, Yamamoto Y: *Phys Rev A* 1985, **32**:2287.
55. Sanders BC, Milburn GJ: *Phys Rev A* 1989, **39**:694.
56. Santamore DH, Goan H-S, Milburn GJ, Roukes ML: *Phys Rev A* 2004, **70**:052105.
57. Buks E, Yurke B: *Phys Rev A* 2006, **73**:023815.
58. Helmer F, Mariantoni M, Solano E, Marquardt F: *Phys Rev A* 2009, **79**:052115.
59. Johnson BR, Reed MD, Houck AA, Schuster DI, Bishop LS, Ginossar E, Gambetta JM, DiCarlo L, Frunzio L, Girvin SM, Schoelkopf RJ: *Nat Phys* 2010, **6**:663.
60. Suchoi O, Abdo B, Segev E, Shtempluck O, Blencowe MP, Buks E: *Phys Rev B* 2010, **81**:174525.
61. Kirchmair G, Vlastakis B, Leghtas Z, Nigg SE, Paik H, Ginossar E, Mirrahimi M, Frunzio L, Girvin SM, Schoelkopf RJ: *Nature* 2013, **495**:205.
62. Koch J, Yu TM, Gambetta J, Houck AA, Schuster DI, Majer J, Blais A, Devoret MH, Girvin SM, Schoelkopf RJ: *Phys Rev A* 2007, **76**:042319.
63. Ferguson DG, Houck AA, Koch J: *Phys Rev X* 2013, **3**:011003.

2

Gain, noise and intermodulation in a nonlinear superconducting resonator

Erik A Tholén[1], Adem Ergül[2], David Schaeffer[2] and David B Haviland[2*]

*Correspondence: haviland@kth.se
[2]Nanostructure Physics, Royal Institute of Technology (KTH), Stockholm, Sweden
Full list of author information is available at the end of the article

Abstract

A superconducting microwave resonator is modified with several weak links to make it nonlinear and operated as a phase-insensitive microwave amplifier. Signal gain is demonstrated by intermodulation with a strong pump. The gain is sharply frequency dependent, and we demonstrate phase dependence by examining correlations between the signal and one idler which is a 3rd order intermodulation product of the pump and signal tones. A calibration procedure is described which is based on measurement of both thermal and quantum noise, revealing that the following HEMT amplifier adds noise at 15 times the quantum limit. When operated as a phase-insensitive amplifier the nonlinear resonator added noise at 2.5 times the quantum limit. Significant power is found at intermodulation products beyond 3rd order, which may be responsible for the inability to reach the quantum limit.

PACS Codes: 74.78.-w; 42.65.Yj; 85.25.Cp

Keywords: intermodulation; parametric amplifier; quantum noise

1 Introduction

Minimal dissipation and a designable nonlinearity are the important attributes of superconductors which are enabling the development of quantum electrodynamic circuits. These attributes are manifested in distributed microwave structures with eigenmodes that are suitably modified by tunnel junctions or weak links to make them nonlinear. Such circuits find use in: single photon detection [1], generating [2] and routing [3] single microwave photons, detecting the state of individual quantum bits [4], and even as the qubit itself [5]. An important general purpose application is microwave signal amplification at the quantum limit of added noise [6]. Here we present measurements of gain, noise and intermodulation in a superconducting nonlinear resonator operated as phase-insensitive signal amplifier. Good agreement is achieved between measurement and a theory that includes one parameter describing a cubic nonlinear term. However, we observe strong response at high-order intermodulation frequencies which is not included in the theory. We believe that power lost to high-order intermodulation is responsible for the inability of the amplifier to squeeze noise at the quantum limit.

2 Amplification at the quantum limit

A fundamental theorem of signal amplification states that any narrow-band linear amplifier providing significant power gain to a signal of arbitrary phase applied to its input, must add noise at its output, where the added noise power per unit bandwidth, or added noise energy, is at least one half the zero-point fluctuation energy at the signal frequency [7].

$$E_N^{\text{added}} \geq \frac{1}{2} h f_s \tag{1}$$

It is however possible to create amplifiers that are not subject to this minimum added noise theorem. The phase-sensitive parametric amplifier operates with a pump that periodically modulates a parameter of a linear system. Signals which drive the system at half the pump frequency are amplified when they are in phase with the pump, whereas quadrature signals are de-amplified. This squeezing of a signal results in overall power gain, where power is transferred from the pump to the signal, and it is in principle a reversible process that can be performed with no added noise. Squeezing at the quantum limit produces a squeezed vacuum state, as recently demonstrated [8].

The utility of a phase-sensitive amplifier is however limited to applications where one has prior knowledge of the phase of the input signal, which is often not the case. A phase-insensitive amplifier is more useful as a general purpose measurement tool and there is growing interest in phase-insensitive amplification with noise performance at the quantum limit [9–15]. In its simplest form, phase-insensitive amplification is realized when a nonlinear system mixes two drive tones, a strong pump tone at frequency f_p and a much weaker signal tone at frequency f_s, to generate an idler tone at $f_i = f_p - f_s$. This mixing can provide gain for arbitrary phase of the signal and squeezing becomes apparent in the correlation between the externally applied signal tone and the internally generated idler tone, which are both locked to the same pump. This type of squeezing is also in principle a reversible amplification of the signal which can be undone if both the signal and idler are applied to another identical nonlinear resonator with an appropriately phase-shifted pump. The reversibility of the process implies that it can be performed without added noise.

The idler in this most simple case is generated by intermodulation of the applied signal and pump tones and it can be understood as resulting from a quadratic nonlinearity. Consider a standing wave in a transmission line with a current-dependent inductance $L(i)$. The flux $\Phi(t)$, which is defined in terms of the potential at one end of the line, $V = \frac{d\Phi}{dt}$, obeys a nonlinear equation of motion.

$$\frac{1}{(2\pi f_0)^2}\frac{d^2\Phi}{dt^2} + \frac{1}{2\pi f_0 Q}\frac{d\Phi}{dt} + \Phi + \sum_{n=2}^{N} g_n \Phi^n = \Phi_{\text{drive}}(t) \tag{2}$$

The coefficients g_n describe an arbitrary conservative nonlinearity of degree N and f_0 and Q are the characteristic frequency and quality factor of the linear resonator. To first order in the nonlinear perturbation, a drive consisting of two tones will generate a correction to the linear response of the resonator containing terms of the form,

$$\sim g_n\left[A_s e^{i2\pi f_s t} + A_p e^{i2\pi f_p t}\right]^n \tag{3}$$

Expanding this binomial, we see that the quadratic nonlinearity generates terms $\sim g_2 \times e^{i2\pi(f_p \pm f_s)t}$ which contains the idler discussed above.

Generally, to any order in a perturbative expansion (so long as it converges) and for a nonlinearity of arbitrary degree, response to a two-tone drive is generated only at frequencies which are intermodulation products of the drive tones,

$$f_{\mathrm{IMP}} = n_s f_s + n_p f_p \tag{4}$$

where n_s and n_p are integers. The order of the intermodulation product is given by $|n_s| + |n_p|$. In nonlinear optics second-order intermodulation is referred to as three-wave mixing, describing the interaction of three electromagnetic waves travelling through a medium having a χ_2 or quadratic nonlinearity. This type of nonlinearity occurs in crystals which lack a point of inversion symmetry [16]. For superconducting resonators the Josephson junction or weak link nonlinearity has a current-flux relation which is an odd function, $i(+\Phi) = -i(-\Phi)$, leading to a nonlinear oscillator equation with only odd powers of Φ, *i.e.* Eq. (2) with $g_{\mathrm{even}} = 0$. Therefore second order intermodulation is typically absent in passive Josephson devices. However, three-wave mixing or second order intermodulation can be realized in actively driven, or parametric Josephson devices, which are non-reciprocal in nature [17].

In the nonlinear resonator studied here we realized phase-insensitive amplification by third-order intermodulation of the signal and pump (4 wave mixing) where the idler is at frequency $f_i = 2f_p - f_s$. This intermodulation product can be generated by a nonlinear oscillator with a g_3 coefficient only, known as the Duffing oscillator or Kerr nonlinearity. A detailed theory of the cavity parametric amplifier perturbed by a Kerr nonlineary has been worked out by Yurke and Buks [9] and extended to the two port system measured here [18]. When comparing our measurements to this theory we could produce excellent agreement between the theory and experiment by adjusting the parameters of the theory [19, 20]. Here we extend this analysis to measurements of noise. We also present measurements of higher order intermodulation products forming a frequency comb of response near resonance. These products are typically not looked for in experiments on parametric amplification, but we found significant power at these frequencies, especially when measuring de-amplification of a signal tone. We speculate that this higher order intermodulation is responsible for not being able to efficiently squeeze with our amplifier.

3 Experimental details and calibration

The nonlinear resonator is a thin-film Nb coplanar waveguide (CPW) fabricated with a photo lithography process by Star Cryoelectronics [21]. Input and output ports were coupled to the resonator by interdigitating the center line of the resonator with tapered CPW feed lines. We attempted to make asymmetric coupling with smaller interdigitating length at the input, but we found that both ports had nearly the same coupling, indicating the coupling is primarily due to fringing fields. The distributed resonator has a series of eigenmodes when its length is close to an integer multiples of the electromagnetic half-wavelength $L \simeq \frac{n\lambda}{2}$. Here we present measurements of the response in the vicinity of the fundamental eigenmode ($n = 1$) with $f_0 = 7.60845$ GHz. A nonlinear inductance was created by milling out a series of 10 weak links with a focused ion beam, each approximately 30 nm × 30 nm in cross section, placed close together in the middle of the resonator, where the oscillating suppercurrent has an anti-node. The sample is flip-chip mounted on to a PC board which transforms the coplanar geometry of the chip to coax input and output connectors. For further details and images see Ref. [20].

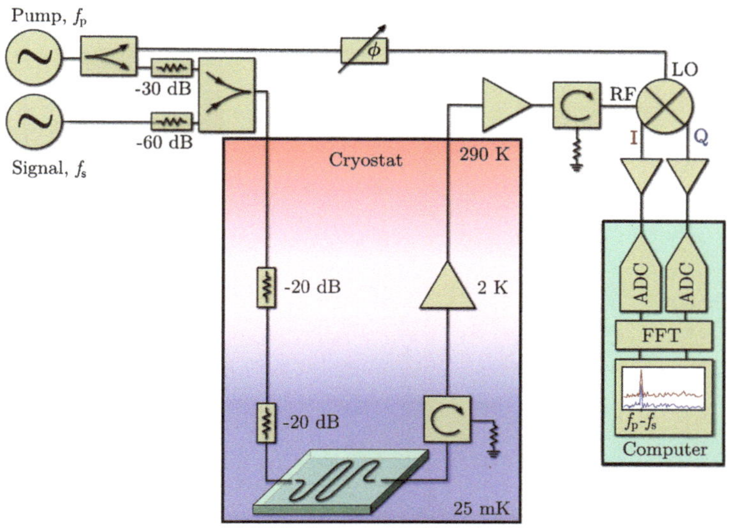

Figure 1 Schematic diagram of the microwave setup used to realize phase-insensitive amplification and homodyne detection.

A schematic diagram of the measurement setup is shown in Figure 1. The sample is mounted in a dilution refrigerator with base temperature 25 mK. The pump and signal tones are combined at room temperature and applied via a coax transmission line to the resonator input. Two stages of attenuation at low temperatures are used on the input side to bring the thermal radiation at the input into thermal equilibrium at the base temperature of the cryostat. Signals emerging from the output are sent through a circulator to a cryogenic pre-amplifier mounted at the 2 K stage of the cryostat. The circulator, with the third terminal connected to a 50 Ω termination acts as an isolator, protecting the resonator from the back-action noise of the cryogenic preamplifier. The signal at the output of the preamplifier is amplified again by a room-temperature microwave amplifier followed by another isolator, and then an IQ mixer for homodyne detection using a phase-shifted copy of the pump signal as the local oscillator. In this arrangement, both the signal and idler tones will be coherently down-converted to the same frequency $\Delta f = f_s - f_p$. A variable delay shifts the phase of the pump signal, allowing one to probe the phase dependence of the signal and idler gains, that is the squeezing.

The cryogenic HEMT amplifier is the dominant source of noise in our setup. We tested amplifiers from Miteq, Amplitech, and Low Noise Factory [22], the later being far superior to the others with $\simeq$ 40 dB of gain in the band 3–8 GHz with a DC power consumption of only 6 mW. The back-action noise from at input stage of the HEMT amplifier is isolated from the resonator and brought in to thermal equilibrium at the base temperature of the cryostat by the 50 Ω termination at one port of the cold circulator. If the 50 Ω termination on the circulator is in equilibrium with the cryostat's calibrated restive thermometer, the noise emitted from this 50 Ω source can used to calibrate the measurement system. Noise from this 50 Ω source is measured in a very narrow frequency band (effective bandwidth 0.5 kHz) centred at the frequency f_c. The measurement band does not overlap with the resonance of the cavity so the noise is reflected at the output of the resonator, re-enters the circulator, and is transported to input of the HEMT amplifier. Thus the total noise power per unit bandwidth measured at the output of the mixer comprises the noise from

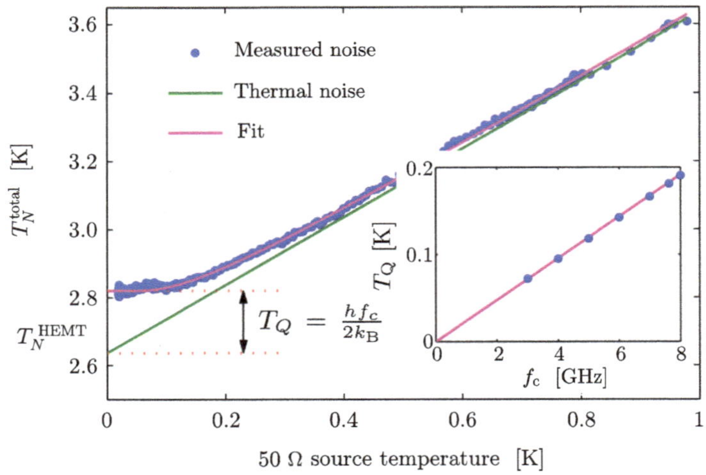

Figure 2 Microwave noise power is calibrated in a narrow band centered at the frequency $f_c = 7.62$ GHz, by varying the temperature of the 50 Ω termination on the cold circulator.
A one-parameter fit to Eq. (5), adjusting only the overall gain factor, allows for determination of the HEMT amplifiers noise temperature $T_N^{\text{HEMT}} = 2.62$ K and observation of the quantum noise $T_Q = 0.18$ K. The data points in the inset are the quantum noise at different center frequencies in the band of the HEMT amplifier, and the straight line is the expected behavior $T_Q = \frac{hf_c}{2k_B}$.

this 50 Ω source, plus the added noise of the HEMT amplifier.

$$E_N^{\text{total}}(T) = E_N^{\text{HEMT}} + \frac{1}{2}hf_c \coth\left(\frac{hf_c}{2k_B T}\right), \tag{5}$$

where we have used the quantum expression for the fluctuations from the 50 Ω source. This simple expression assumes that all inputs and outputs are matched to 50 Ω, which is the case.

We measured the noise as a function of temperature and fit the theory Eq. (5) to the measured data as shown in Figure 2, where the noise is expressed as an effective temperature $T_N = \frac{E_N}{k_B}$. The least-squared fit was obtained by adjusting only one parameter, the overall scale factor on the vertical axis. This scale factor contains the gains and attenuations from all of the various components in our measurement chain, which would not be easy to calibrate independently. At high temperatures $k_B T > hf_c$ the data approach the Johnson-Nyquist thermal noise from the 50 Ω source, which the fit determines as the straight green line in Figure 2. The $T = 0$ intersection of this line with the noise axis gives the added noise of the HEMT amplifier, $T_N^{\text{HEMT}} = 2.62$ K. The difference between the amplifiers added noise and the measured noise extrapolated to $T = 0$, is the quantum noise, T_Q. We performed this measurement at several center frequencies f_c to verify that the quantum noise determined by this fitting procedure behaved as expected $T_Q = \frac{hf_c}{2k_B}$ (see inset of Figure 2).

Thus it is possible to use noise to calibrate the microwave measurement chain against the cryostat's calibrated restive thermometer. We are also able to directly measure the quantum noise with the HEMT amplifier, which adds only $T_N^{\text{HEMT}} \simeq 15T_Q$ to the total noise. In contrast to two-point calibrations involving low temperature switches [23], or exotic hot quasiparticale noise sources [12], this straight-forward calibration method is rather simple to perform. Our confidence in the accuracy of this method was bolstered

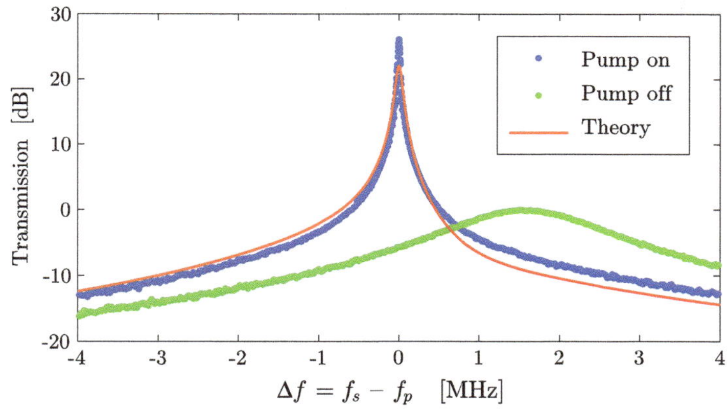

Figure 3 Frequency response of the parametric amplifier with the pump turned off and the pump turned on. The pump frequency is f_p = 7.6069 GHz which is 1.55 MHz below the resonant frequency of the resonator in the linear regime. The pump power is –83.8 dBm at the input of the sample box. The red line is a fit to the theory made by adjusting the strength of the Kerr nonlinearity.

when we observed that problems with the 50 Ω source not being in thermal equilibrium with the cryostats thermometer were clearly evident in the poor quality of the fit to Eq. (5) to the data [20]. This calibration method has been use for Planck spectroscopy of thermal microwave states [24].

4 Measurements and analysis

Signal amplification by the nonlinear resonator is demonstrated in Figure 3 which shows a measurement of the signal transmission when the pump power and frequency are fixed at $P_p = -83.8$ dBm and $f_p = 7.6069$ GHz, very close to bifurcation point of the nonlinear resonator. In this measurement the signal source and mixer are replaced by the output and input ports of a network analyser respectively, and the signal power is fixed to $P_s =$ -133.7 dBm while its frequency is swept. The measurement was calibrated so that 0 dB corresponds to unity transmission between the input and output connectors of the board on which the chip is mounted. The transmission curve at this signal power (green data in Figure 3) in the absence of the pump is also shown in Figure 3 which is well described by the Lorentzian lineshape of a driven, damped harmonic oscillator with resonant frequency $f_0 = 7.60845$ GHz and width $\gamma = 900$ kHz ($Q = \frac{2\pi f_0}{\gamma} = 8{,}454$).

When the pump is turned on we observe gain in a rather narrow band which is sharply peaked at zero detuning between the pump and signal, with a maximum gain of 28 dB (blue data points in Figure 3). Using the values of f_0 and Q determined from the transmission curve in the linear regime, and the value of a cubic nonlinear term determined by fitting transmission curves at many different signal powers, when the resonator is in the nonlinear regime (data not shown) we could calculate the theoretically expected frequency dependent gain [9, 18] (red line in Figure 3). Good correspondence with the measured signal gain is seen over many times the resonator bandwidth. We found that including the nonlinear damping parameter in the theory did not improve the quality of the fit.

In a previous publication [19] we reported a detailed study of the pump power and phase dependence of the signal and idler gain, where excellent agreement with theory [9, 18] was found. We were not able to achieve equal magnitude of the amplification and deamplification of a signal, which we understood as most likely being the result of too weak coupling

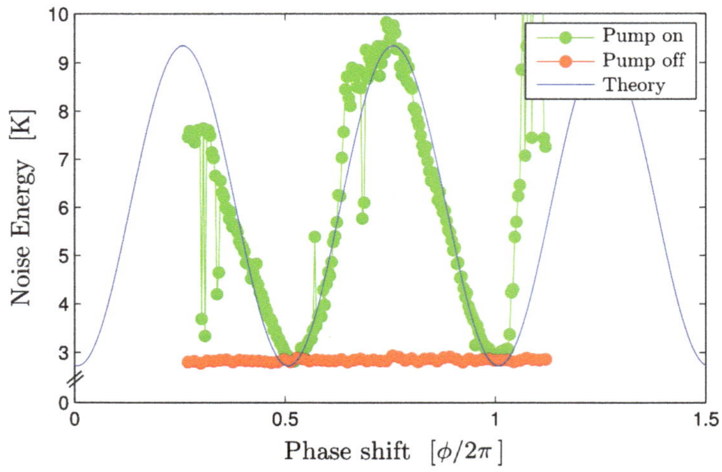

Figure 4 Phase dependence of the noise power in a 30 kHz bandwidth, with the pump on (green) and pump off (red). The solid line is the theoretical expectation taking in to account the frequency dependence of the gain. Instabilities in the amplifier cause jumps in the measurements.

of the input and output ports. Here we report noise and intermodulation measurements which were made in a subsequent cool-down of the sample. The thermal cycling of the sample caused a slight shift of the bifurcation point and for the following measurements the working point of the pump was $f_p = 7.6029$ GHz and $P_p = -84$ dBm.

With the signal turned off we measured only the noise in a 30 kHz bandwidth as the phase shifter is varied. Note that here we are measuring noise close to resonance, so the noise measured is inside the cavity and it is effected by the noise entering from both the input and output ports. In Figure 4 we see that noise at the output of the mixer is clearly phase-dependent, verifying that intermodulation with the pump is causing a correlation of the frequency components of a random signal that are equally spaced from pump, $f_p \pm \Delta f$. We see that when the phase shifter is adjusted for minimum noise (maximum de-amplification) we recover the independently measured value of the HEMT amplifiers added noise 2.6 K. This value is the same as the noise measured when the pump is off (red data in Figure 4). When the phase shifter is adjusted for maximum gain we find that the nonlinear cavity amplifier adds and additional 6.8 K of noise to the total noise power. Taking in to account the frequency dependence and phase dependence of the gain, we can estimate the phase dependence of the total measured noise as,

$$T_N^{\text{total}}(\phi) = T_N^{\text{HEMT}} + \int_0^{30\text{ kHz}} G(\phi,f)\ T_Q + T_N^{\text{NLO}}\ df \tag{6}$$

where T_N^{NLO} is the temperature of the added noise of the nonlinear oscillator. This theoretical curve is shown in Figure 4 with $T_N^{\text{NLO}} = 0.44$ K. The jumps observed in the measured data are due to instabilities of the nonlinear oscillator which is biased close to the bifurcation point.

In another measurement we applied a signal with fixed frequency with $\Delta f = f_s - f_p = 10$ kHz and measured the power spectrum at the output of the mixer over a much wider frequency band corresponding to $\sim 10\Delta f$. This measurement was made by Fast Fourier Transform (FFT) of the output of the mixer with a 1 kHz resolution bandwidth, where

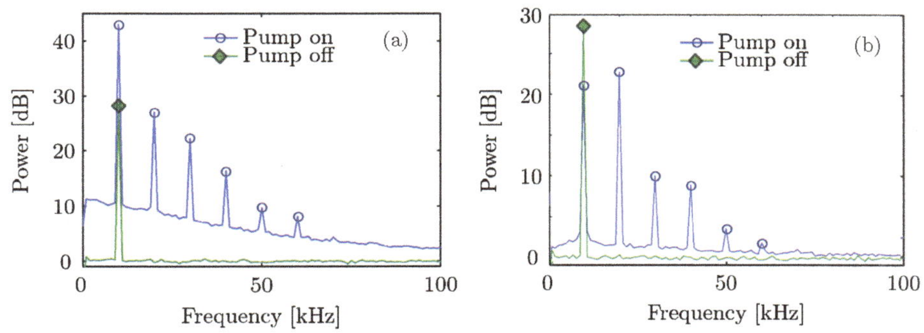

Figure 5 Power spectrum measured at the output of the mixer when a signal is applied at $\Delta f = f_s - fp = 10$ kHz. Measurements are shown when the pump is on (blue) and when the pump is off (green), for two cases: (**a**) the phase shifter is tuned for maximum amplification, and (**b**) the phase shifter is tuned for maximum de-amplification. The measurement reveals both signal and noise gain, and the additional peaks at integer multiples of Δf are due to higher-order intermodulation products of the signal and pump.

1,000 power spectra were averaged. To avoid Fourier leakage between different components in the spectrum, we synchronized the microwave generators and the sampling clock on the ADC, and we chose both pump and signal frequencies to be integer multiples of the sampling frequency.

Figure 5 shows the averaged power spectra when the phase shifter is adjusted for maximum amplification (Figure 5(a)) and maximum de-amplification (Figure 5(b)). Measurements with the pump both on and off are shown in each figure. For these measurements, the power is normalized so that 0 dB corresponds to the pump off noise level. Comparing the height of the signal peak at 10 kHz for the pump on and pump off case, we find 14 dB of amplification and 7 dB of de-amplification of the signal tone. Also evident is a large number of peaks at integer multiples of Δf, corresponding to higher order intermodulation products of the signal and pump. Between the peaks we can see the noise level, which reveals the frequency dependence of the amplifiers gain. In Figure 5(a) we find that the signal-to-noise ratio (SNR) at 10 kHz improves by a factor of 5.9 when the pump is on. Since the pump off noise is measured to be that of the HEMT amplifier, we can determine the added noise of the nonlinear oscillator to be $T_N^{\mathrm{NLO}} = T_N^{\mathrm{HEMT}}/5.9 = 0.44$ K, consistent with the previous measurement.

5 Discussion and conclusions

The peaks in Figure 5(a) and Figure 5(b) at integer multiples of Δf correspond to odd order intermodulation products (*e.g.* $2f_p \pm f_s, f_p \pm 2f_s$, $3f_p \pm 2f_s$, $2f_p \pm 3f_s$ *etc.*). It is difficult to pin-point the source of these higher order intermodulation products: They may be generated in the nonlinear resonator as a result of expansion coefficients g_n with odd $n > 3$ in Eq. (3), or from the g_3 term in higher order perturbation theory. Nonlinearity of the mixer may also generate these higher order intermodulation products. In any case, participation of both signal and pump is required, and power is taken from the signal band in their creation. The amplitude of the higher order intermodulation products is quite large, with the 5th order product exceeding the signal amplitude at maximum de-amplification and intermodulation up 15th order can be resolved above the noise. These higher order products are clearly diminishing our ability to measure gain and deamplification of the signal, and we speculate that they also inhibiting the measurement of noise squeezing at the quantum limit.

One way in which one might control this loss of power to higher order intermodulation is to filter the higher-order IMPs by designing a system where the linear response is negligible at undesirable intermodulation products. Another strategy would be to design a drive consisting of a frequency comb with the amplitude and phase of each component appropriately tuned to cancel the undesired response. In fact, these IMPs can be measured in a phase-coherent way and used for accurate reconstruction of the actual nonlinearity [25]. Development is under way to make system which will allow for driving and measuring response with arbitrary, phase-coherent frequency combs [26].

In this paper we measured intermodulation response near one resonance and analyzed the data within the context of a basic theory of a nonlinear system with only two degrees of freedom, *i.e.* one oscillator equation describing one eigenmode of the transmission line [9]. A distributed system actually has many standing waves and the ideal transmission line has resonances at integer multiples of the fundamental resonant frequency, nf_0. Deviations from this ideal behaviour occur due to finite coupling to input and output ports, as well as the frequency shift due to the linear part of the weak-link inductance. Because we observe IMPs of 15th order near the fundamental resonance, one really should examine the first 15 eigenmodes to determine if they are excited by harmonics or intermodulation products of the signal and pump. If so, we expect additional loss of signal power, diminishing the performance of the amplifier. Since the frequencies of these higher eigenmodes are out of the band of our HEMT amplifier, we are unable to measure their response and typically they are not measured in other experiments with cavity parametric amplifiers. Note however, that we do not expect excitation near even multiples of f_0 because $g_{\text{even}} = 0$ for the nonlinear oscillator considered here. The excitation of higher eigenmodes by harmonics and intermodulation might be avoided by designing lumped-element resonators with a more complicated mode structure [27], but it is a potential problem in designs which use separate transmission line resonators to store the signal and idler [12]. Filtering response at higher harmonics and higher order intermodulation products is generally more difficult to achieve when the quality factor of the eigenmodes is decreased to achieve larger bandwidth of the amplifier.

In conclusion, we have realized phase-insensitive amplification of a microwave signal at 7.6 GHz using a nonlinear superconducting resonator pumped close to bifurcation. A peak gain of 28 dB was realized, but it was strongly frequency-dependent in a narrow band with a 3 dB bandwidth of only 1 MHz. We used a calibrated resistive thermometer to calibrate the microwave noise measurement and with the following HEMT amplifier we measured the standard quantum limit on noise, $T_Q = \frac{hf}{2k_B}$, where the noise temperature of the HEMT amplifier was only $T_N^{\text{HEMT}} = 2.6$ K, corresponding to $15T_Q$. When operated as an amplifier the nonlinear resonator achieved 14 dB of signal gain in a 10 kHz bandwidth with noise temperature $T_N^{\text{NLO}} = 0.44$ K, corresponding to $2.5T_Q$. We were not able to demonstrate squeezing of vacuum fluctuations by examining correlations between the signal and 3rd order idler. Correlations involving higher order idlers were not examined.

Competing interests

The authors declare that they have no competing interests.

Authors' contributions

EAT and DBH conceived the experiment. EAT and AE performed the experiments and analysis. All authors took part in preparation of the manuscript.

Author details

[1]Intermodulation Products AB, Solna, Sweden. [2]Nanostructure Physics, Royal Institute of Technology (KTH), Stockholm, Sweden.

Acknowledgements

We acknowledge stimulating discussions with Steve Girvin, Jack Lidmar, Hans Hansson, Per Delisng and Chris Wilson. This work was funded by the EU project SCOPE, FET-Open grant number 218783, the Swedish Research Council VR, the Kunt and Allice Wallenberg Foundation and the Olle Engkvist Foundation.

References

1. Gao J, Vissers MR, Sandberg MO, da Silva FCS, Nam SW, Pappas DP, Wisbey DS, Langman EC, Meeker SR, Mazin BA, Leduc HG, Zmuidzinas J, Irwin KD: *Appl Phys Lett* 2012, **101**(14):142602.
2. Houck AA, Schuster DI, Gambetta JM, Schreier JA, Johnson BR, Chow JM, Frunzio L, Majer J, Devoret MH, Girvin SM, Schoelkopf RJ: *Nature* 2007, **449**(7160):328.
3. Hoi IC, Wilson CM, Johansson G, Palomaki T, Peropadre B, Delsing P: *Phys Rev Lett* 2011, **107**:073601.
4. Boissonneault M, Doherty AC, Ong FR, Bertet P, Vion D, Esteve D, Blais A: *Phys Rev A* 2012, **85**:022305.
5. Paik H, Schuster DI, Bishop LS, Kirchmair G, Catelani G, Sears AP, Johnson BR, Reagor MJ, Frunzio L, Glazman LI, Girvin SM, Devoret MH, Schoelkopf RJ: *Phys Rev Lett* 2011, **107**:240501.
6. Clerk AA, Devoret MH, Girvin SM, Marquardt F, Schoelkopf RJ: *Rev Mod Phys* 2010, **82**:1155.
7. Caves CM: *Phys Rev D* 1982, **26**(8):1817.
8. Mallet F, Castellanos-Beltran MA, Ku HS, Glancy S, Knill E, Irwin KD, Hilton GC, Vale LR, Lehnert KW: *Phys Rev Lett* 2011, **106**:220502.
9. Yurke B, Buks E: *J Lightwave Technol* 2006, **24**(12):5054.
10. Tholén EA, Ergül A, Doherty EM, Weber FM, Grégis F, Haviland DB: *Appl Phys Lett* 2007, **90**(25):253509.
11. Abdo B, Suchoi O, Segev E, Shtempluck O, Blencowe M, Buks E: *Europhys Lett* 2009, **85**(6):68001.
12. Bergeal N, Schackert F, Metcalfe M, Vijay R, Manucharyan VE, Frunzio L, Prober DE, Schoelkopf RJ, Girvin SM, Devoret MH: *Nature* 2010, **465**(7294):64.
13. Abdo B, Kamal A, Devoret M: *Phys Rev B* 2013, **87**:014508.
14. Yaakobi O, Friedland L, Macklin C, Siddiqi I: *Phys Rev B* 2013, **87**:144301.
15. Wustmann W, Shumeiko V: *Phys Rev B* 2013, **87**:184501.
16. Franken PA, Hill AE, Peters CW, Weinreich G: *Phys Rev Lett* 1961, **7**:118.
17. Kamal A, Clarke J, Devoret MH: *Phys Rev B* 2012, **86**:144510.
18. Stannigel K: **Parametric amplification in superconducting resonators.** *Master's thesis*. KTH; 2007.
19. Tholén EA, Ergül A, Stannigel K, Hutter C, Haviland DB: *Phys Scr* 2009, **T137**:014019.
20. Tholén E: **Intermodulation in microresonators: for microwave amplification and nanoscale surface analysis.** *Ph.D. thesis*. KTH, Nanostructure Physics; 2009. http://kth.diva-portal.org/smash/get/diva2:277877/FULLTEXT01.pdf.
21. STAR Cryoelectronics, Santa Fe, New Mexico. http://www.starcryo.com/.
22. Low Noise Factory, Göteborg, Sweden. http://www.lownoisefactory.com/.
23. Castellanos-Beltran MA, Irwin KD, Hilton GC, Vale LR, Lehnert KW: *Nat Phys* 2008, **4**(12):929.
24. Mariantoni M, Menzel EP, Deppe F, Araque Caballero MA, Baust A, Niemczyk T, Hoffmann E, Solano E, Marx A, Gross R: *Phys Rev Lett* 2010, **105**:133601.
25. Platz D, Forchheimer D, Tholén EA, Haviland DB: *Nanotechnology* 2012, **23**(26):265705.
26. Tholén EA, Platz D, Forchheimer D, Tholén MO, Hutter C, Haviland DB: *Rev Sci Instrum* 2011, **82**:026109. arXiv:1008.2722v1.
27. Levenson-Falk EM, Vijay R, Siddiqi I: *Appl Phys Lett* 2011, **98**:123115.

3

Creation of two-dimensional Coulomb crystals of ions in oblate Paul traps for quantum simulations

Bryce Yoshimura[1*], Marybeth Stork[2], Danilo Dadic[3], Wesley C Campbell[3] and James K Freericks[1]

*Correspondence: bty4@georgetown.edu
[1]Department of Physics, Georgetown University, 37th and O St. NW, Washington, DC 20007, USA
Full list of author information is available at the end of the article

Abstract

We develop the theory to describe the equilibrium ion positions and phonon modes for a trapped ion quantum simulator in an oblate Paul trap that creates two-dimensional Coulomb crystals in a triangular lattice. By coupling the internal states of the ions to laser beams propagating along the symmetry axis, we study the effective Ising spin-spin interactions that are mediated via the axial phonons and are less sensitive to ion micromotion. We find that the axial mode frequencies permit the programming of Ising interactions with inverse power law spin-spin couplings that can be tuned from uniform to r^{-3} with DC voltages. Such a trap could allow for interesting new geometrical configurations for quantum simulations on moderately sized systems including frustrated magnetism on triangular lattices or Aharonov-Bohm effects on ion tunneling. The trap also incorporates periodic boundary conditions around loops which could be employed to examine time crystals.

Keywords: ion trap; quantum simulation; Ising model

1 Introduction

Using a digital computer to predict the ground state of complex many-body quantum systems, such as frustrated magnets, becomes an intractable problem when the number of spins becomes too large. The constraints on the system's size become even more severe if one is interested in the (nonequilibrium) quantum dynamics of the system. Feynman proposed the use of a quantum-mechanical simulator to efficiently solve these types of quantum problems [1]. One successful platform for simulating lattice spin systems is the trapped ion quantum simulator, which has already been used to simulate a variety of scenarios [2–13]. In one realization [14], ions are cooled in a trap to form a regular array known as a Coulomb crystal and the quantum state of each simulated spin can be encoded in the internal states of each trapped ion. Laser illumination of the entire crystal then can be used to program the simulation (spin-spin interactions, magnetic fields, etc.) via coupling to phonon modes, and readout of the internal ion states at the end of the simulation corresponds to a projective measurement of each simulated spin on the measurement basis.

To date, the largest number of spins simulated in this type of device is about 300 ions trapped in a rotating approximately-triangular lattice in a Penning trap [15]. In that experiment, a spin-dependent optical dipole force was employed to realize an Ising-type spin-spin coupling with a tunable power law behavior. Although the Penning trap can implement a transverse magnetic field to rotate the ions, the Penning trap simulator was not able to apply a time-dependent transverse magnetic field to perform certain desirable tasks such as the adiabatic state preparation of the ground state of a frustrated magnet. The reason why the transverse field has not been tried yet is that the Penning trap cannot be cooled to the ground state for the phonons, and the presence of phonons causes problems with the quantum coherence of the system. Improvements of the experiment are currently in progress which may allow for a transverse field Ising model simulation in the near future. The complexity of the Penning trap apparatus also creates a barrier to adoption and therefore does not seem to be as widespread as radio-frequency (RF) Paul traps.

Experiments in linear Paul traps have already performed a wide range of different quantum simulations within a one-dimensional linear crystal [2–9]. The linear Paul trap is a mature architecture for quantum information processing, and quantum simulations in linear chains of ions have benefited from a vast toolbox of techniques that have been developed over the years. Initially, the basic protocol [14] was illustrated in a two-ion trap [2], which was followed by a study of the effects of frustration in a three-ion trap [3]. These experiments were scaled up to larger systems first for the ferromagnetic case [4] and then for the antiferromagnetic case [5]. Effective spin Hamiltonians were also investigated using a Trotter-like stroboscopic approach [6]. As it became clear that the adiabatic state preparation protocol was difficult to achieve in these experiments, ion trap simulators turned to spectroscopic measurements of excited states [7] and global quench experiments to examine Lieb-Robinson bounds and how they change with long-range interactions (in both Ising and XY models) [8, 9].

It is therefore desirable to be able to use the demonstrated power of the Paul trap systems to extend access to the 2D physics that is native to the Penning trap systems. However, extension of this technology to higher dimensions is hampered by the fact that most ions in 2D and 3D Coulomb crystals no longer sit on the RF null. This leads to significant micromotion at the RF frequency and can couple to the control lasers through Doppler shifts if the micromotion is not perpendicular to the laser-illumination direction, leading to heating and the congestion of the spectrum by micromotion sidebands. However, the micromotion may not cause as many problems as previously thought. It has recently been shown that a robust quantum gate can be implemented even in the presence of large micromotion [16].

In an effort to utilize the desirable features of the Paul trap system to study the 2D physics, arrays of Paul traps are being pursued [17–22]. It has also been shown that effective higher-dimensional models may be programmed into a simulator with a linear crystal if sufficient control of the laser fields can be achieved [23]. In this article, we study an alternative approach to applying Paul traps to the simulation of frustrated 2D spin lattices. We consider a Paul trap with axial symmetry that forms an oblate potential, squeezing the ions into a 2D crystal. The micromotion in this case is purely radial due to symmetry, and lasers that propagate perpendicular to the crystal plane will therefore not be sensitive to Doppler shifts from micromotion. We study the parameter space of a particular model trap geometry to find the crystal structures, normal modes, and programmable spin-spin interactions for 2D triangular crystals in this trap. We find a wide parameter space for making such crystals, and an Ising spin-spin interaction with widely-tunable range, suitable for studying spin frustration and dynamics on triangular lattices with tens of ions.

In the future, we expect the simulation of larger systems to be made possible within either the Penning trap, or in linear Paul traps that can stably trap large numbers of ions. It is likely that spectroscopy of energy levels will continue to be examined, including new theoretical protocols [24]. Designing adiabatic fast-passage experiments along the lines of what needs to be done for the nearest-neighbor transverse field Ising model [25] might improve the ability to create complex quantum ground states. Motional effects of the ions are also interesting, such as tunneling studies for motion in the quantum regime [26]. It is also possible that novel ideas like time crystals [27, 28] could be tested (although designing such experiments might be extremely difficult).

The organization of the remainder of the paper is as follows: In Section 2, we introduce and model the potential energy and effective trapping pseudopotential for the oblate Paul trap. In Section 3, we determine the equilibrium positions and the normal modes of the trapped ions, with a focus on small systems and how the geometry of the system changes as ions are added in. In Section 4, we show representative numerical results for the equilibrium positions and the normal modes. We then show numerical results for the effective spin-spin interactions that can be generated by a spin-dependent optical dipole force. In Section 5, we provide our conclusions and outlook.

2 Oblate Paul trap

The quantum simulator architecture we study here is based on a Paul trap with an azimuthally symmetric trapping potential that has significantly stronger axial confinement than radial confinement, which we call an 'oblate Paul trap' since the resulting effective potential resembles an oblate ellipsoid. As we show below, this can create a Coulomb crystal of ions that is a (finite) two-dimensional triangular lattice. 2D Coulomb crystals in oblate Paul traps have been studied since the 1980's and were used, for instance, by the NIST Ion Storage Group [29] to study spectroscopy [30, 31], quantum jumps [32], laser absorption [33, 34] and cooling processes [35]. 2D crystals in oblate Paul traps have also been studied by other groups both experimentally [36, 37] and theoretically [38–44].

The particular oblate Paul trap we study has DC 'end cap' electrodes above and below a central radio-frequency (RF) ring, as depicted in Figure 1. The trap we propose uses modern microfabrication and lithography technology (manufactured by Translume, Ann Arbor, MI) to realize the DC end cap electrodes as surface features on a monolithic fused silica substrate, providing native mechanical indexing and easier optical access to the ions than discrete end cap traps. Similar to Penning traps, oblate Paul traps can be used to study frustration effects when the lattice of ions has multiple rings. However in a Penning trap, the lattice of ions is rotating and the ions are in a strong magnetic field, which can add significant complications. In an oblate Paul trap, the lattice of ions is stationary except for the micromotion of each ion about its equilibrium positions (which is confined to the plane of the crystal by symmetry) and the qubits can be held in nearly zero magnetic field, permitting the use of the $m = 0$ 'clock state' used in linear Paul trap quantum simulators [3]. For trapped ions in a crystal that is a single polygon ($N = 3, 4,$ or 5), we can study periodic boundary conditions applied to a linear chain of trapped ions in the linear Paul quantum simulators. Oblate Paul traps can also potentially be used to perform experiments that are similar to those recently exploring the Aharonov-Bohm effect [26] with more ions.

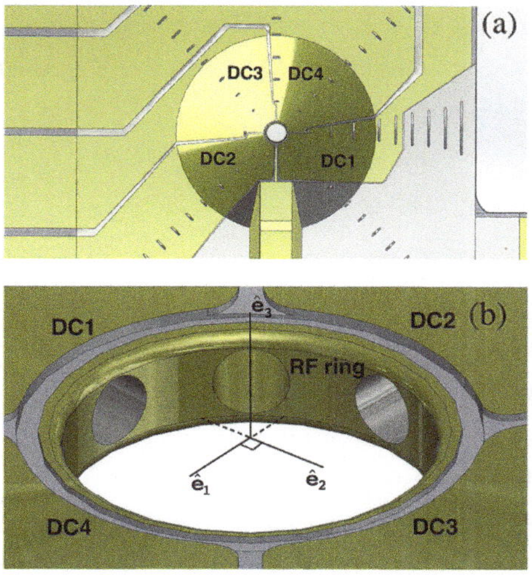

Figure 1 Top view of the proposed oblate Paul trap. The ions are trapped in the through hole in the center, which is magnified in a three-dimensional image in panel (b). The diameter and height of the RF ring electrodes is 500 μm and 140 μm, respectively. Radial optical access tunnels are visible in (b) and will contribute to the breaking of rotational symmetry, but play no other role in this analysis. Electrodes 1-4 are labeled in both panels and the RF ring is shown in panel (b). The origin is defined to be at the center of the trap. For our calculations, we hold all four electrodes on either the top or bottom face at the same potential as the segmenting is for compensation of stray fields in the experiment and plays no role in the ion crystal structure.

It is well known that Maxwell's equations forbid the possibility to use a static electric field to trap charged particles in free space through Earnshaw's theorem. However, a static electric field can create a saddle-point, which confines the charged particles in some directions and deconfines them in the other directions. A static electric field that provides a saddle-point is

$$\mathbf{E}(\tilde{x}_1, \tilde{x}_2, \tilde{x}_3) = A(\tilde{x}_1\hat{e}_1 + \tilde{x}_2\hat{e}_2 - 2\tilde{x}_3\hat{e}_3), \tag{1}$$

where A is a constant and $\hat{e}_i$ are the perpendicular unit vectors with $i = 1, 2, 3$. Using a static electric field with a saddle-point, both Penning and radio-frequency (RF) Paul traps have successfully trapped charged particles in free space by applying an additional field. In the Penning trap, one applies a strong uniform magnetic field, such that the charged particles are confined to a circular orbit via the Lorentz force, $q\mathbf{v} \times \mathbf{B}/c$. The RF Paul trap applies a time-varying voltage to its electrodes, which produce a saddle potential that oscillates sinusoidally as a function of time. This rapid change of sign allows for certain ions to be trapped because for particular charge to mass ratios, the effective focusing force is stronger than the defocussing force.

If the ions remain close to the nulls of the potential, then the micromotion of the ions is small, and it is a good approximation to describe the system via a static pseudopotential that approximates the trapping effect of the time-varying potential. We use the numerical modeling software Comsol to simulate this effective pseudopotential that arises from applying a time-varying voltage to the RF ring and additional DC voltages on the other electrodes. The effective total potential energy, $\tilde{V}(\tilde{x}_1, \tilde{x}_2, \tilde{x}_3)$, of an ion in the oblate Paul trap can be approximated by

$$\tilde{V}(\tilde{x}_1, \tilde{x}_2, \tilde{x}_3) = \psi(\tilde{x}_1, \tilde{x}_2, \tilde{x}_3) + q\phi(\tilde{x}_1, \tilde{x}_2, \tilde{x}_3), \tag{2}$$

where $\psi(\tilde{x}_1, \tilde{x}_2, \tilde{x}_3)$ is the effective pseudopotential due to the RF fields and $\phi(\tilde{x}_1, \tilde{x}_2, \tilde{x}_3)$ is the additional potential due to the DC voltage applied on the top and bottom electrodes and the RF ring. The resulting pseudopotential at a certain point in space will depend upon the RF frequency, Ω_{RF}, and the RF electric field amplitude, $E_{o,\mathrm{RF}}(\tilde{x}_1, \tilde{x}_2, \tilde{x}_3)$, at that point [45] and is given by

$$\psi(\tilde{x}_1, \tilde{x}_2, \tilde{x}_3) = \frac{q^2}{4m\Omega_{\mathrm{RF}}^2} \left|\mathbf{E}_{o,\mathrm{RF}}(\tilde{x}_1, \tilde{x}_2, \tilde{x}_3)\right|^2, \tag{3}$$

which depends on the charge, q, and the mass, m, of the particular ion being trapped. After simulating the field using Comsol, we find that the electric field amplitude from the RF field near the trap center can be approximated by

$$\mathbf{E}_{o,\mathrm{RF}} \approx -\frac{2V_{o,\mathrm{RF}}}{r_o^2}(\tilde{x}_1\hat{e}_1 + \tilde{x}_2\hat{e}_2 - 2\tilde{x}_3\hat{e}_3), \tag{4}$$

where $V_{o,\mathrm{RF}}$ is the amplitude of the RF voltage. Plugging this into Eq. (3) yields

$$\psi(\tilde{x}_1, \tilde{x}_2, \tilde{x}_3) = \frac{q^2 V_{o,\mathrm{RF}}^2}{m\Omega_{\mathrm{RF}}^2 r_o^4}(\tilde{x}_1^2 + \tilde{x}_2^2 + 4\tilde{x}_3^2), \tag{5}$$

where $r_o = 512\ \mu\mathrm{m}$ is a fitting parameter, that is determined by grounding the top and bottom electrodes and numerically modeling the square of the RF electric field amplitude, as shown in Figure 2. We calculate the DC electric field as having 3 contributions: one from the voltage applied to the RF ring, $\phi_r(\tilde{x}_1, \tilde{x}_2, \tilde{x}_3)$, one from the voltage applied to the top electrodes, $\phi_t(\tilde{x}_1, \tilde{x}_2, \tilde{x}_3)$ and one from the bottom electrodes, $\phi_b(\tilde{x}_1, \tilde{x}_2, \tilde{x}_3)$. The DC voltage on the RF ring, $\phi_r(\tilde{x}_1, \tilde{x}_2, \tilde{x}_3)$, is given by

$$\phi_r(\tilde{x}_1, \tilde{x}_2, \tilde{x}_3) = \frac{V_r}{r_o^2}(\tilde{x}_1^2 + \tilde{x}_2^2 - 2\tilde{x}_3^2), \tag{6}$$

where V_r is the DC voltage on the ring.

We numerically model the electrostatic potential due to the DC voltage applied to the either the top or bottom electrodes, as shown in Figure 3. We find that near the trap center, the numerical results for the electrostatic potential produced by a voltage of $V_{t,b}$ the top or bottom electrodes is reasonably modeled by the polynomial

$$\phi_{t,b}(\tilde{x}_1, \tilde{x}_2, \tilde{x}_3) = V_{t,b}\left(\frac{\tilde{x}_3^2}{a^2} + \frac{\tilde{x}_3}{b_{t,b}} - \frac{\tilde{x}_1^2 + \tilde{x}_2^2}{c^2} + d\right), \tag{7}$$

with fitting parameters $a = 524\ \mu\mathrm{m}$, $b_t = 761\ \mu\mathrm{m}$, $b_b = -761\ \mu\mathrm{m}$, $c = 704\ \mu\mathrm{m}$, $d = 0.812$. Due to the symmetry of the oblate Paul trap, the parameters satisfy $b_b = -b_t$.

We can use the results from Eqs. (5)-(7) in Eq. (2) to yield the final effective potential energy of an ion in this oblate Paul trap

$$\begin{aligned}\tilde{V}(\tilde{x}_1, \tilde{x}_2, \tilde{x}_3) = {} & \frac{q^2 V_{o,\mathrm{RF}}^2}{m\Omega_{\mathrm{RF}}^2 r_o^4}(\tilde{x}_1^2 + \tilde{x}_2^2 + 4\tilde{x}_3^2) + q\frac{V_r}{r_o^2}(\tilde{x}_1^2 + \tilde{x}_2^2 - 2\tilde{x}_3^2) \\ & + qV_t\left(\frac{\tilde{x}_3^2}{a^2} + \frac{\tilde{x}_3}{b_t} - \frac{\tilde{x}_1^2 + \tilde{x}_2^2}{c^2} + d\right) \\ & + qV_b\left(\frac{\tilde{x}_3^2}{a^2} + \frac{\tilde{x}_3}{b_b} - \frac{\tilde{x}_1^2 + \tilde{x}_2^2}{c^2} + d\right).\end{aligned} \tag{8}$$

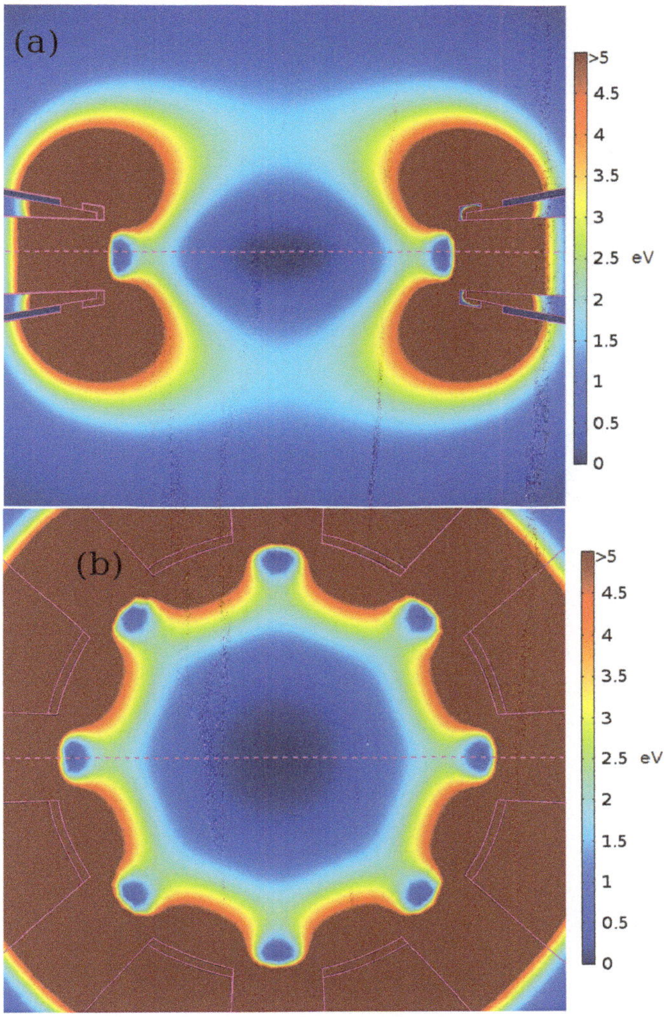

Figure 2 Numerical results for the pseudopotential produced by applying our anticipated operating parameters of a 500 V RF amplitude and a RF frequency of 35 MHz, and grounding the top and bottom electrodes. The oblate Paul trap's edges are shown as solid magenta lines. For our anticipated parameters the pseudopotential well depth is 1.6 eV. This numerical result is used to calculate $r_o = 512\ \mu$m. (**a**) The pseudopotential is shown in the $\hat{e}_1 - \hat{e}_3$ plane. The dashed magenta line identifies the plane of panel (b). (**b**) The pseudopotential is shown in the $\hat{e}_1 - \hat{e}_2$ plane and the dashed magenta line shows the plane of panel (a).

Since the effective potential energy is just a function of $\tilde{x}_1^2 + \tilde{x}_2^2$, it is rotationally symmetric around the $\hat{e}_3$-axis and we would expect there to be a zero frequency rotational mode in the phonon eigenvectors. That mode can be lifted from zero by breaking the symmetry, which can occur by adding additional fields that do not respect the cylindrical symmetry, and probably occur naturally due to imperfections in the trap, the optical access ports, stray fields, etc.

3 Equilibrium structure and normal modes

Using Eq. (8) (the calculated pseudopotential), we solve for the equilibrium structure of the crystal in the standard way. We first construct an initial trial configuration for the ions and then minimize the total potential energy of the oblate Paul trap (including the trap

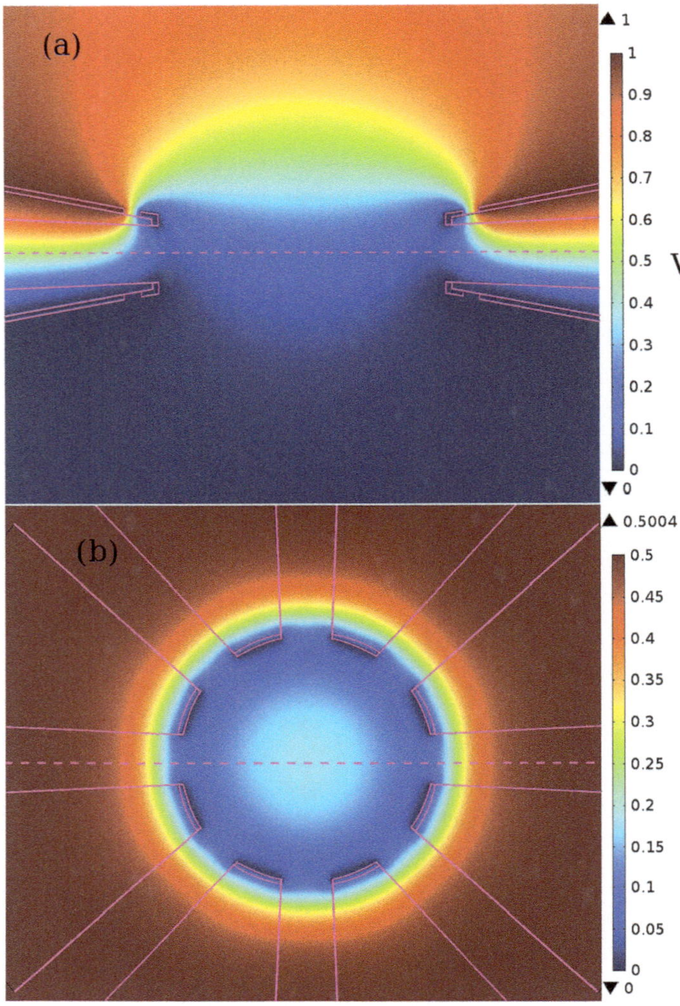

Figure 3 Grounding the RF ring and applying 1 V to the top electrodes, we plot the numerical electric potential. The oblate Paul trap's electrodes are shown as the solid magenta lines. (**a**) In the $\hat{e}_1 - \hat{e}_3$ plane the electrostatic potential due to the top electrodes has a gradient along the $\hat{e}_3$ direction. This gradient produces an electrostatic force that confines the ions in the center of the trap. (The dashed magenta line shows the plane of panel (b).) (**b**) The electrostatic potential is shown in the $\hat{e}_1 - \hat{e}_2$ plane, where the dashed magenta line shows the plane of panel (a). For the actual DC voltage applied to the trap, we multiply the potential by actual voltage difference.

potential and the Coulomb repulsion between ions), as summarized in Eq. (9); MatLab is used for the nonlinear minimization with a multidimensional Newton's method. We rewrite the total potential energy of the oblate Paul trap in a conventional form (up to a constant) via

$$\tilde{V}(\tilde{x}_1, \tilde{x}_2, \tilde{x}_3) = \frac{1}{2} m \left[\sum_{i=1}^{2} (\omega_{\psi,i}^2 + \omega_{r,i}^2 - \omega_{t,i}^2 - \omega_{b,i}^2) \tilde{x}_i^2 + \omega_{\psi,3}^2 \tilde{x}_3^2 - \omega_{r,3}^2 \tilde{x}_3^2 + \omega_{t,3}^2 \left(\tilde{x}_3 + \frac{a^2}{2b_t} \right)^2 + \omega_{b,3}^2 \left(\tilde{x}_3 + \frac{a^2}{2b_b} \right)^2 \right] + \frac{1}{2} \sum_{\substack{m,n=1 \\ m \neq n}}^{N} \frac{k_e e^2}{\tilde{r}_{nm}}, \quad (9)$$

where $k_e = 1/4\pi\epsilon_o$. Here $\tilde{x}_{in}$ is the ith component of the nth ion's location and $\tilde{r}_{nm} = \sqrt{(\tilde{x}_{1n}-\tilde{x}_{1m})^2+(\tilde{x}_{2n}-\tilde{x}_{2m})^2+(\tilde{x}_{3n}-\tilde{x}_{3m})^2}$. The frequencies in Eq. (9) are defined via

$$\omega_{\psi,1} = \frac{\sqrt{2}qV_{0,\mathrm{RF}}}{m\Omega_{\mathrm{RF}}r_0^2}, \qquad \omega_{\psi,1} = \omega_{\psi,2} = \frac{\omega_{\psi,3}}{2}, \tag{10}$$

$$\omega_{r,1} = \sqrt{\frac{2qV_r}{mr_0^2}}, \qquad \omega_{r,1} = \omega_{r,2} = \frac{\omega_{r,3}}{\sqrt{2}}, \tag{11}$$

$$\omega_{t,1} = \sqrt{\frac{2qV_t}{mc^2}}, \qquad \omega_{t,1} = \omega_{t,2} = \frac{a}{c}\omega_{t,3}. \tag{12}$$

We will express all distances in terms of a characteristic length, l_o, which satisfies

$$l_o^3 = \frac{k_e e^2}{m\omega_{\psi,3}^2} \tag{13}$$

and we will work with dimensionless coordinates $x = \tilde{x}/l_o$ when calculating the equilibrium positions. Furthermore, we measure all frequencies relative to $\omega_{\psi,3}$. The normalized frequencies are

$$\beta_i = \sqrt{\left(\omega_{\psi,i}^2+\omega_{r,i}^2-\omega_{t,i}^2-\omega_{b,i}^2\right)}/\omega_{\psi,3} \tag{14}$$

for $i = 1,2$, $\beta_{r,3} = \omega_{r,3}/\omega_{\psi,3}$, $\beta_{t,3} = \omega_{t,3}/\omega_{\psi,3}$, and $\beta_{b,3} = \omega_{b,3}/\omega_{\psi,3}$. The dimensionless total potential energy becomes

$$\begin{aligned} V &= \frac{\tilde{V}}{k_e e^2/l_o} \\ &= \frac{1}{2}\sum_{n=1}^{N}\left[\beta_1^2 x_{1n}^2+\beta_2^2 x_{2n}^2+x_{3n}^2-\beta_{r,3}^2 x_{3n}^2+\beta_{t,3}^2(x_{3n}+x_{o,t})^2+\beta_{b,3}^2(x_{3n}+x_{o,b})^2\right] \\ &\quad + \frac{1}{2}\sum_{\substack{m,n=1\\ m\neq n}}^{N}\frac{1}{r_{nm}}, \end{aligned} \tag{15}$$

where we have defined $x_{o,t} = a^2/(2l_o b_t)$ and $x_{o,b} = a^2/(2l_o b_r)$.

To find the equilibrium positions, we use the gradient of the total potential energy and numerically minimize the total potential energy using a multidimensional Newton's method. The gradient of Eq. (9) is

$$\begin{aligned} \vec{\nabla}V &= \sum_{i=1}^{3}\sum_{m=1}^{N}\frac{\partial V}{\partial x_{im}}\hat{e}_{im} \\ &= \sum_{m=1}^{N}\left[\sum_{i=1}^{2}\hat{e}_{im}\beta_i^2 x_{im}+\hat{e}_{3m}\left[x_{3m}-\beta_{r,3}^2 x_{3m}+\beta_{t,3}^2(x_{3m}+x_{o,t})+\beta_{b,3}^2(x_{3m}+x_{o,b})\right]\right. \\ &\quad \left.+\sum_{\substack{n=1\\ n\neq m}}^{N}\sum_{i=1}^{3}\hat{e}_{im}\frac{x_{in}-x_{im}}{r_{nm}^3}\right]. \end{aligned} \tag{16}$$

The force on ion m in the $\hat{i}$ direction will be $-\vec{\nabla}V \cdot \hat{e_{im}}$. We seek the solution in which all ions lie in a plane parallel to the $\hat{e}_1 - \hat{e}_2$ plane, such that $x_{3m} = \overline{x}_3$ for all $m \in [1, N]$. As a result of this condition, $x_{3n} = x_{3m}$, and there is no x_3 contribution to the Coulomb potential term. The value of $\overline{x}_3$ is determined by setting the $\hat{e}_{3m}$ term equal to zero in Eq. (16) and is given by the condition

$$\overline{x}_3 = \frac{-\beta_{t,3}^2 x_{o,t} - \beta_{b,3}^2 x_{o,b}}{1 - \beta_{r,3}^2 + \beta_{t,3}^2 + \beta_{b,3}^2}. \tag{17}$$

Using $\overline{x}_3$, the ion equilibrium positions are numerically obtained when all $3N$ components of the force on each ion are zero, which is given by $\vec{\nabla}V|_{equilib.} = 0$.

After the equilibrium positions $\{\overline{x}_{in}, n = 1, \ldots, N, i = 1, 2, 3\}$ are found, we expand the total potential about the equilibrium positions up to quadratic order

$$V = V^{(0)} + \sum_{i=1}^{3}\sum_{m=1}^{N} q_{im} \frac{\partial}{\partial x_{im}} V|_{eq} + \frac{1}{2}\sum_{i,j=1}^{3}\sum_{\substack{n=1\\m=1}}^{N} q_{im}q_{jn} \left.\frac{\partial^2 V}{\partial x_{im}\,\partial x_{jn}}\right|_{eq}. \tag{18}$$

The nonzero terms of the expansion are the zeroth order and the quadratic terms; the first order term is zero because the equilibrium position is defined to be where the gradient of the total potential is zero, however the zeroth term is also neglected since it is a constant. We calculate the Lagrangian of the trapped ions using the quadratic term of the Taylor expanded total potential, with q_{in} being the dimensionless displacement from the equilibrium position for the nth ion in the ith direction. The expanded Lagrangian becomes

$$L = \frac{1}{2\omega_{\psi,3}^2}\sum_{i=1}^{3}\sum_{m=1}^{N} \dot{q}_{im}^2 - \frac{1}{2}\sum_{i,j=1}^{3}\sum_{m,n=1}^{N} q_{im}K_{mn}^{ij}q_{jn}, \tag{19}$$

where K_{mn}^{ij} represents the elements of the effective spring constant matrices which are given by

$$(i = 1, 2) \quad K_{mn}^{ii} = \begin{cases} \beta_i^2 - \sum_{\substack{n'=1\\n'\neq m}}^{N} \left[\frac{1}{\overline{r}_{n'm}^3} - 3\frac{(\overline{x}_{in'} - \overline{x}_{im})^2}{\overline{r}_{n'm}^5}\right] & \text{if } m = n, \\ \frac{1}{\overline{r}_{mn}^3} - 3\frac{(\overline{x}_{in} - \overline{x}_{im})^2}{\overline{r}_{mn}^5} & \text{if } m \neq n, \end{cases} \tag{20}$$

$$K_{mn}^{12} = K_{mn}^{21} = \begin{cases} 3\sum_{\substack{n'=1\\n'\neq m}}^{N} \frac{(\overline{x}_{1n'} - \overline{x}_{1m})(\overline{x}_{2n'} - \overline{x}_{2m})}{\overline{r}_{n'm}^5} & \text{if } m = n, \\ -3\frac{(\overline{x}_{1n} - \overline{x}_{1m})(\overline{x}_{2n} - \overline{x}_{2m})}{\overline{r}_{mn}^5} & \text{if } m \neq n, \end{cases} \tag{21}$$

$$K_{mn}^{33} = \begin{cases} \beta_3^2 - \sum_{\substack{n'=1\\n'\neq m}}^{N} \frac{1}{\overline{r}_{n'm}^3} & \text{if } m = n, \\ \frac{1}{\overline{r}_{mn}^3} & \text{if } m \neq n, \end{cases} \tag{22}$$

where $\beta_3 = \sqrt{1 - \beta_{r,3}^2 + \beta_{t,3}^2 + \beta_{b,3}^2}$ and $\overline{r}_{mn} = \sqrt{(\overline{x}_{1m} - \overline{x}_{1n})^2 + (\overline{x}_{2m} - \overline{x}_{2n})^2}$ is the planar interparticle distance between ions n and m. Note that motion in the 3-direction (axial direction) is decoupled from motion in the $\hat{e}_1 - \hat{e}_2$ plane.

After applying the Euler-Lagrange equation to Eq. (19) and substituting the eigenvector solution $q_{im} = Re(b^{\alpha}_{im} e^{i\omega_\alpha t})$, we are left to solve the standard eigenvalue equation

$$-b^{\alpha}_{im}\left(\frac{\omega_\alpha}{\omega_{\psi,3}}\right)^2 + \sum_{j=1}^{3}\sum_{n=1}^{N} K^{ij}_{mj} b^{\alpha}_{jn} = 0. \tag{23}$$

There are two sets of normal modes: eigenvectors of the $N \times N$ matrix K^{33} yield the 'axial' modes (those corresponding to motion perpendicular to the crystal plane) and eigenvectors of the $2N \times 2N$ matrix K^{ij}, $i,j \in [1,2]$, yield the 'planar' modes (those corresponding to ion motion in the crystal plane).

4 Results

Now that we have constructed the formal structure to determine the equilibrium positions, phonon eigenvectors, and phonon frequencies, and we have determined the total pseudopotential of the trap, we are ready to solve these systems of equations to determine the expected behavior of the trapped ions. We present several numerical examples to illustrate the equilibrium structure, eigenvalues of the normal modes, and the effective spin-spin interaction J_{mn} for the axial modes with a detuning of the spin-dependent optical dipole force above the axial center-of-mass phonon frequency, ω_{CM}. We use an ytterbium ion with mass $m = 171$ u, where u is the atomic mass unit, and a positive charge $q = e$. For the frequency of the RF voltage, we use $\Omega_{RF} = 2\pi \times 35$ MHz and the amplitude of the potential applied to the RF ring is $V_{o,RF} \approx 500$ V. The DC voltage applied to the RF ring and to the top and bottom electrodes will be $|V_{DC}| \leq 100$ V. We work in a region where the trapped ion configurations are stable. The ion crystal is stable only when both β_1 and β_2 are real and nonzero, as defined in Eq. (14) and this region is shown in Figure 4, which depends on the voltages applied to the RF ring and to the top and bottom electrodes. We work with ion crystals that contain up to 20 ions.

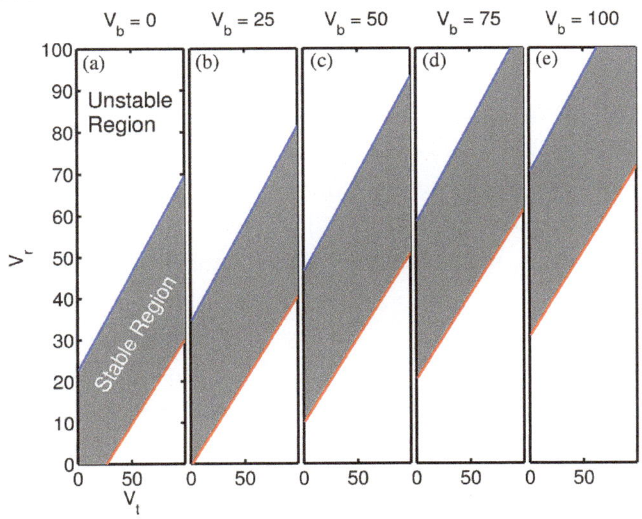

Figure 4 β_1 and β_3 are both real when the ion configuration is stable (shaded grey). We use Eq. (14) to find a region where the ion configuration is stable. The white regions are unstable.

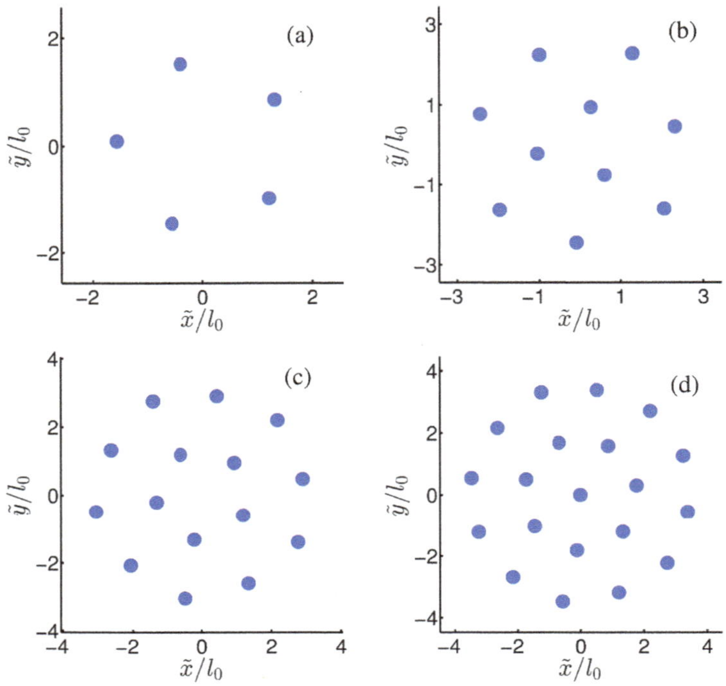

Figure 5 Equilbrium positions (blue dots) calculated for $N = 5$ (a), 10 (b), 15 (c), 20 (d) with $V_r = 46.3$ V and $V_t = V_b = 50$ V. (**a**) $N = 5$ is the maximum number of ions for a single ring. (**b**) $N = 10$ is the first instance where there is a second ring in the center of the configuration. (**c**) Similar to panel (a), $N = 15$ is the maximum number of ions to have two rings. (**d**) $N = 20$ is the maximum number ions expected to operate within our trap. As previously noted, the additional ions when added to the $N = 15$ equilibrium configuration occupy the outer rings instead of the center.

4.1 Equilibrium configurations

A single ion will sit in the center of the trap. As more ions are added, because the ions repel each other, a 'hard core' like structure will form, starting with rings of ions until it becomes energetically more stable to have one ion in the center of the ring, and then additional shells surrounding it, and so on. We find that as we increase the number of ions, the single ring is stable for $N = 3, 4,$ and 5. Increasing N further creates more complex structures. We show the common equilibrium configurations for $N = 5, 10, 15, 20$ with DC voltages of $V_r = 46.3$ V and $V_t = V_b = 50$ V, in Figure 5. As mentioned above, $N = 5$ is the last configuration that is comprised of a single ring of ions, as depicted in Figure 5(a). The $N = 5$ configuration is ideal to use in order to study when periodic boundary conditions are applied to the linear chain, this is due to the configuration being in a single ring. For configurations with $N = 6$ through 8, the additional ions are added to an outer ring. When $N = 9$ the additional ions are added to the center. In Figure 5(b), $N = 10$ is the first configuration that forms a ring in the center, with three ions. The equilibrium configuration of $N = 15$ is the maximum number to have two rings, as shown in Figure 5(c). Ion configurations with $N > 15$ have a single ion at the center, as an example of this, we show $N = 20$ in Figure 5(d). The common configurations for $N > 5$ are nearly formed from triangular lattices (up to nearest neighbor) and this could be used to study frustration in the effective spin models (except, of course, that due to the finite number of ions there are many cases where the coordination number of an interior ion is not equal to 6, as seen in Fig-

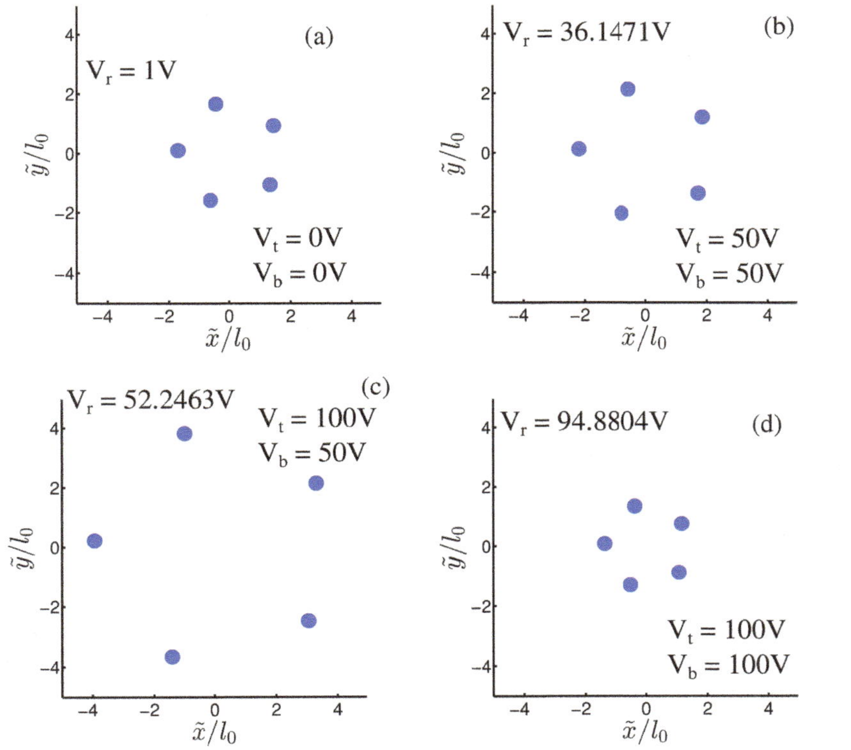

Figure 6 Equilibrium positions (blue dots) for $N = 5$ with various DC voltages on the RF ring and independently applied to the top and bottom electrodes. The equilibrium shape remains the same for the four cases shown: (**a**) $V_r = 1$ V and $V_t = V_b = 0$ V, (**b**) $V_r = 36.147$ V and $V_t = V_b = 50$ V, (**c**) $V_r = 52.225$ V, $V_t = 100$ V, and $V_b = 50$ V and (**d**) $V_r = 94.880$ V and $V_t = V_b = 100$ V.

ure 5). The shape of all of these clusters for small N agree with those found in Ref. [39], except for $N = 10, 12$, and 14, which have small differences due to the different potential that describes the oblate Paul trap from the potential used in [39].

We next show the dependence of the equilibrium positions on the DC voltages applied to the RF ring and independently applied to the top and bottom electrodes. We fix $N = 5$. As each DC voltage is independently varied, the shape of the equilibrium configuration for $N = 5$ remains the same and only the distances between ions change, as shown in the four cases in Figure 6. The equilibrium positions are robust to when $V_t \neq V_b$ as shown in Figure 6(c), although this voltage setting will not be used due to the effects of micromotion. The enhanced micromotion is due to the plane of the trapped ions are no longer at the center of the trap but at some offset. This offset of the plane will expose the ions to a stronger restoring force.

4.2 Normal modes

After determining the equilibrium positions, we can find the spring constants and then solve the eigenvalue problem to find the normal modes. Note that due to rotational symmetry, there always is a zero frequency planar mode corresponding to the free rotation of the crystal. In an actual experiment, however, we expect that the rotational symmetry of the trap will be broken by stray fields, the radial optical access tunnels, imperfections in the electrodes, etc., so that mode will be lifted from zero.

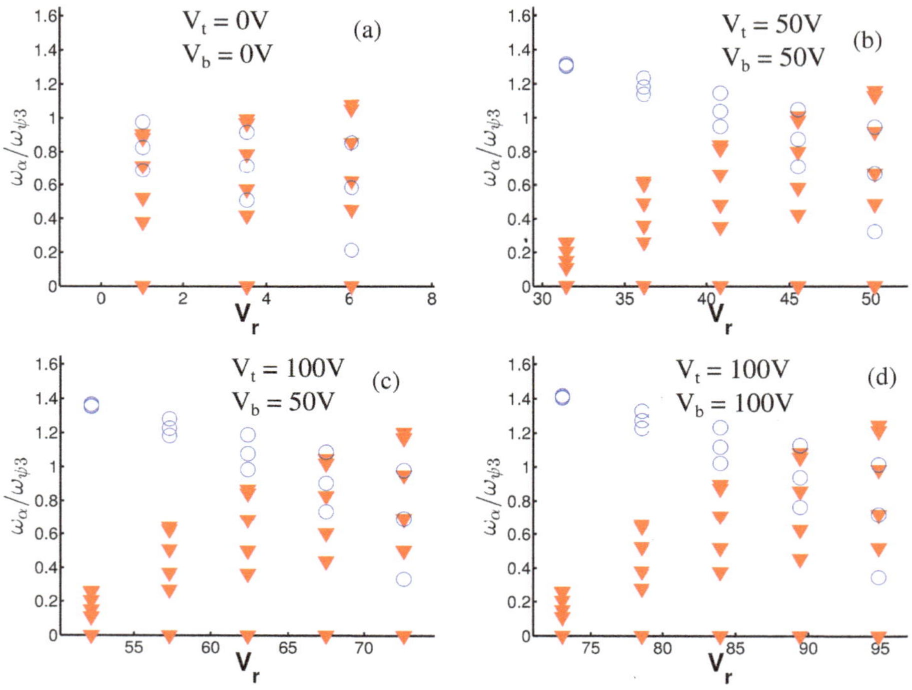

Figure 7 Eigenvalues of the normal modes as a function of V_r for $N = 5$. The axial mode frequencies (blue circles) decrease and planar mode frequencies (red triangles) increase as V_r increases, as shown in the four cases illustrated: (**a**) $V_t = V_b = 0$ V, (**b**) $V_t = V_b = 50$ V, (**c**) $V_t = 100$ V and $V_b = 50$ V and (**d**) $V_t = V_b = 100$ V. For (b)-(d) the eigenvalues of the normal modes also are closer together at low V_r and separate as V_r increases. The axial frequencies have two degeneracies and the planar frequencies have four degeneracies as discussed in the main text.

We show the eigenvalues of the normal modes for $N = 5$ in Figure 7. In Figure 7 for a given V_t, V_b, and V_r there are 5 axial phonon frequencies, however the axial phonon frequencies has two pairs of degenerate modes. One degenerate pair are the two orthogonal tilt modes and the other pair is the two orthogonal zig-zag modes. Similarly there are 10 planar phonon frequencies, however there are 4 pairs that are degenerate. The axial phonon frequencies decrease as the DC voltage on the RF ring increases and the planar phonon frequencies decrease as the DC voltage on the RF ring decreases. For the majority of the combinations of V_t and V_b the axial phonon frequencies lie in a narrow band that is separated from the planar mode frequencies, which also lie in a narrow band. As V_r increases the axial band broadens and eventually overlaps the planar band, which is also broadening. This behavior of the axial phonon frequencies and planar phonon frequencies occurs when $V_t \neq V_b$ as shown in Figure 7(c), although this voltage setting will not be used due to the micromotion which is larger in this case. When the axial band has an eigenvalue that goes soft, the system is no longer stable within one plane (which is the equivalent of the zig-zag transition in the linear Paul trap). When $V_t = V_b = 0$ the initial clustering of the axial modes and planar modes is not present, as shown in Figure 7(a).

4.3 Ising spin-spin interaction

The ions in our trap have two hyperfine states that are separated by a frequency ω_0. Three laser beams with two beatnotes at frequencies $\omega_0 \pm \mu$ will illuminate the ions, selectively exciting phonon modes as described in [3]. In this case, we choose the laser beams to propagate along the $\pm\hat{e}_3$ direction, as defined in Figure 1(b), such that the laser beams are insensitive to the micromotion which is entirely radial. The phonon modes are excited in a spin-dependent way to generate effective spin-spin interactions which can be described by the Ising spin coupling matrix, J_{mn}

$$\mathcal{H} = \sum_{mn}^{N} J_{mn}\sigma_m^z\sigma_n^z, \tag{24}$$

where σ_i^z is the Pauli spin matrix of ion i in the $\hat{e}_3$-direction and we have neglected the time-dependent terms of the spin couplings J_{mn}. The explicit formula for J_{mn} is [46]

$$J_{m,n} = J_0 \sum_{\alpha} \frac{b^*_{m,\alpha} b_{n,\alpha}}{(\frac{\mu}{\omega_{\text{CM}}})^2 - (\frac{\omega_m}{\omega_{\text{CM}}})^2}, \tag{25}$$

where the coefficient J_0, depends on the carrier transition Rabi frequency (Ω), the difference in wavevector between the laser beams (δk), the ion mass (m), and the frequency of the center-of-mass mode, (ω_{CM}), and is given by

$$J_0 = \frac{\Omega^2 \hbar (\delta k)^2}{2m\omega_{\text{CM}}^2}.$$

We calculate the spin-spin coupling, using Eq. (25), for a small number of ions in order to neglect the micromotion and assume the ions are at their average position. The micromotion will be approximately equal to the displacement of the ion from the RF null times the Mathieu q-parameter [47]. The ions with the largest micromotion will occur in the outermost 'ring' of the crystal. For N ions and an average inter-ion spacing s, the ions in the outermost ring are a distance $s\sqrt{N}$ from the center, which means that the amplitude of the micromotion is approximately $qs\sqrt{N}$, and the ratio of this micromotion amplitude to the inter-ion spacing is therefore $q\sqrt{N}$.

It is expected that if one detunes, μ, to be larger than the center-of-mass mode frequency, then one can generate long-range spin-spin couplings that vary as a power law from 0 to 3, as they decay with distance [14, 15, 48]. Hence, we fit the spin-spin couplings to a power law in distance as a function of detuning in Figure 8 for $N = 10$ and 20. Note

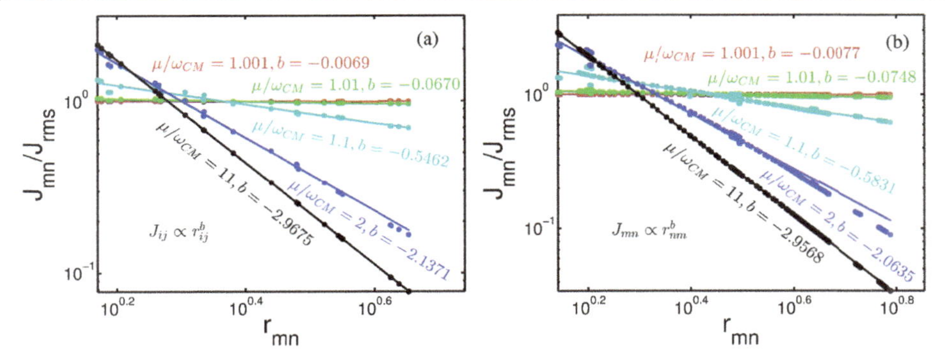

Figure 8 Spin-spin coupling versus distance for different detunings with DC voltages of $V_r = 94.9$ V and $V_t = V_b = 100$ V and different number of ions (different panels; (a) is $N = 10$ and (b) is $N = 20$) on a log-log plot. The colored lines are the fit to a power-law behavior $J_{mn} \propto r_{mn}^{-b}$. The detunings are $\mu = 1.001, 1.01, 1.1, 2,$ and 11 to the blue of ω_{CM} in units of ω_{CM}.

that our system is still rather small, so there are likely to be finite size effects that modify the simple power law behavior.

The trap could also be use the recent work on a controlled phase flip (CPF) quantum gate that includes the corrections due to the micromotion and is optimized to attain high fidelity [16]. The high fidelity is obtained by dividing the time of the gate into m equal segments, during each segment of equal time a particular Rabi frequency is applied, Ω_β ($\beta = 1, 2, \ldots, m$), that has been optimized for the βth segment. The optimized Ω_β are determined by solving the motional dynamics, that explicitly take into account all the micromotion contributions. The motion of the ions are described by a set of coupled time-dependent Mathieu equations.

4.4 Quantum motional effects

The trap could also be used to examine different types of quantum motional effects of ions, similar to the recent work on the Aharonov-Bohm effect [26]. In order to examine such effects, one would need to cool the system to nearly the ground state. This can be accomplished by including Raman side-band cooling after Doppler cooling the system for all modes except the soft rotational mode, at least when the potential is large enough that the mode frequencies are sizeable. To cool the rotational mode, one would need to add a perturbation to the system that lifts the phonon mode frequency, side-band cool it, and then adiabatically reduce the frequency by removing the perturbation. This procedure will cool off that phonon mode, which can yield quite small quanta in it [26]. Once the system has been prepared in this state, then quantum tunneling effects, or coherent motional effects could be studied in the trap for a range of different ion configurations. It might also be interesting to extend these types of studies to cases where the ions no longer lie completely in one plane, but have deformed into a full three-dimensional structure (as long as the larger micromotion does not cause problems). Finally, many of these ideas would need to be used if one tried to examine time crystals, especially the cooling of the rotational mode to be able to see quantum effects [27, 28].

5 Conclusion

In this work, we have studied 2D ion crystals in an oblate Paul trap for use in quantum simulations. With this system, one can trap a modest number of ions in 2D planar structures that are likely to be highly frustrated without needing a Penning trap, providing a controlled way to study the onset of frustration effects in quantum simulations. We calculated the equilibrium positions and the phonon frequencies for the proposed oblate Paul trap over its stable region. The equilibrium positions with $N \leq 5$ form a single ring configuration and could potentially be used to study periodic boundary conditions and the Aharonov-Bohm effect when $N = 4$ or 5 (and possibly time crystals). Once $N > 5$, the equilibrium configurations have multiple rings that are nearly formed from triangular lattices. One can generate an effective Ising Hamiltonian by driving axial modes with a spin-dependent optical dipole force. When detuning is to the blue of the axial center-of-mass mode, the spin-spin coupling, J_{mn}, has an approximate power law that is within the expected range of 0 to 3. In the future, as this trap is tested and performs simulations of spin models with ions, the work presented here will be critical to determining the parameters of the Hamiltonian and for selecting the appropriate configurations to use in the different simulations.

Competing interests
The authors declare that they have no competing interests.

Authors' contributions
The idea from this work came from Wes Campbell and Jim Freericks. Danilo Dadic and Wes Campbell developed the electric field potentials for the pseudopotential description of the trap as well as designing the trap parameters. Bryce Yoshimura, Marybeth Stork and Jim Freericks performed all of the theoretical calculations. The paper manuscript was first drafted by Bryce Yoshimura and then all authors contributed to revisions.

Author details
[1]Department of Physics, Georgetown University, 37th and O St. NW, Washington, DC 20007, USA. [2]Department of Physics and Astronomy, Washington University, Campus Box 1105, One Brookings Dr., St. Louis, Missouri 63130, USA.
[3]Department of Physics and Astronomy, University of California Los Angeles, 475 Portola Plaza, Los Angeles, CA 90095, USA.

Acknowledgements
We thank Dr. Philippe Bado, Dr. Mark Dugan and Dr. Christopher Schenck of Translume (Ann Arbor, MI) for valuable discussions. B.Y. acknowledges the Achievement Rewards for College Scientists Foundation for supporting this work. M.S. acknowledges the National Science Foundation under grant number DMR-1004268 for support. J.K.F. and B.Y. acknowledge the National Science Foundation under grant number PHY-1314295 for support. D.D. and W.C.C. acknowledge support from the U.S. Air Force Office of Scientific Research Young Investigator Program under grant number FA9550-13-1-0167 and support from the AFOSR STTR Program. J.K.F. also acknowledges support from the McDevitt bequest at Georgetown University.

References
1. Feynman R: **Simulating physics with computers.** *Int. J. Theor. Phys.* 1982, **21**(6/7):467-488.
2. Friedenauer A, Schmitz H, Glueckert J, Porras D, Schaetz T: **Simulating a quantum magnet with trapped ions.** *Nat. Phys.* 2008, **4**(10):757-761.
3. Kim K, Chang M-S, Korenblit S, Islam R, Edwards E, Freericks J, Lin G-D, Duan L-M, Monroe C: **Quantum simulation of frustrated Ising spins with trapped ions.** *Nature* 2010, **465**(7268):590-593.
4. Islam R, Edwards E, Kim K, Korenblit S, Noh C, Carmichael H, Lin G-D, Duan L-M, Wang C-C, Freericks J, Monroe C: **Onset of a quantum phase transition with a trapped ion quantum simulator.** *Nat. Commun.* 2011, **2**:377.
5. Islam R, Senko C, Campbell W, Korenblit S, Smith J, Lee A, Edwards E, Wang C-C, Freericks J, Monroe C: **Emergence and frustration of magnetism with variable-range interactions in a quantum simulator.** *Science* 2013, **340**(6132):583-587.
6. Lanyon B, Hempel C, Nigg D, Müller M, Gerritsma R, Zähringer F, Schindler P, Barreiero J, Rambach M, Kirchmair G, Hennrich M, Zoller P, Blatt R, Roos C: **Universal digital quantum simulation with trapped ions.** *Science* 2011, **334**(6052):57-61.
7. Senko C, Smith J, Richerme P, Lee A, Campbell W, Monroe C: **Coherent imaging spectroscopy of a quantum many-body spin system.** *Science* 2014, **345**:430.
8. Richerme P, Gong Z-X, Lee A, Senko C, Smith J, Foss-Feig M, Michalakis S, Gorshkov A, Monroe C: **Non-local propagation of correlations in long-range interacting quantum systems.** *Nature* 2014, **511**:198.
9. Jurcevic P, Lanyon B, Hauke P, Hempel C, Zoller P, Blatt R, Roos C: **Observation of entanglement propagation in a quantum many-body system.** *Nature* 2014, **511**:202.
10. Wuerker RF, Shelton H, Langmuir RV: **Electrodynamic containment of charged particles.** *J. Appl. Phys.* 1959, **30**(3):342.
11. Casdroff R, Blatt R: **Ordered structures and statistical properties of ion clouds stored in a Paul trap.** *Appl. Phys. B* 1988, **45**:175-182.
12. Walther H: **From a single ion to a mesoscopic system - crystallization of ions in Paul traps.** *Phys. Scr.* 1995, **59**:360-368.
13. Landa H, Drewsen M, Reznik B, Retzker A: **Classical and quantum modes of coupled Mathieu equations.** *J. Phys. A, Math. Theor.* 2012, **45**:455305.
14. Porras D, Cirac J: **Effective quantum spin systems with trapped ions.** *Phys. Rev. Lett.* 2004, **92**(20):207901.
15. Britton J, Sawyer B, Keith A, Wang C-C, Freericks J, Uys H, Biercuk M, Bollinger J: **Engineered two dimensional Ising interactions in a trapped-ion quantum simulator with hundreds of spins.** *Nature* 2012, **484**(7395):489-492.
16. Shen C, Duan L-M: **High fidelity quantum gates for trapped ions under micromotion.** *Phys. Rev. A* 2014, **90**:022332.
17. Clark R, Lin T, Brown K, Chuang I: **A two-dimensional lattice ion trap for quantum simulation.** *J. Appl. Phys.* 2009, **105**(1):013114.
18. Sterling R, Rattanasonti H, Weidt S, Lake K, Srinivasan P, Webster S, Kraft M, Hensinger W: **Fabrication and operation of a two-dimensional ion-trap lattice on a high-voltage microchip.** *Nat. Commun.* 2014, **5**:3637.
19. Chiaverini J, Lybarger W: **Laserless trapped-ion quantum simulations without spontaneous scattering using microtrap arrays.** *Phys. Rev. A* 2008, **77**(2):022324.
20. Kumph M, Brownnutt M, Blatt R: **Two-dimensional arrays of radio-frequency ion traps with addressable interactions.** *New J. Phys.* 2011, **13**:073043.
21. Siverns J, Weidt S, Lake K, Lekitsch B, Hughes M, Hensinger W: **Optimization of two-dimensional ion trap arrays for quantum simulation.** *New J. Phys.* 2012, **14**:085009.
22. Schmied R, Wesenberg J, Leibfried D: **Optimal surface-electrode trap lattices for quantum simulation with trapped ions.** *Phys. Rev. Lett.* 2009, **102**(23):233002.
23. Korenblit S, Kafri D, Campbell W, Islam R, Edwards E, Gong Z-X, Lin G-D, Duan L-M, Kim J, Kim K, Monroe C: **Quantum simulation of spin models on an arbitrary lattice with trapped ions.** *New J. Phys.* 2012, **14**:095024.

24. Yoshimura B, Campbell W, Freericks J: **Diabatic ramping spectroscopy of many-body excited states for trapped-ion quantum simulators.** *Phys. Rev. A* 2014 (to appear).
25. del Campo A, Rams M, Zurek W: **Assisted finite-rate adiabatic passage across a quantum critical point: exact solution for the quantum Ising model.** *Phys. Rev. Lett.* 2012, **109**(11):115703.
26. Noguchi A, Shikano Y, Toyoda K, Urabe S: **Aharonov-Bohm effect in the tunnelling of a quantum rotor in a linear Paul trap.** *Nat. Commun.* 2014, **5**:3868.
27. Wilcek F: **Quantum time crystals.** *Phys. Rev. Lett.* 2012, **109**(16):160401.
28. Li T, Gong Z-X, Yin Z-Q, Quan H, Yin X, Zhang P, Duan L-M, Zhang X: **Space-time crystals of trapped ions.** *Phys. Rev. Lett.* 2012, **109**(16):163001.
29. Itano W, Bergquist J, Wineland D: **Coulomb clusters of ions in a Paul trap.** In *Proceedings of the Workshop on Crystalline Ion Beams: 4-7 October 1988; Wertheim, Germany.* Edited by Hasse R, Hofmann I, Liese D; 1989:241.
30. Bergquist J, Wineland D, Itano W, Hemmati H, Daniel H, Leuchs G: **Energy and radiative lifetime of the $5d^9\ 6s^2\ {}^2d_{5/2}$ state in Hg II by Doppler-free two-photon laser spectroscopy.** *Phys. Rev. Lett.* 1985, **55**(15):1567-1570.
31. Wineland D, Itano W, Bergquist J, Hulet R: **Laser-cooling limits and single-ion spectroscopy.** *Phys. Rev. A* 1987, **36**(5):2220-2231.
32. Bergquist J, Hulet R, Itano WM, Wineland D: **Observation of quantum jumps in a single atom.** *Phys. Rev. Lett.* 1986, **57**(14):1699-1702.
33. Wineland D, Itano W, Bergquist J: **Absorption spectroscopy at the limit: detection of a single atom.** *Opt. Lett.* 1987, **12**(6):389-391.
34. Bergquist J, Itano W, Wineland D: **Recoilless optical absorption and Doppler sidebands of a single trapped ion.** *Phys. Rev. A* 1987, **36**(1):428-431.
35. Heinzen D, Wineland D: **Quantum-limited cooling and detection of radio-frequency oscillations by laser-cooled ions.** *Phys. Rev. A* 1990, **42**(5):2977-3004.
36. Block M, Drakoudis A, Leuthner H, Seibert P, Werth G: **Crystalline ion structures in a Paul trap.** *J. Phys. B* 2000, **33**(11):375-382.
37. Kaufmann H, Ulm S, Jacob G, Poschinger U, Landa H, Retzker A, Plenio MB, Schmidt-Kaler F: **Precise experimental investigation of eigenmodes in a planar ion crystal.** *Phys. Rev. Lett.* 2012, **109**:263003.
38. Schiffer J: **Phase transitions in anisotropically confined ionic crystals.** *Phys. Rev. Lett.* 1993, **70**(6):818.
39. Bedanov V, Peeters F: **Ordering and phase transitions of charged particles in a classical finite two-dimensional system.** *Phys. Rev. B* 1994, **49**(4):2667-2676.
40. Buluta I, Kitaoka M, Georgescu S, Hasegawa S: **Investigation of planar coulomb crystals for quantum simulation and computation.** *Phys. Rev. A* 2008, **77**(6):062320.
41. Buluta I, Hasegawa S: **The structure of planar coulomb crystals in rf traps.** *J. Phys. B, At. Mol. Opt. Phys.* 2009, **42**(15):154004.
42. Clossen T, Roussel M: **The flattening phase transition in systems of trapped ions.** *Can. J. Chem.* 2009, **87**(10):1425-1435.
43. Landa H, Drewsen M, Reznik B, Retzker A: **Modes of oscillation in radiofrequency Paul traps.** *New J. Phys.* 2012, **14**:093023.
44. Landa H, Retzker A, Schaetz T, Reznik B: **Entanglement generation using discrete solitons in Wigner crystals.** e-print 2014. arXiv:1308.2943.
45. Dehmelt H: **Radiofrequency spectroscopy of stored ions I: storage.** In *Advances in Atomic and Molecular Physics.* Edited by Bates D, Estermann I. *Volume* 3. New York: Academic Press; 1967:53-72.
46. Zhu S-L, Monroe C, Duan L-M: **Trapped ion quantum computation with transverse phonon modes.** *Phys. Rev. Lett.* 2006, **97**(5):050505.
47. Berkeland DJ, Miller JD, Bergquist JC, Itano WM, Wineland DJ: **Minimization of ion micromotion in a Paul trap.** *J. Appl. Phys.* 1998, **83**:5025.
48. Senko C, Smith J, Richerme P, Lee A, Campbell WC, Monroe C: **Coherent imaging spectroscopy of a quantum many-body spin system.** *Science* 2014, **345**:430.

4

The circuit quantum electrodynamical Josephson interferometer

Robert Jirschik[1,2] and Michael J Hartmann[1*]

*Correspondence: mh@tum.de
[1] Physik Department, Technische Universitaet Muenchen, James Franck Str., Garching, 85748, Germany
Full list of author information is available at the end of the article

Abstract

Arrays of circuit cavities offer fascinating perspectives for exploring quantum many-body systems in a driven dissipative regime where excitation losses are continuously compensated by coherent input drives. Here we investigate a system consisting of three transmission line resonators, where the two outer ones are driven by coherent input sources and the central resonator interacts with a superconducting qubit. Whereas a low excitation number regime of such a device has been considered previously with a numerical integration, we here specifically address the high excitation density regime. We present analytical approximations to these regimes in the form of two methods. The first method is a Bogoliubov or linear expansion in quantum fluctuations which can be understood as an approximation for weak nonlinearities. As the second method we introduce a combination of mean-field decoupling for the photon tunneling with an exact approach to a driven Kerr nonlinearity which can be understood as an approximation for low tunneling rates. In contrast to the low excitation regime we find that for high excitation numbers the anti-bunching of output photons from the central cavity does not monotonously disappear as the tunnel coupling between the resonators is increased.

1 Introduction

In recent years, a new direction of research in cavity quantum electrodynamics (cavity-QED) has developed, in which multiple cavities that are coupled via the exchange of photons are considered. Such setups are particularly intriguing if the cavities are connected to form regular arrays and if the strong coupling regime is achieved in each cavity of the array. Such devices would give rise to novel structures where coherent light matter interactions exceed dissipative processes simultaneously in multiple locations of the array [1–6].

Whereas it is rather challenging to build mutually resonant high finesse cavities in the optical regime, it is for microwave photons perfectly feasible to engineer large arrays of mutually resonant cavities on one chip in an architecture known as circuit-QED [6, 7]. Here multiple superconducting transmission line resonators with virtually identical lengths in the centimeter range can be coupled via capacitors or inductive links [6]. The individual transmission line resonators can feature strong optical nonlinearities by coupling them to superconducting qubits such as transmons [8, 9] or phase qubits [7].

Yet in all experiments that involve light-matter interactions, some photons will inevitably be lost from the structure due to imperfect light confinement or emitter relaxation. To compensate for such losses, cavity arrays are thus most naturally studied in a regime where an input drive continuously replaces the dissipated excitations. This mode of operation eventually gives rise to a driven dissipative regime, where the dynamical balance of loading and loss processes leads to the emergence of stationary states [10–13]. The properties of these stationary states may however vary significantly if one changes system parameters such as the photon tunneling rate between cavities, the light-matter interaction in a cavity or even the relative phase between a pair of coherent input drives [12, 14].

A device that is ideally suited for studying the effect of relative phases between input drives is the so called quantum optical Josephson interferometer [14], which consists of two coherently driven linear cavities connected through a central cavity with a single-photon nonlinearity. Here, the interplay between tunneling and interactions in the steady state of the system has been analyzed for regimes of weak input drives where the number of excitations in each cavity is rather small by Gerace *et al.* [14]. For opposite phases of the input drives one finds a destructive interference which suppresses population of the central cavity, whereas for input drives in phase the central cavity is populated with anti-bunched excitations due to its strong nonlinearity.

For the considered regime of low excitation numbers the approach [14] employed a full numerical analysis relying on an excitation number truncation of the Hilbert spaces. For high input drives however, the theoretical description of the system poses a challenge as conventional numerical methods quickly become computationally infeasible. Hence, such a numerical approach cannot describe regimes with more intense input drives where excitation numbers begin to grow and is therefore unable to explore a possible transition from a quantum regime with anti-bunched output photons to a classical regime with uncorrelated output photons.

Here we present approaches that are capable of describing this transition. For a parameter regime with a weak nonlinearity in the central resonator we expand the intra cavity fields to linear order in the quantum fluctuations around their coherent parts. This Bogoliubov-type expansion provides a good approximation provided the density of quantum fluctuations around the coherent background is small compared to the excitation number in the latter. This regime is realized for weak nonlinearities in the central resonator.

In turn for strong nonlinearities but moderate photon tunneling rates between the resonators we combine an exact solution for a driven Kerr nonlinearity by Drummond and Walls [15] with a mean-field decoupling of the photon tunneling between resonators. As the solution is based on a description of the quantum state in terms of a P-function we will refer to it as P-function mean-field [16]. We find that this approach provides an accurate description in a large parameter range provided the photon tunneling is low enough such that the fields in the outer resonators do not deviate significantly from coherent ones. An overview of the discussed approaches and the parameter regimes where they provide an accurate description is sketched in Figure 1.

The remainder of the paper is organized as follows, in Section 2 we describe the setup and model we consider and in Section 3 we revisit the numerical approach by Gerace *et al.* [14] to present results for time resolved correlation functions which so far have not been considered. In the next Section 4 we discuss the approach based on a Bogoliubov expansion and the results it yields. In Section 5 we then introduce our mean-field approach based on an exact solution for the central resonator. Finally Section 6 presents conclusions and a discussion of the parameter regimes covered by each of the discussed descriptions.

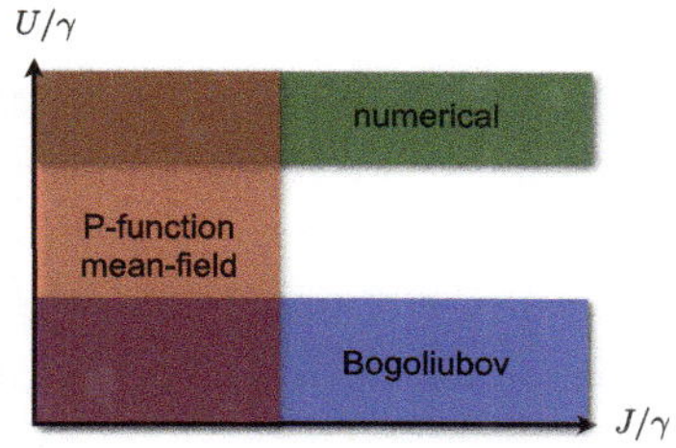

Figure 1 Sketch of parameter regions where the discussed approaches provide accurate descriptions. With increasing driving strength Ω, the validity range of the numerical approach shrinks to larger nonlinearities whereas the validity ranges for the Bogoliubov and *P*-function mean-field approaches grow. For definitions of U, J, Ω and γ see equation (1). The validity boundaries of the various approaches are discussed further in Figures 8, 9 and 12.

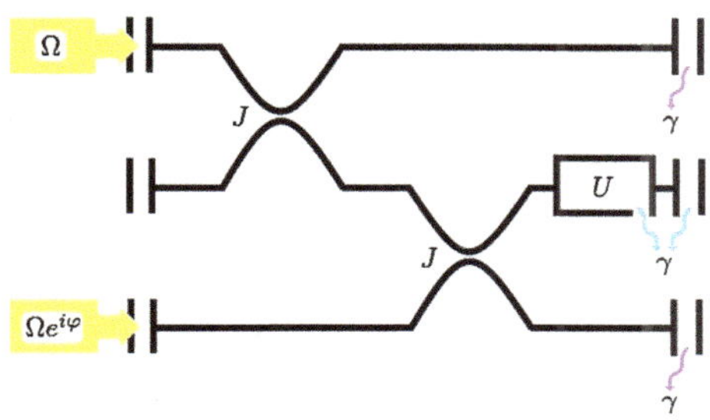

Figure 2 Schematic of a possible realization of the quantum-optical Josephson interferometer with transmission line resonators in circuit QED. Here only the central conductors of the resonators are drawn. The three-resonator chip allows for a coupling J between two neighboring resonators [17]. A voltage-node in the center is the reason for an off-center implementation of the nonlinearity U that is capacitively coupled to the middle resonator. The two outer transmission line resonators are driven by coherent input sources with amplitudes Ω. Each transmission line resonator has an individual dissipation-rate, which depends on the resonator's coupling capacitors at both ends. We consider the case where the combined decay rate of the central resonator and the transmon is equal to the decay rate of the two outer resonators γ.

2 Setup

The circuit QED Josephson interferometer we consider is a multicavity driven-dissipative system consisting of three coupled transmission line resonators where the two outer ones are coherently driven and the central transmission line resonator interacts with a superconducting qubit, see Figure 2 for a sketch. We consider here a transmon as the qubit and assume a regime with strong interactions between the central resonator and this qubit so that polaritons, combined light matter excitations, become a useful description of the nonlinear central resonator. For large enough qubit-resonator coupling one can focus the description only on one excitation species in the central resonator [18]. Employing this approach, the Hamiltonian for the circuit QED Josephson interferometer can be written in a frame rotating at the frequency ω_{in} of the input drives as

$$\hat{H} = \sum_{k=1}^{3} \Delta_k \hat{a}_k^\dagger \hat{a}_k - \frac{U}{2}\hat{a}_2^\dagger \hat{a}_2^\dagger \hat{a}_2 \hat{a}_2 - J\left(\hat{a}_1^\dagger \hat{a}_2 + \hat{a}_2^\dagger \hat{a}_1 + \hat{a}_2^\dagger \hat{a}_3 + \hat{a}_3^\dagger \hat{a}_2\right)$$
$$+ \frac{\Omega}{2}\left[\hat{a}_1^\dagger + \hat{a}_1 + \hat{a}_3^\dagger e^{i\varphi} + \hat{a}_3 e^{-i\varphi}\right]. \tag{1}$$

Here $\hat{a}_k$ annihilates an excitation in transmission line resonator k. For the two outer resonators the excitations are photons whereas for the central resonators they are excitations of the considered polariton mode. $\Delta_k = \omega_k - \omega_{\text{in}}$ is the detuning between the input drives and resonance frequency of excitations in the kth transmission line resonator. For simplicity, we only consider the on-resonance case with $\Delta_k = 0$. The parameter U describes the effect of the Kerr nonlinearity due to the transmon in the central resonator and J is the tunneling rate between neighboring resonator sites, Ω is the drive amplitude, assumed to be equal for the two inputs, and φ describes the phase difference between them. At this point it is important to note that the nonlinearity of a transmon qubit is attractive (corresponding to $U > 0$ in our notation).

The circuit QED Josephson interferometer is an open quantum system due to experimental imperfections such as the finite reflectivity of the coupling capacitors at each end of the transmission line resonators and relaxation of the transmon. We consider here the case where excitations in all three transmission line resonators have the same dissipation rates. As excitations in the second resonator can also decay via transmon relaxation, equal dissipation rates can be achieved for example by choosing suitable capacitances at the ends of the transmission lines. Since the couplings to the environments are sufficiently small [19, 20] we can make use of the master equation approach [21]. As a simplification we assume the reservoir state to be in vacuum, which is justified by the fact that circuit QED experiments are typically performed at a temperature in the millikelvin range [22]. These considerations lead to the master equation

$$\dot{\rho} = \mathcal{L}\rho, \tag{2}$$

where the Liouvillian $\mathcal{L}$ is given by

$$\mathcal{L}\rho = i[\rho, \hat{H}] + \frac{\gamma}{2}\sum_{k=1}^{3}(2\hat{a}_k\rho\hat{a}_k^\dagger - \hat{a}_k^\dagger\hat{a}_k\rho - \rho\hat{a}_k^\dagger\hat{a}_k), \tag{3}$$

the dot denotes a time derivative and γ describes the excitation loss rate of an individual resonator site. Since the excitation losses are continuously compensated by a coherent input drive, a dynamical equilibrium leading to a steady state will be established. We are here interested in properties of this steady state, such as the mean excitation number and its correlation functions, mainly for the central resonator.

In cavity or circuit quantum electrodynamics experiments the output fields of the investigated resonators are typically accessible by measurements. Hence in our device the quantity of interest in measurements is the output field of the central resonator. It is linked to the field in the resonator via the input-output relation [23],

$$\hat{a}_{\text{out}}(t) = \sqrt{\kappa}\hat{a}_{\text{res}}(t) + \hat{a}_{\text{in}}(t), \tag{4}$$

where $\hat{a}_{\text{out}}$ and $\hat{a}_{\text{in}}$ are the output and input fields and a_{res} the field in the resonator which leaks out at a rate κ. In cavity and circuit QED the photons in the cavity usually couple to the photon modes in the output channel so that one can identify the cavity field $\hat{a}_{\text{res}}$ with the annihilation operator of photons in the cavity. In the central cavity of our setup however, two polaritonic excitations build up due to the strong coupling to the qubit and the photon field in the central resonator can be expressed as a superposition of them, $\hat{a}_{\text{res}} = \eta\hat{a}_2 + \eta'\tilde{a}_2$ with coefficients η and η'. (Note that η can always be chosen to be positive and that $|\eta|^2 + |\eta'|^2 = 1$.) In the regime of strong coupling we are interested in, the frequencies of the polariton modes $\hat{a}_2$ and $\tilde{a}_2$ differ strongly enough so that the input drives selectively only excite $\hat{a}_2$ and we can neglect the $\tilde{a}_2$ excitations in the input-output relation since they only contribute vacuum output. Hence for calculating the output fields of the central resonator we can use the modified input-output relation,

$$\hat{a}_{\text{out}}(t) = \sqrt{\gamma}\hat{a}_2(t) + \hat{a}_{\text{in}}(t), \tag{5}$$

where $\gamma = \eta\kappa$. For vacuum input fields, we thus obtain an out photon flux $\langle \hat{a}^\dagger_{out}\hat{a}_{out}\rangle = \gamma\langle \hat{a}^\dagger_2\hat{a}_2\rangle$ for the central resonator.

The quantumness or non-classicality of light fields can be characterized by the statistics of their excitations. Here a quantity of central interest is the so-called (time-resolved) second order correlation function, here for resonator 2,

$$g^{(2)}(\tau) = \frac{\langle \hat{a}_2(t)^\dagger \hat{a}_2^\dagger(t+\tau)\hat{a}_2(t+\tau)\hat{a}_2(t)\rangle}{|\langle \hat{a}_2(t)^\dagger \hat{a}_2(t)\rangle|^2}. \tag{6}$$

The significance of the physical interpretation is to have a value for the probability to measure a second particle at time $t + \tau$ after a particle has been measured at time t and compare this to a coherent field where $g^{(2)} = 1$. Therefore, if $g^{(2)} < 1$ it is less likely to measure a second particle. In this case we speak of an anti-bunched light field that is necessarily non-classical, whereas for $g^{(2)} > 1$ we have bunched light meaning a higher probability than for a coherent field. We note that the $g^{(2)}$-function for the output fields is identical to the $g^{(2)}$-function of intra-cavity fields provided the input fields are in vacuum.

Let us finally stress that although we refer to an implementation in circuit electrodynamics here, our calculations do in most parts not make use of details of this technology and are therefore applicable to other realizations as well. Yet we prefer to discuss them in the context of circuit QED as this technology offers excellent perspectives for realizing a coupling between resonators as we envision it here.

3 The numerical approach in the weak driving regime

In the regime where the laser intensity is rather low, we are able to numerically solve the master equation directly. Here, an entire regime of rather high nonlinearities $U/\gamma > 1$, which lessen the probability of high intra-cavity excitations, can be covered for arbitrary values of the tunneling rate J. We introduce a cut-off in the excitation number to give a matrix expression for the excitation ladder operators in a Fock basis of bosonic number states. This reduces the infinitely high dimensional Hilbert space to a small number of dimensions and hence allows for the numerical calculation of all operator expectation-values. To corroborate our results we test their convergence as we increase the excitation number cut-off.

For the steady states we consider, $g^{(2)}(\tau)$ as given in equation (6) is independent of the time t and only depends on the time delay τ. In the weak driving regime the results for no delay $\tau = 0$ have already been covered in [14] and we therefore focus for this regime on the time-resolved second order correlation function for $\tau \neq 0$.

Applying the quantum regression theorem (QRT) [24] we find the results presented in Figure 3. Here, as expected, the $g^{(2)}(\tau)$-function converges to 1 for a long time delay $\tau \to \infty$. The amplitudes of the flucutations around $g^{(2)}(\tau) = 1$ increase with increasing nonlinearities and increasing tunneling rates J, however at $\tau = 0$ the system is more excitation anti-bunched if the tunneling strength is weaker and the nonlinearity higher.

With the used approach of directly solving the master-equation numerically in a truncated Hilbert-space one can cover a regime for arbitrary values of J if the nonlinearity fulfills $U/\gamma > 1$ in the weak driving regime of $|\Omega|/\gamma < 2$. Nevertheless, this method is unable to show a clear crossover between anti-bunched and uncorrelated excitations because the weak drive and high nonlinearity restrict it to low numbers of excitations, implying anti-bunched statistics. Therefore we will present a method to extend the investigation of this

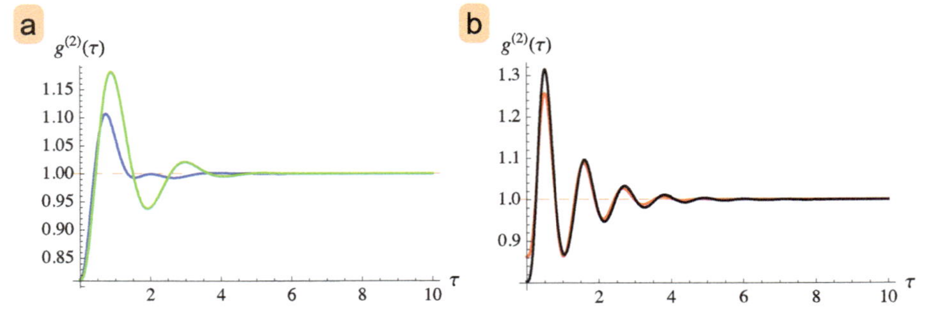

Figure 3 The second order correlation function $g^{(2)}(\tau)$ as found from a fully numerical solution. (**a**) For $\varphi = \Delta = 0$, $J/\gamma = 2$, $U/\gamma = 2$ and $\Omega/\gamma = 1$ (green) or $\Omega/\gamma = 2$ (blue). (**b**) For $\varphi = \Delta = 0$, $J/\gamma = 4$, $U/\gamma = 2$ and $\Omega/\gamma = 1$ (black) or $\Omega/\gamma = 2$ (red).

set-up to the strongly driven regime, motivated by the possibility of high intra-resonator excitations.

4 The quantum fluctuations approach in the strong driving regime

For higher laser intensities, the number of excitations increases rapidly. Since the two outer transmission line resonators do not directly couple to qubits, their only deviation from a coherent state stems from the coupling to the central resonator, which features a nonlinearity due to its coupling to a transmon qubit. Therefore it is for strong inputs no longer computationally feasible to numerically solve the full master equation. Assuming a weak nonlinearity, we thus use a generalized Bogoliubov approach to non-equilibrium steady states [12], which linearizes the equations for the quantum fluctuations around the classical fields and can be understood as an approximation for low nonlinearities. We expand the creation and annihilation operators around the coherent state by writing

$$\hat{a}_j = \alpha_j + \widehat{\delta a}_j, \tag{7}$$

$$\hat{a}_j^\dagger = \alpha_j^* + \widehat{\delta a}_j^\dagger, \tag{8}$$

where the complex number $\alpha_j = \langle \hat{a}_j \rangle$ represents the coherent background and the operators $\widehat{\delta a}_j$ and $\widehat{\delta a}_j^\dagger$ describe the quantum fluctuations around it. Using equations (7) and (8) we expand the master equation in powers of the operators $\widehat{\delta a}_j$ and $\widehat{\delta a}_j^\dagger$,

$$\dot{\rho} = \left(\mathcal{L}^{(1)} + \mathcal{L}^{(2)} + \mathcal{L}^{(3)} + \mathcal{L}^{(4)}\right)[\rho], \tag{9}$$

where $\mathcal{L}^{(n)}[\rho]$ denotes the nth order of quantum fluctuations $\widehat{\delta a}_j$ in the Liouvillian $\mathcal{L}$.

4.1 Steady state coherent background

To determine the coherent background in terms of the α_j, we require that the first order Liouvillian vanishes for all quantum fluctuations, $\mathcal{L}^{(1)} \stackrel{!}{=} 0$. The result is equivalent to the solution for the steady state coherent background. With the above requirement we get a cubic equation for the steady state mean excitation number $n_2 = |\alpha_2|^2$ of the coherent background in the second cavity

$$4\gamma^2 U^2 n_2^3 + \left(\gamma^2 + 8J^2\right)^2 n_2 = 8J^2\Omega^2(1 + \cos\varphi) \tag{10}$$

which has only one physically meaningful solution with $n_2 \geq 0$:

$$n_2 = \frac{\eta^2 - 3^{1/3} U^2 \gamma^2 (\gamma^2 + 8J^2)^2}{2 \times 3^{2/3} U^2 \gamma^2 \eta}, \tag{11}$$

where

$$\eta = \sqrt[3]{72\gamma^4 J^2 U^4 \Omega^2 (\cos(\varphi) + 1) + \sqrt{3} \sqrt{\gamma^6 U^6 \left(6912 \gamma^2 J^4 U^2 \Omega^4 \cos^4\left(\frac{\varphi}{2}\right) + (\gamma^2 + 8J^2)^6 \right)}}. \tag{12}$$

The values of α_1 and α_3 can be determined from the solution (11) via the additional relations

$$\alpha_1 = \frac{i(2J\alpha_2 - \Omega)}{\gamma}, \tag{13}$$

$$\alpha_3 = \frac{i(2J\alpha_2 - \Omega e^{i\varphi})}{\gamma}. \tag{14}$$

The solution (11) is plotted in Figure 4. One can clearly see that the coherent background decreases for increasing nonlinearities. This feature appears since the frequencies of the input drives are resonant with the transition between zero- and single-excitation states in the resonators. Hence with increasing nonlinearity it becomes less probable that higher excitation states become populated as well. Moreover the population of the coherent part of the intra-resonator fields shows a maximum for specific values of $J \geq 0$. These maxima can be understood when writing the Hamiltonian (1) in terms of Bloch modes [12], which have frequencies that depend on the tunneling rate J. Thus with increasing J a balance establishes between the higher detuning between Bloch modes and input drives leading to less excitations and an increased tunneling leading to more excitations in the central cavity. Here, it can already be seen that the number of polaritons in the second resonator exceeds the computational feasibility of conventional numerical methods to calculate the 3-site system.

Figure 5 shows the effects of varying the phase-difference φ of the two coherent input laser sources. One immediately notices the emptying of the central cavity for a phase-difference $\varphi = (2m + 1)\pi$ with integer m. The oscillation of the excitation number in the central cavity is a result of the interference between the two coherent input drives. On the other hand, changing the tunneling-rate does not affect the minimum, but changes the severeness of the interference effect in agreement with the previously visualized results in Figure 4. We further find (not plotted) that for an increase in driving strength the oscillation due to the phase-difference φ is steeper, changes the value of J for which the background occupation number has a maximum and decreases the incline around it. Also the sharpness of the peak of maximal background occupation is less pronounced for higher nonlinearities.

4.2 Steady state quantum fluctuations

We are mainly interested in the statistics of output photons from the central resonator. These are described by the mean excitation number $\langle \hat{n}_2 \rangle$ and the second-order correlation

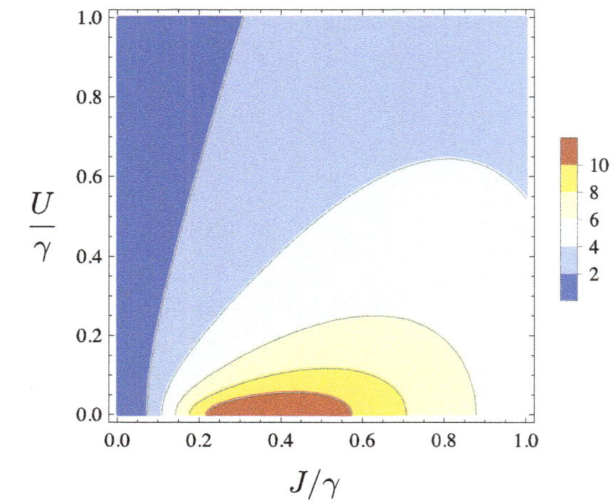

Figure 4 The coherent excitation number $n_2 = |\alpha_2|^2$ in the second transmission line resonator as a function of the nonlinearity U/γ and the tunneling rate J/γ for in-phase laser input $\varphi = 0$ and $\Omega/\gamma = 5$. Stronger drives lead to even higher excitation numbers n_2.

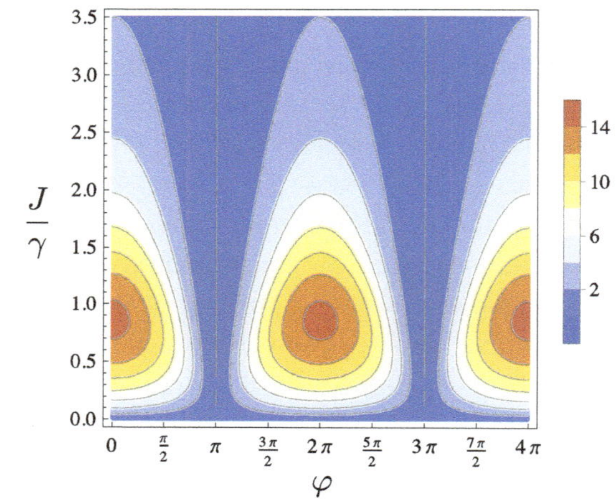

Figure 5 Background excitation number n_2 as a function of the phase difference φ and tunneling J/γ. Due to a destructive interference, excitations are suppressed for $\varphi = (2m + 1)\pi$ with integer m. The plot shows the effect for a low nonlinearity $U/\gamma = 0.2$ and a strong drive $\Omega/\gamma = 10$.

function $g_2^{(2)}$. To calculate the mean excitation number in the steady state, we can write in the expansion approach,

$$\langle \hat{a}^\dagger \hat{a} \rangle = |\alpha|^2 + \langle \widehat{\delta a}^\dagger \widehat{\delta a} \rangle, \tag{15}$$

where we have dropped the site index as we are only focusing on resonator 2. The second order correlation function has been introduced in Eq. (6). The case with zero time-delay $\tau = 0$ is important in order to determine whether the output light field exhibits Poissonian or sub-Poissonian statistics. We expand the $g^{(2)}$-function to get

$$g^{(2)}(\tau = 0) = \frac{|\alpha|^4 + 4|\alpha|^2 \langle \widehat{\delta a}^\dagger \widehat{\delta a} \rangle + (\alpha^*)^2 \langle \widehat{\delta a} \widehat{\delta a} \rangle + (\alpha)^2 \langle \widehat{\delta a}^\dagger \widehat{\delta a}^\dagger \rangle}{|\alpha|^4 + 2|\alpha|^2 \langle \widehat{\delta a}^\dagger \widehat{\delta a} \rangle}. \tag{16}$$

Therefore, in order to calculate the important statistical properties of our system we need to combine the previously derived results of the coherent background with the mean correlation values of the quantum fluctuations. We thus solve the master equation (2) to second order in the quantum fluctuations for ρ,

$$\mathcal{L}^{(2)}[\rho] = 0, \tag{17}$$

using $\mathcal{L}^{(1)}[\rho] = 0$ and plugging the solution (11) into $\mathcal{L}^{(2)}[\rho]$. For the considered case of $\Delta = 0$, the eigenvalues of $\mathcal{L}^{(2)}[\rho]$ all have real parts $-\gamma/2$ so that the solutions are stable. Neglecting higher order terms in the master equation, *i.e.* $\mathcal{L}^{(3),(4)}[\rho] \approx 0$, is a good approximation provided that quantum fluctuations are small compared to the coherent background,

$$\langle \widehat{\delta a}_j^\dagger \widehat{\delta a}_j \rangle \ll |\alpha_j|^2 \quad \text{for } j = 1, 2, 3. \tag{18}$$

To analyze the validity of the solutions we find, we thus check whether they fulfill this property.

For the central transmission line resonator, the solution for the mean quantum fluctuation number reads

$$\langle \widehat{\delta a}_2^\dagger \widehat{\delta a}_2 \rangle = \frac{2n_2^2 U^2(\gamma^4 + 16J^4 + 3\gamma^2(2J^2 + n_2^2 U^2))}{(\gamma^2 + 3n_2^2 U^2)(\gamma^4 + 64J^4 + 4\gamma^2(4J^2 + 3n_2^2 U^2))}. \tag{19}$$

Considering the fact that we neglected the terms of third and fourth order in quantum fluctuations, which contain additional pre-factors of the nonlinearity U, we come to the conclusion that our approach can also be considered an expansion in U. As a consequence, we can deduce the results represent a good approximation in a regime of weak nonlinearities U. Moreover, the approach becomes much better for higher laser-input intensities because the coherent background excitation number in Eq. (11) scales much stronger with the laser input amplitude Ω than the number of quantum fluctuations in Eq. (19).

In contrast to the mean excitation number in the second resonator, the quantum fluctuations do play a significant role for the $g^{(2)}$-function, see Eq. (16). The second order correlation function of the central resonator in the strong driving regime is plotted in Figure 6. Depending on the tunneling rate J we observe smooth transitions from a coherent to an anti-bunched excitation field as the nonlinearity is increased beyond specific values. Figure 7 shows that the oscillations in the excitation number due to interference effects can be observed in the $g^{(2)}$-function as well.

4.3 Limitations of the Bogoliubov approach

The major limitation of the Bogoliubov approach is obviously the validity of the assumption (18). In order to determine the regime where this is fulfilled, we plot the ratio $\langle \widehat{\delta a}_i^\dagger \widehat{\delta a}_i \rangle / |\alpha_i|^2$ $(i = 1, 2, 3)$. For parameters where this ratio is sufficiently small, the approximation works fairly well. The ratio $\langle \widehat{\delta a}_2^\dagger \widehat{\delta a}_2 \rangle / |\alpha_2|^2$ is plotted in Figure 8, whereas Figure 9 depicts the same ratio for the two outer resonator sites, *i.e.* $\langle \widehat{\delta a}_1^\dagger \widehat{\delta a}_1 \rangle / |\alpha_1|^2 = \langle \widehat{\delta a}_3^\dagger \widehat{\delta a}_3 \rangle / |\alpha_3|^2$.

As a conclusion, we can safely assume that our approximation works well in a regime for strong input drives $\Omega/\gamma > 5$ and low nonlinearities $U/\gamma < 0.2$ for arbitrary values of the tunneling rate. For larger drives, such as $\Omega/\gamma > 10$ the range of validity extends to

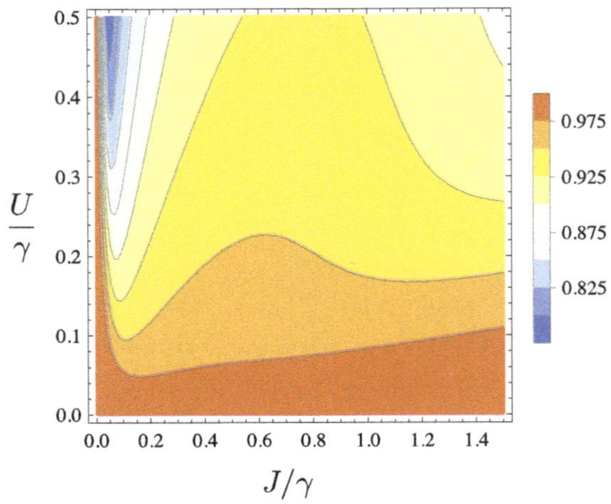

Figure 6 The $g^{(2)}$-function as given by equation (16) as a function of J/γ and U/γ in the in-phase strong driving regime for $\Omega/\gamma = 5$. The red color indicates a coherent field, where $g^{(2)} \approx 1$, whereas the blue area marks excitation-antibunching with $g^{(2)} < 1$.

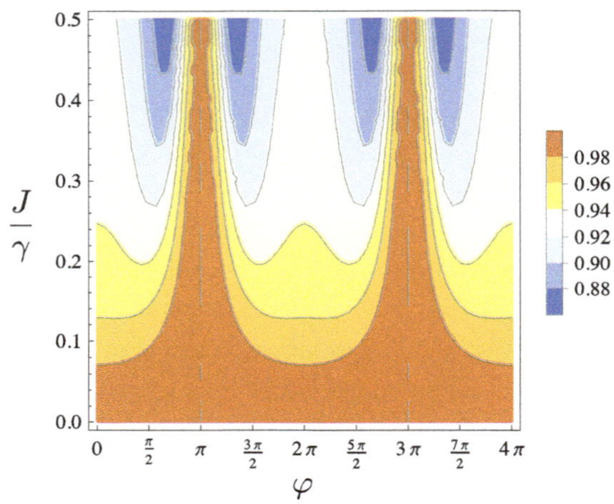

Figure 7 The $g^{(2)}$-function as given by equation (16) as a function of φ and U/γ for $J/\gamma = 1$ and $\Omega/\gamma = 5$. The red color indicates a coherent field, where $g^{(2)} \approx 1$, whereas the blue area marks excitation-antibunching with $g^{(2)} < 1$.

larger U/γ. Thus we are motivated to find access to the regime of larger nonlinearities U/γ because, as is evident from Figure 8, the assumption $\langle \widehat{\delta a}_i^\dagger \widehat{\delta a}_i \rangle \ll |\alpha_i|^2$ breaks down in that regime, primarily in the second resonator. We present an approach to this parameter regime in the next section.

5 *P*-Function mean-field approach to the low tunnelling regime

As one can see from Figures 8 and 9, the main limitation of a Bogoliubov approximation is the ratio of the mean quantum fluctuation number to the mean excitation number in the second cavity which significantly increases for large nonlinearities U and weak drives Ω. Hence if we are able to take higher quantum fluctuations in resonator 2 into account the results should become more exact and be applicable to a larger parameter range. On the other hand the deviation from a coherent state for the first and third transmission line

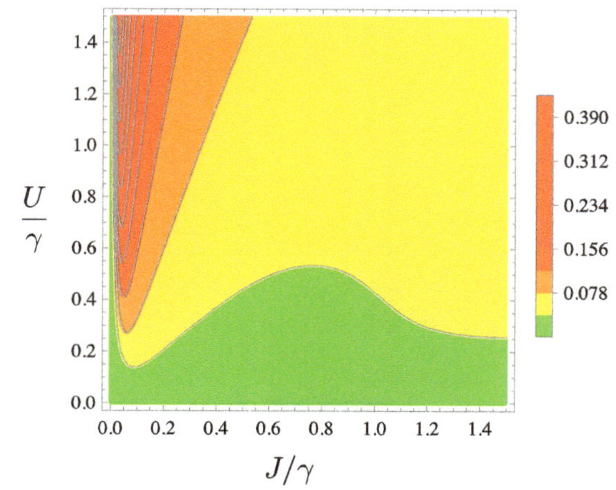

Figure 8 Ratio of quantum fluctuations to the coherent background excitation number for the mid-resonator, $\langle \widehat{\delta a}_2^\dagger \widehat{\delta a}_2 \rangle / |\alpha_2|^2$ as a function of J/γ and U/γ in the strong driving regime with $\Omega/\gamma = 5$. For higher drives Ω the ratio becomes smaller and also stays small up to larger nonlinearities U.

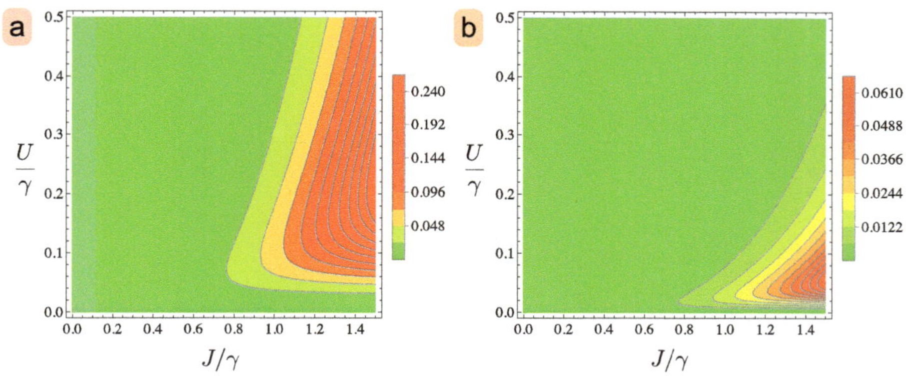

Figure 9 Ratio of quantum fluctuations to the coherent background excitation number for the two outer resonators, $\langle \widehat{\delta a}_1^\dagger \widehat{\delta a}_1 \rangle / |\alpha_1|^2 = \langle \widehat{\delta a}_3^\dagger \widehat{\delta a}_3 \rangle / |\alpha_3|^2$ as a function of J/γ and U/γ, (a) $\Omega/\gamma = 5$, (b) $\Omega/\gamma = 10$.

resonator is extremely small because they do not contain any nonlinearity themselves and the nonlinearity of the center cavity only slightly affects the two outer cavities. Thus, if we are able to find a way to solve the second transmission line resonator's mean values exactly, our results should become better. This will be the ansatz of our mean-field approach where we ignore quantum fluctuations in the two outer resonators and substitute $\hat{a}_{1,3} \rightarrow \alpha_{1,3}$. With this substitution, the Hamiltonian for resonator 2 takes the form

$$\hat{H}_{MF} = \Delta\omega \hat{a}_2^\dagger \hat{a}_2 - \frac{U}{2} \hat{a}_2^\dagger \hat{a}_2^\dagger \hat{a}_2 \hat{a}_2 - J(\alpha_1 + \alpha_3)\hat{a}_2^\dagger - J\left(\alpha_1^* + \alpha_3^*\right)\hat{a}_2 \tag{20}$$

which can be written in the form of a coherently driven Kerr nonlinearity [15],

$$\hat{H}_{MF} = \Delta\omega \hat{a}_2^\dagger \hat{a}_2 - \frac{U}{2} \hat{a}_2^\dagger \hat{a}_2^\dagger \hat{a}_2 \hat{a}_2 + \frac{\tilde{\Omega}}{2} \hat{a}_2^\dagger + \frac{\tilde{\Omega}^*}{2} \hat{a}_2 \tag{21}$$

by introducing the drive $\tilde{\Omega} = -2J(\alpha_1 + \alpha_3)$. To determine α_1 and α_3 self consistently, we use equations (13) and (14) to obtain,

$$\tilde{\Omega} = \frac{2iJ}{\gamma}\left(\Omega + \Omega e^{i\varphi} - 4J\langle\hat{a}_2\rangle\right) \tag{22}$$

in the $\Delta = 0$ case. Our approach can thus be understood as a mean-field decoupling of the three cavities. With the consistency condition (22), the Hamiltonian (21) becomes a single site model which can be solved exactly using a P-function based method introduced by Drummond and Walls [15].

What we have done is basically a trade-off: We allow for a more crude approximation of the quantum tunneling effects in the two outer cavities by completely neglecting their quantum fluctuations. The latter are however expected to be very small, see Figure 9. In turn we gain an exact solution for the resulting one-site problem, giving a better approximation of the nonlinear effects in the second transmission line resonator. In contrast to the Bogoliubov expansion that has been performed in Section 4, we here do not expand in U but in the tunneling rate J [25]. With the approach introduced here we thus cover a regime with arbitrary values for the nonlinearity U but only for moderate values of J. The solution for the normal-ordered mean values of the middle transmission line resonator in the $\Delta = 0$ case reads

$$\langle(\hat{a}_2^\dagger)^k(\hat{a}_2)^l\rangle = \left(\frac{\tilde{\Omega}^*}{U}\right)^k\left(\frac{\tilde{\Omega}}{U}\right)^l \frac{\Gamma(-i\frac{\gamma}{U})\Gamma(i\frac{\gamma}{U})}{\Gamma(k-i\frac{\gamma}{U})\Gamma(l+i\frac{\gamma}{U})} \frac{{}_0\mathcal{F}_2(k-i\frac{\gamma}{U}, l+i\frac{\gamma}{U}, 2|\frac{\tilde{\Omega}}{U}|^2)}{{}_0\mathcal{F}_2(-i\frac{\gamma}{U}, i\frac{\gamma}{U}, 2|\frac{\tilde{\Omega}}{U}|^2)}, \tag{23}$$

where

$${}_0\mathcal{F}_2(c,d,z) = \sum_{n=0}^{\infty}\left(\frac{z^n\Gamma(c)\Gamma(d)}{\Gamma(c+n)\Gamma(d+n)n!}\right) \tag{24}$$

is the generalized Gauss hypergeometric series, defined in terms of the Γ-function. Note that $\tilde{\Omega}$ is a function of $\langle\hat{a}_2\rangle$, see Eq. (22). The starting point of our approach is thus to find the solution of the expectation value $\langle\hat{a}_2\rangle$ from Eq. (23) which, after inserting equation (22), becomes a nonlinear algebraic equation for $\langle\hat{a}_2\rangle$,

$$\begin{aligned}\langle\hat{a}_2\rangle &= \frac{2J}{\gamma^2}\left(\Omega(1+e^{i\varphi}) - 4J\langle\hat{a}_2\rangle\right)\\ &\times \frac{{}_0\mathcal{F}_2(-\frac{i\gamma}{U}, 1+\frac{i\gamma}{U}, 2|\frac{2iJ}{\gamma U}(\Omega(1+e^{i\varphi}) - 4J\langle\hat{a}_2\rangle)|^2)}{{}_0\mathcal{F}_2(-\frac{i\gamma}{U}, \frac{i\gamma}{U}, 2|\frac{2iJ}{\gamma U}(\Omega(1+e^{i\varphi}) - 4J\langle\hat{a}_2\rangle)|^2)}.\end{aligned} \tag{25}$$

The result for $\langle\hat{a}_2\rangle$ can be put into Eq. (22) allowing for an expression of the 'new' drive $\tilde{\Omega}$. Based on this solution the correlation values for all higher normal-ordered momenta in Eq. (23) can be subsequently derived in a self-consistent manner.

Figure 10 shows the behaviour of the second order correlation function for excitations in the second cavity depending on the external experimental parameters. As expected, for low nonlinearities the state in the middle transmission line resonator is close to a coherent state. An increase of the nonlinearity leads to a state becoming more and more anti-bunched. In contrast to the low excitation regime [14], we here find that $g^{(2)}(0)$ does

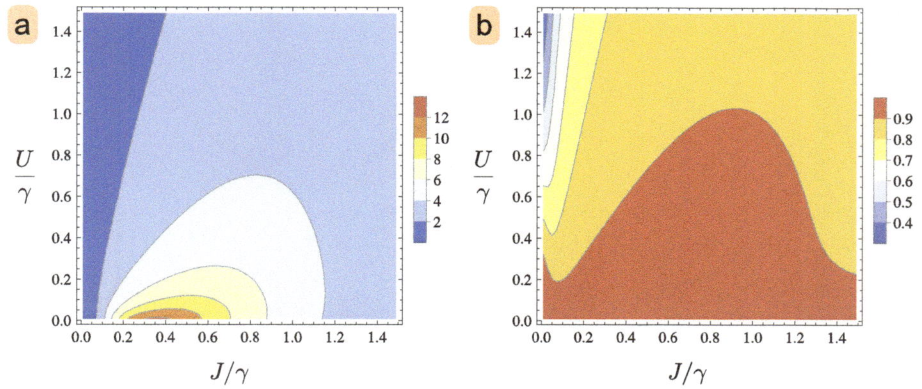

Figure 10 The mean polariton number (a) and $g_2^{(2)}$-function (b) of the central resonator in the strong driving regime, $\Omega/\gamma = 5$, as a function of J/γ and U/γ. The plots show the results gained from equation (23). Due to the previously explained nature of our approach being a good approximation for lower tunneling strength, we are able to cover an additional regime compared to the Bogoliubov expansion, which broke down for higher values of the nonlinearity. (Compare with Fig. 8.)

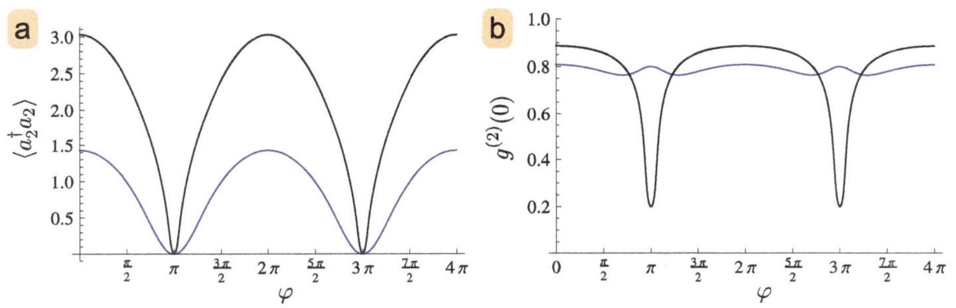

Figure 11 Dependence of the second resonator's excitation statistics on the phase difference φ of the input laser drives for $\Omega/\gamma = 5$, $U/\gamma = 0.5$ and $J/\gamma = 0.1$ (blue) or $\Omega/\gamma = 10$, $U/\gamma = 2$ and $J/\gamma = 0.5$ (black) as obtained from equation (23). (**a**) Mean excitation number and (**b**) Second order correlation function. The oscillation in the mean excitation number can be attributed to the interference of the two input laser fields.

not monotonously approach unity as the tunneling J is increased. Instead anti-bunching can also become more pronounced. We attribute this non-monotonous behavior to the competition of two effects. On the one hand, an increased tunneling rate means that more photons from the coherent fields in the outer resonators flow into the central resonator and weaken the anti-bunching. On the other hand, an increased tunneling rate detunes the normal modes of the three-cavity system from the laser drives so that the outer cavities are less populated. If this second effect starts to dominate over the previous one, the photon flow into the central resonator decreases and its excitations show more tendency to anti-bunch. This signature is also visible, albeit less pronounced, in Figure 8.

In Figure 11 we investigate the influence of the phase-difference φ on the excitation statistics of the central transmission line resonator. The mean excitation number fluctuates with the phase-difference φ, exhibiting maxima at $\varphi = 2n\pi$ and minima at $\varphi = 2(n+1)\pi$, where n is an integer, due to constructive and destructive interference of the input laser drives respectively. An increase in the tunneling rate from $J/\gamma = 0.1$ to $J/\gamma = 0.5$ enables a higher number of excitations to tunnel into, and therefore excite, the central

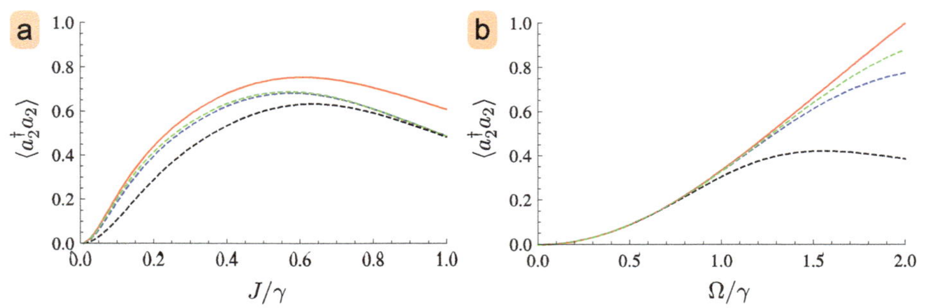

Figure 12 Comparison of the results for mean polariton number in the central cavity obtained by the *P*-function mean-field method, red line, and by the conventional numerical method, dashed lines, with cut-offs at three (black), four (blue) and five (green) excitations. (**a**) Deviation due to an increase in the tunneling rate for $\Omega/\gamma = 1.5$ and $U/\gamma = 2$. (**b**) Deviations due to an increase in driving strength for $J/\gamma = 0.2$ and $U/\gamma = 0.5$.

transmission line resonator, leading to a higher number of excitations altogether. The $g^{(2)}$-function also shows oscillations, however, their behavior is very different for different tunneling rates. Whereas maxima appear for both tunneling rates at a phase-difference $\varphi = 2n\pi$, the second order correlation function for $J/\gamma = 0.1$ shows additional local maxima at $\varphi = (2n+1)\pi$. These may be attributed to the low excitation number at these points which induce a tendency for $g^{(2)}$-functions to increase.

5.1 Comparison to the numerical solution

Finally we quantify the parameter range where the P-function based mean-field approach we introduce here is in good agreement with a full numerical solution for regimes where the latter can be obtained. Figure 12(a) compares the mean-field solution (red) and full numerical solutions (dashed lines) for $\Omega/\gamma = 1.5$ and $U/\gamma = 2$ as the tunneling rate J is increased. Whereas the numerical solutions for cut-offs at four (blue) and five (green) excitations agree very well indicating that the cut-offs are high enough, one can see deviations between the numerical and the mean-field solutions as J increases. These should be interpreted as a failure of the mean-field approach for larger J. Figure 12(b) in turn compares the mean-field solution (red) and full numerical solutions (dashed lines) for $J/\gamma = 0.2$ and $U/\gamma = 0.5$ as the drive strength Ω is increased. Here the numerical solutions converge to the mean-field solution as the cut-off number is increased, indicating that the mean-field approach is a very accurate approximation throughout. The agreements and disagreements displayed in Figures 12(a) and (b) are in agreement with the intuition that enhancing the coherent drives at the outer cavities should make the fields in those cavities more coherent and hence the mean-field approximation more accurate whereas an increase in the tunneling rate enhances the influence of the nonlinearity on the outer cavities and hence drives them away from a coherent state which renders the mean-field approach inaccurate.

6 Conclusion and discussion of accessible regimes

In summary we have complemented the discussion of a quantum optical Josephson interferometer with two methods. One is a linearization or Bogoliubov expansion of the intracavity fields around their coherent parts. The second method, which we have introduced here, combines an exact solution for the central resonator with a mean-field decoupling of the tunneling terms to the outer resonators.

With the proposed methods the circuit QED Josephson interferometer can be described for a significant range of all experimentally adjustable parameters, such as the input drive Ω, the nonlinearity U and the tunneling strength J for a given dissipation rate γ. If one is faced with low laser intensities and thus low intra-cavity excitations, the solution can be obtained by solving the master-equation numerically via an excitation cut-off in the matrix representation of the ladder operators as has been shown in Section 3. In a strong driving regime with intermediate to high tunneling rates but low nonlinearities a good approximation to the excitation output statistics can be given by a Bogoliubov expansion in the quantum fluctuations around the coherent background parts of the intra-cavity fields, which we presented in Section 4. However, for a strong input drive, low tunneling rates and intermediate to high values of the nonlinearity the Bogoliubov expansion breaks down. Here we derived a method in Section 5, which we termed P-function mean-field approach, that works extremely well in said regime. The only regime for which a good approximation could not be gained was one where the drive is strong enough to produce high intra-cavity excitations, but the tunneling rate and nonlinearity are still relatively high. As a result they are responsible for non-neglectible quantum effects in comparison to the drive or to each other and therefore neither an expansion in U nor in J would be valid. A summary of the validity ranges for each of the presented methods is sketched in Figure 1.

Competing interests
The authors declare that they have no competing interests.

Authors' contributions
RJ performed the calculations and analysis, MH conceived and supervised the project, both authors discussed the results and wrote the manuscript.

Author details
[1]Physik Department, Technische Universitaet Muenchen, James Franck Str., Garching, 85748, Germany. [2]Department of Physics and Astronomy, University College London, Gower Street, London, WC1E 6BT, United Kingdom.

Acknowledgements
The authors thank Peter Degenfeld-Schonburg for fruitful discussions. This work is part of the Emmy Noether project HA 5593/1-1 and the CRC 631, both funded by the German Research Foundation, DFG.

References

1. Hartmann MJ, Brandao FGSL, Plenio MB: **Strongly interacting polaritons in coupled arrays of cavities.** *Nat Phys* 2006, **2**(12):849-855.
2. Angelakis DG, Santos MF, Bose S: **Photon-blockade-induced Mott transitions and XY spin models in coupled cavity arrays.** *Phys Rev A* 2007, **76**(3):031805.
3. Greentree AD, Tahan C, Cole JH, Hollenberg LCL: **Quantum phase transitions of light.** *Nat Phys* 2006, **2**(12):856-861.
4. Hartmann MJ, Brandao FG, Plenio MB: **Quantum many-body phenomena in coupled cavity arrays.** *Laser Photonics Rev* 2008, **2**(6):527-556.
5. Tomadin A, Fazio R: **Many-body phenomena in QED-cavity arrays [Invited].** *J Opt Soc Am B, Opt Phys* 2010, **27**(6):A130-A136.
6. Houck AA, Türeci HE, Koch J: **On-chip quantum simulation with superconducting circuits.** *Nat Phys* 2012, **8**(4):292-299.
7. Lucero E, Barends R, Chen Y, Kelly J, Mariantoni M, Megrant A, O'Malley P, Sank D, Vainsencher A, Wenner J, White T, Yin Y, Cleland AN, Martinis JM: **Computing prime factors with a Josephson phase qubit quantum processor.** *Nat Phys* 2012, **8**(10):719-723.
8. Koch J, Terri MY, Gambetta J, Houck AA, Schuster D, Majer J, Blais A, Devoret MH, Girvin SM, Schoelkopf RJ: **Charge-insensitive qubit design derived from the Cooper pair box.** *Phys Rev A* 2007, **76**(4):042319.
9. Fedorov A, Steffen L, Baur M, Da Silva M, Wallraff A: **Implementation of a Toffoli gate with superconducting circuits.** *Nature* 2011, **481**(7380):170-172.

10. Carusotto I, Gerace D, Tureci H, De Liberato S, Ciuti C, Imamoğlu A: **Fermionized photons in an array of driven dissipative nonlinear cavities.** *Phys Rev Lett* 2009, **103**(3):033601.
11. Nissen F, Schmidt S, Biondi M, Blatter G, Türeci HE, Keeling J: **Nonequilibrium dynamics of coupled qubit-cavity arrays.** *Phys Rev Lett* 2012, **108**(23):233603.
12. Hartmann MJ: **Polariton crystallization in driven arrays of lossy nonlinear resonators.** *Phys Rev Lett* 2010, **104**(11):113601.
13. Jin J, Rossini D, Fazio R, Leib M, Hartmann MJ: **Photon solid phases in driven arrays of nonlinearly coupled cavities.** *Phys Rev Lett* 2013, **110**(16):163605.
14. Gerace D, Türeci HE, Imamoglu A, Giovannetti V, Fazio R: **The quantum-optical Josephson interferometer.** *Nat Phys* 2009, **5**(4):281-284.
15. Drummond PD, Walls DF: **Quantum theory of optical bistability. I. Nonlinear polarisability model.** *J Phys A, Math Gen* 1980, **13**(2):725.
16. Le Boité A, Orso G, Ciuti C: **Steady-state phases and tunneling-induced instabilities in the driven dissipative Bose-Hubbard model.** *Phys Rev Lett* 2013, **110**:233601. doi:10.1103/PhysRevLett.110.233601.
17. Peropadre B, Zueco D, Wulschner F, Deppe F, Marx A, Gross R, García-Ripoll JJ: **Tunable coupling engineering between superconducting resonators: from sidebands to effective gauge fields.** *Phys Rev B* 2013, **87**:134504. doi:10.1103/PhysRevB.87.134504.
18. Leib M, Hartmann MJ: **Bose-Hubbard dynamics of polaritons in a chain of circuit quantum electrodynamics cavities.** *New J Phys* 2010, **12**(9):093031.
19. Fink J, Göppl M, Baur M, Bianchetti R, Leek P, Blais A, Wallraff A: **Climbing the Jaynes-Cummings ladder and observing its nonlinearity in a cavity QED system.** *Nature* 2008, **454**(7202):315-318.
20. Bishop LS, Chow J, Koch J, Houck A, Devoret M, Thuneberg E, Girvin S, Schoelkopf R: **Nonlinear response of the vacuum Rabi resonance.** *Nat Phys* 2008, **5**(2):105-109.
21. Carmichael H: *An Open Systems Approach to Quantum Optics*. Lectures presented at the Université Libre de Bruxelles, October 28-November 4, 1991. Berlin: Springer; 1993. [*Lecture Notes in Physics Monographs*, vol. 18.]
22. Wallraff A, Schuster D, Blais A, Frunzio L, Huang RS, Majer J, Kumar S, Girvin S, Schoelkopf R: **Strong coupling of a single photon to a superconducting qubit using circuit quantum electrodynamics.** *Nature* 2004, **431**(7005):162-167.
23. Collett MJ, Gardiner CW: **Squeezing of intracavity and traveling-wave light fields produced in parametric amplification.** *Phys Rev A* 1984, **30**:1386.
24. Lax M: **Formal theory of quantum fluctuations from a driven state.** *Phys Rev* 1963, **129**(5):2342.
25. Degenfeld-Schonburg P, Hartmann MJ: **Self-consistent projection operator approach to quantum many-body systems.** arXiv:1307.7027 2013.

5

Topological zero-dimensional photonic modes in chiral coupled-cavity arrays

Vassilios Yannopapas*

*Correspondence:
vyannop@mail.ntua.gr
Department of Physics, National Technical University of Athens, Athens, 157 80, Greece

Abstract

We show theoretically that a properly chosen one-dimensional array of coupled photonic resonators (cavities) may possess localized zero-dimensional topological modes bearing resemblance with the corresponding edge modes/Majorana states in semiconductor nanowires atop a superconducting substrate. These modes constitute a manifold of degenerate states and are robust to the geometrical characteristics of the array, to the dielectric properties of the cavities and of the host medium. Such arrays can be realized in the laboratory as chains of microwave cavities within a metallic wire-network or as lattices of sinusoidally curved dielectric waveguides in the optical regime.

PACS Codes: 73.43.-f; 81.05.Xj; 42.70.Qs

Keywords: chirality; topological states; tight-binding; plasmons; cavities

1 Introduction

The topological properties of the quantum states define a new paradigm in the description and classification of condensed matter. Namely, atomic crystals possessing a topologically nontrivial electronic band structure constitute a new class of materials whose salient properties are robust to phase transitions which modify the symmetry order of the atomic solid. Typical examples of such topological solids are the integer/fractional quantum Hall (I/FQHE) systems and the topological insulators. Well-known examples of topological properties are the existence of chiral edge states in QHE systems and the presence of gapless surface states in topological insulators which are both insensitive to the presence of randomness/disorder.

Over the years there has been a continuous transfer of ideas and phenomena from atomic solids to their man-made electromagnetic (EM) counterparts, i.e., photonic crystals and metamaterials. Among several phenomena in traditional condensed matter physics which have found their analogues in artificial electromagnetic systems, the topological properties of matter have been the focus of intensive research in the past years motivated by two main scopes: the investigation of new states of photons in the context of quantum simulation and the design of disorder-immune integrated photonic devices. Examples of topological electronic systems which have been simulated in photonics are 2D lattices of gyromagnetic cylinders simulating the integer QHE [1–12], metamaterials simulating topological insulators [13–18], metamaterial-based microwave networks simulating the fractional QHE [19], and many others.

The main application in topological condensed matter comes from the presence of electron states at the boundaries of finite solids, i.e., chiral edge states in 2D systems (e.g., quantum spin Hall effect) and gapless surface states in 3D systems (topological insulators), which are robust to disorder. In finite one-dimensional (1D) systems with non-trivial topology, zero-dimensional localized states may appear in model Hamiltonians such as Kitaev's model [20]. In particular, Kitaev predicted that when a semiconductor nanowire is placed atop a superconducting substrate, Majorana-type fermionic states emerge at the edges of the nanowire; this is a phenomenon with important application in topologically fault-tolerant quantum computing [21–23]. Unfortunately, a proof-of-principle experiment for these zero-dimensional edge states requires a highly sophisticated laboratory setup. Unlike electron systems, topological zero-dimensional states can more easily simulated with photons either in the microwave regime with coupled-cavity arrays [24] or in the optical regime with arrays of metallic nanoparticles [25, 26]. Namely, the EM Green's tensor describing the coupling among the electric dipoles either in EM cavities [24] or nanoparticles [25] as well as the derived photonic frequency bands are in analogy with Bogoliubov-de Gennes equations of Kitaev's model of Majorana edge states in semiconductor nanowires. Besides classical electrodynamics, quantum simulators for Majorana states have been proposed in the context of cold atoms [27] and trapped ions [28].

In the present work we show that an array of coupled cavities immersed in a chiral host medium may support a *manifold* of zero-dimensional EM modes similarly to the fractional QHE, with obvious application in fault-tolerant quantum computing which is a non-local, decoherence-free type of quantum computing [21–23]. We note that we are working in a regime of high EM-radiation intensity in which case the classical theory of light (Maxwell's equations) is applicable. However, for low intensities where the quantum nature of light (photons) is employed, additional phenomena based on quantum correlations may arise which can possibly enrich the range of applications of the presented system such as, e.g., in multipartite entanglement [29]. The paper is structured as follows. Section 2 presents a tight-binding coupled-dipole formalism which, for suitable EM designs, transforms a classical electrodynamic problem described by Maxwell's equations to an eigenvalue problem equivalent to an electronic tight-binding Hamiltonian of ordinary atomic solids. Section 3 applies the formalism of Section 2 to a coupled-cavity array design which exhibits nontrivial topological frequency bands. Section 4 discusses the emergence of zero-dimensional edges modes in a coupled-cavity array. Section 5 assesses possible experimental realizations of topological coupled-cavity arrays and Section 6 concludes the paper.

2 Tight-binding coupled-dipole method

The EM crystals under study here are amenable to a photonic tight-binding description within the framework of the coupled-dipole method [30]. The latter is an exact means of solving Maxwell's equations in the presence of nonmagnetic scatterers. We consider a lattice of cavities within a lossless metallic host. The ith cavity is represented by a dipole of moment $\mathbf{P}_i = (P_{x;i}, P_{y;i}, P_{z;i})$ which stems from an incident electric field $\mathbf{E}^{inc}$ and the field which is scattered by all the other cavities of the lattice. This way the polarization vectors of all the cavities are coupled to each other and to the external field leading to the coupled-dipole equation

$$\mathbf{P}_i = \alpha_i(\omega)\left[\mathbf{E}^{inc} + \sum_{i'\neq i} \mathbf{G}_{ii'}(\omega)\mathbf{P}_{i'}\right]. \tag{1}$$

$\mathbf{G}_{ii'}(\omega)$ is the electric part of the free-space Green's tensor and $\alpha_i(\omega)$ is the polarizability of the ith cavity. Equation (1) is a $3N \times 3N$ linear system of equations where N is the number of cavities of the system.

For a particle/cavity of electric permittivity ϵ embedded within a material host of permittivity ϵ_h, the polarizability α is provided by the Clausius-Mossotti formula $\alpha = (3V/4\pi)(\epsilon - \epsilon_h)/(\epsilon + 2\epsilon_h)$, where V is the volume of the particle/cavity. For a lossless Drude-type (metallic) host i.e., $\epsilon_h(\omega) = 1 - \omega_p^2/\omega^2$ (where ω_p is the bulk plasma frequency), the polarizability α exhibits a pole at $\omega_0 = \omega_p\sqrt{2/(\epsilon + 2)}$ (surface plasmon resonance). By making a Laurent expansion of α around ω_0 and keeping the leading term [13], we may write $\alpha = F/(\omega - \omega_0) \equiv F/\Omega$ where $F = (27V/8\pi)\omega_0\epsilon/(2\epsilon + 4)$. For sufficiently high value of the permittivity of the dielectric cavity the electric field of the surface plasmon is much localized at the surface of the cavity. As a result, in a periodic lattice of cavities, the interaction of neighboring surface plasmons is very weak leading to much narrow frequency bands. By treating such a lattice in a tight binding (TB) manner [13], we may assume that the Green's tensor $\mathbf{G}_{ii'}(\omega)$ does not vary much with frequency and therefore, $\mathbf{G}_{ii'}(\omega) \simeq \mathbf{G}_{ii'}(\omega_0)$. In this case, Eq. (1) becomes an eigenvalue problem

$$\sum_{i' \neq i} \mathbf{G}_{ii'}(\omega_0)\mathbf{P}_{i'} = \Omega \mathbf{P}_i, \tag{2}$$

where F has been absorbed within the definition of $\mathbf{G}_{ii'}(\omega_0)$ and we have set $\mathbf{E}^{inc} = \mathbf{0}$ in Eq. (1) as we seek the eigenmodes of the system of cavities. In the following, we will be dealing with cavities lying in the same plane. We can, therefore, treat separately the case where the electric field lies is perpendicular to the chain axis (transverse modes) from the case where the electric field is parallel to the chain axis (longitudinal modes). In what follows, we will only deal with transverse modes.

In this case, $\mathbf{P}_i = (P_{x;i}, P_{y;i})$ and the Green's tensor $\mathbf{G}_{ii'}(\omega_0)$ is given by

$$\mathbf{G}_{ii'}(\omega_0) = Fq_0^3 C(q_0|r_{ii'}|)\mathbf{I}_2 \tag{3}$$

with $\mathbf{r}_{ii'} = (x_{ii'}, y_{ii'}, z_{ii'}) = \mathbf{r}_i - \mathbf{r}_{i'}$, $q_0 = \omega_0\sqrt{\epsilon_h(\omega_0)}/c$ and $\mathbf{I}_2$ is the 2×2 unit matrix. The vectors $\mathbf{r}_i$ denote the position of the cavities. Since we focus our attention around the surface plasmon frequency ω_0, we operate in the subwavelength regime where $q_0|r_{ii'}| \ll 1$. In this regime, the function $C(q_0|r_{ii'}|)$ is written as

$$\begin{aligned} q_0^3 FC(q_0|r_{ii'}|) &\simeq q_0^3 F \exp(\mathrm{i}q_0|r_{ii'}|)/(q_0|r_{ii'}|) \\ &= t_{ii'} \exp(\mathrm{i}\phi_{ii'}), \end{aligned} \tag{4}$$

where $t_{ii'}$ and $\phi_{ii'}$ are real numbers. In what follows, the cavities are connected via coupling elements, i.e., waveguides or transmission lines, in which case the phase factors ϕ_{ij} are not necessarily related with the wavevector of the host medium ϵ_h and can therefore be considered as independent parameters.

For a 1D lattice of cavities, we assume the Bloch ansatz for the polarization field, i.e.,

$$\mathbf{P}_i = (P_{x;i}, P_{y;i}) = \mathbf{P}_{n\beta} = \exp(\mathrm{i}kR_n)\mathbf{P}_{0\beta}. \tag{5}$$

The cavity index i becomes composite, $i \equiv n\beta$, where n enumerates the unit cell and β the positions of inequivalent cavities in the unit cell. Also, $R_n = na$, $n = -\infty \cdots \infty$, denotes the lattice vectors (a is the lattice period) and k is the Bloch wavevector. By substituting Eq. (5) into Eq. (2) we finally obtain

$$\sum_{\beta'} \tilde{\mathbf{G}}_{\beta\beta'}(\omega_0, k)\mathbf{P}_{0\beta'} = \Omega \mathbf{P}_{0\beta}, \tag{6}$$

where

$$\tilde{\mathbf{G}}_{\beta\beta'}(\omega_0, k) = \sum_{n'} \exp\left[ik(R_n - R_{n'})\right]\mathbf{G}_{n\beta;n'\beta'}(\omega_0). \tag{7}$$

Solution of Eq. (6) provides the transverse frequency band structure of a 1D periodic system of cavities.

We note that in the Green's tensor of Eq. (3) the Drude metal within which the cavities are formed is an isotropic medium. Since the case of a isotropic *chiral* host medium [31] will be needed in what follows, we provide the corresponding Green's tensor for the case of a 1D chain,

$$\mathbf{G}_{ii'}(\omega_0) = Fq_0^3 C(q_0|r_{ii'}|)\mathbf{I}_2 - z_{ii'}\mathrm{i}\eta\sigma_y, \tag{8}$$

where η is the chiral parameter of the host medium and σ_y is the corresponding Pauli matrix.

3 Topological frequency bands

Next we derive the EM Green's tensor for the 1D chain of Figure 1. Namely, we assume that the nearest neighbor (NN) hopping strengths are complex with alternating opposite arguments, $t\exp(\pm i\phi)$, i.e., the phase factors in the hopping strengths in the chain alternate as, $+\phi, -\phi, +\phi, -\phi, +\phi, -\phi, \ldots$.

We note that in EM lattices such as those considered here, a negative phase $-\phi$ can be easily achieved when the cavities are connected e.g., by 1D left-handed transmission lines, i.e. transmission lines supporting backward-propagating waves where the phase velocity is opposite to the group velocity [32, 33]. Alternatively, the cavities may be connected by waveguides loaded with a left-handed metamaterial. Obviously, a positive phase $+\phi$ can be achieved by similar means (right-handed transmission lines or waveguides loaded with a right-handed material). For the 1D chain of Figure 1, the Green's tensor of Eq. (8) becomes,

$$\tilde{\mathbf{G}}_{\beta\beta'}(k) = \begin{pmatrix} \mathbf{0}_2 & \tilde{\mathbf{G}}_{AB} \\ \tilde{\mathbf{G}}^*_{AB} & \mathbf{0}_2 \end{pmatrix}, \tag{9}$$

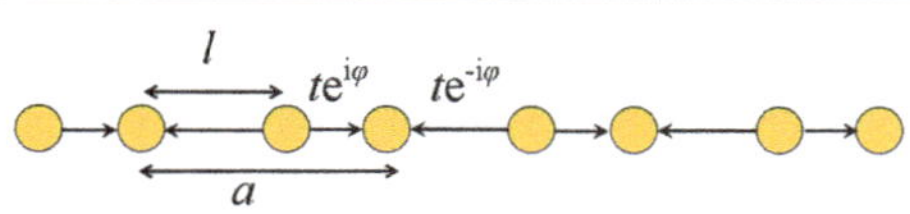

Figure 1 1D chain of cavities coupled with complex NN hopping strengths. The arrows denote the sign of the phase factor of the (complex) hopping strengths. Due to the dissimilar (opposite) phases in the hopping strengths between a cavity and its left/right first neighbours, the unit cell contains a 'diatomic' basis of length a.

Figure 2 Frequency band structure for $\phi = \pi/2$, $l = 0.3a$ of the chain of Figure 1 where the cavities are embedded in a chiral ($\eta = 0.5$ - solid lines) and non-chiral ($\eta = 0$ - dashed lines) host medium. The thick straight line at $\Omega = 0$ corresponds to the manifold of degenerate zero-dimensional modes.

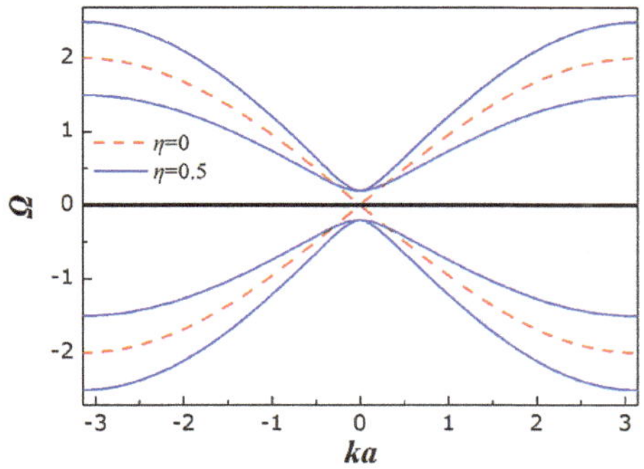

where

$$\tilde{\mathbf{G}}_{AB} = t[\exp(\mathrm{i}\phi) + \exp[-\mathrm{i}(ka + \phi)]]\mathbf{I}_2 - \mathrm{i}\eta[(l - a) + l\exp(-\mathrm{i}ka)]\sigma_y \tag{10}$$

and $\mathbf{0}_2$ is the 2×2 zero matrix.

For a non-chiral medium ($\eta = 0$) the (doubly degenerate) eigenvalues of Eq. (9) (frequency bands) are

$$\Omega_{\pm}(k) = \pm 2t\cos(ka/2 + \phi). \tag{11}$$

For $\eta \neq 0$, the expressions for the frequency bands are much more involved and will not be presented here - we only depict them graphically in what follows.

In Figure 2 we show the frequency bands for $\phi = \pi/2$ and for a nonchiral ($\eta = 0$) and a chiral ($\eta = 0.5$) medium hosting the cavities with $l = 0.3a$. Evidently, a Dirac point [34, 35] is present at $k = 0$ irrespective of the (nonzero) value of ϕ. For a non-chiral host medium, the introduction of chirality ($\eta = 0.5$) opens up a frequency gap in place of the Dirac point.

4 Zero-dimensional topological modes

The topological nature of the EM modes of the 1D chain of cavities under study can be revealed by assessing the modes of finite chains of Figure 1. The modes of a finite chain of N cavities are found by dropping the Bloch anzsatz of Eq. (5) and by diagonalizing instead the corresponding real-space $4N \times 4N$ Green's tensor $\mathbf{G}^r$

$$\mathbf{G}^r = \begin{pmatrix} G_{i,i} & G_{i,i+1} & 0 & 0 & 0 & \cdots & 0 & 0 & 0 \\ G_{i-1,i} & G_{i,i} & G_{i,i+1} & 0 & 0 & \cdots & 0 & 0 & 0 \\ 0 & G_{i-1,i} & G_{i,i} & G_{i,i+1} & 0 & \cdots & 0 & 0 & 0 \\ \vdots & 0 & \ddots & \ddots & \ddots & \cdots & G_{i-1,i} & G_{i,i} & G_{i,i+1} \\ 0 & \cdots & \cdots & \cdots & \cdots & \cdots & 0 & G_{i-1,i} & G_{i,i} \end{pmatrix}, \tag{12}$$

where

$$\mathbf{G}_{i,i} = \begin{pmatrix} \mathbf{0}_2 & \mathbf{B} \\ \mathbf{B}^* & \mathbf{0}_2 \end{pmatrix}, \quad \mathbf{G}_{i,i+1} = \begin{pmatrix} \mathbf{0}_2 & \mathbf{0}_2 \\ \mathbf{B}^* & \mathbf{0}_2 \end{pmatrix}, \quad \mathbf{G}_{i-1,i} = \begin{pmatrix} \mathbf{0}_2 & \mathbf{B} \\ \mathbf{0}_2 & \mathbf{0}_2 \end{pmatrix} \tag{13}$$

Figure 3 The manifold of degenerate zero-dimensional modes for a chain of 4 cavities. The arrows denote the polarization at a particular cavity. Each row represents a different mode of the manifold.

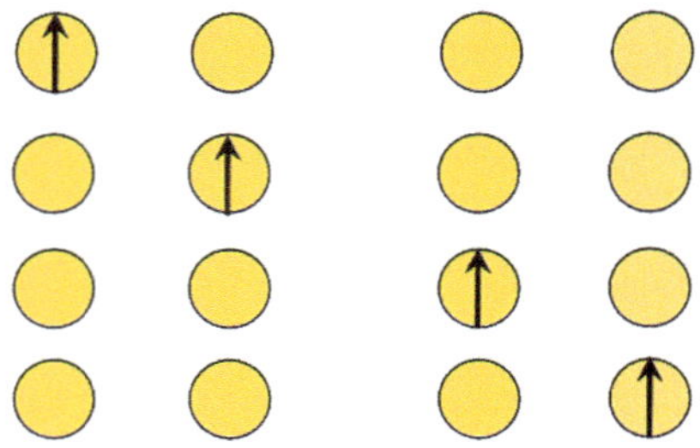

and

$$\mathbf{B} = t\exp(\mathrm{i}\phi)\mathbf{I}_2 + \mathrm{i}\eta(a - l)\sigma_y. \tag{14}$$

By calculating the eigenspectrum of Eq. (12) we find that half of the eigenmodes, i.e., $2N$ modes, are ordinary chain modes spread all over the chain whilst the other $2N$ modes constitute a *manifold* of modes which are degenerate at $\Omega = 0$ (thick straight line at $\Omega = 0$ in Figure 2). A similar manifold of states is met in the ground state of a fractional QHE system [36] and in quantum-optics based simulators of the fractional QHE [37]. Each of the degenerate modes is localized at a *single* cavity as illustrated in Figure 3. Namely, each row of Figure 3 corresponds to a different mode of the chain wherein the polarization vector is zero at all cavities except from one. A similar picture (4 degenerate states) holds when the polarization vectors are normal to the page. All these states constitute a manifold of degenerate states of the chain. The above degenerate zero-dimensional modes may find application in quantum-computing devices with multipartite entangelement [29].

We note that the above zero-dimensional modes are present with ($\eta \neq 0$) or without the presence of chirality ($\eta = 0$) in the host medium. In the non-chiral case, however, these modes may be fragile since they have a infinitesimal frequency difference from the EM modes spreading over the entire chain (and correspond to the Dirac-cone frequency bands). It is therefore mandatory to generate a frequency band gap at $\Omega = 0$ to isolate the degenerate manifold of states from the rest of states. And of course, the band gap should be also topologically nontrivial in order to preserve the degenerate zero-dimensional states. This is achieved by introducing chirality in our system (by assuming that $\eta \neq 0$ in Eqs. (8) and (10)), which is the optical analogue of spin-orbit coupling in electron systems [12]. Due to the non-trivial topological nature of these zero-dimensional EM modes, they are immune to the presence of disorder and/or possible fabrication imperfections [24, 25] facilitating this way their experimental observation. We note that, similarly to the Majorana edge states in Kitaev's model, the observed emergence of zero-dimensional modes in the middle of the frequency gap of Figure 2 signifies the non-trivial topology of the corresponding frequency bands of Figure 2 [26].

5 Physical realisation

In order to realize the proposed coupled-cavity arrays in the laboratory, it is important to create a negative phase (argument) of the hopping strength. As stated in Section 2, a negative phase can be achieved in a medium supporting backward-propagating waves, i.e. waves where the phase velocity is opposite to the group velocity. Such media are the so-called negative refractive-index metamaterials or left-handed metamaterials which are artificial dielectrics supporting backward-propagating waves. Therefore, in order to achieve

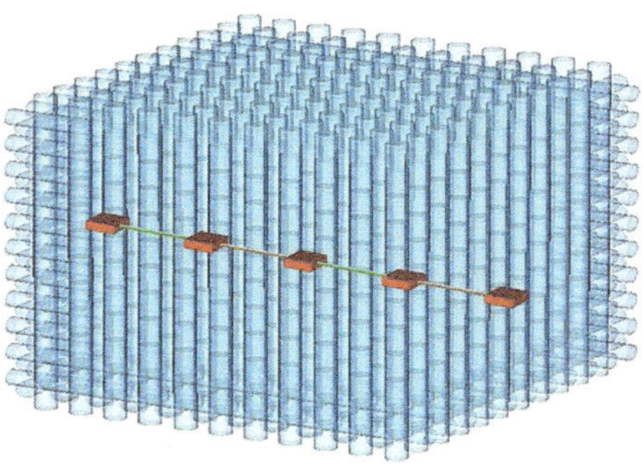

Figure 4 Microwave design for a 1D coupled-cavity array with topological properties: cuboid dielectric cavities joined together with metamaterial-based coupling elements and embedded within a 3D network of metallic wires (artificial plasma). The coupling elements are alternately of positive and negative refractive index.

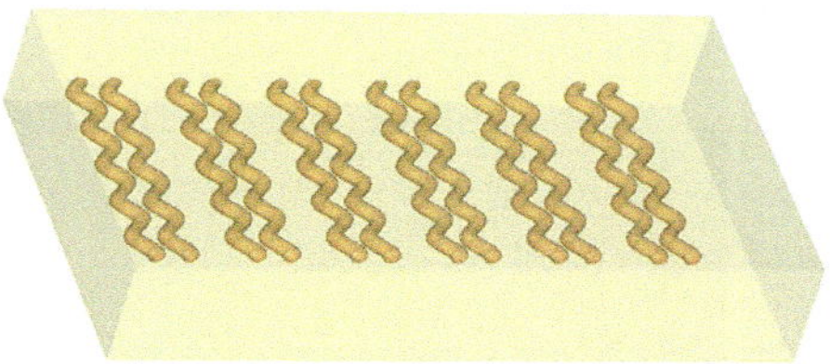

Figure 5 Optical design for a 1D coupled-cavity array with topological properties: a 1D periodic array of sinusoidally curved waveguides.

a negative phase coupling, i.e., $\exp(-i\phi)$, between two NN cavities, the latter must be physically connected via a metamaterial element. Such can be a waveguide loaded with a negative-index metamaterial or a left-handed transmission line [19].

A feasible coupled-cavity array design with desired properties is the one depicted in Figure 4. As our model system requires dielectric cavities within a homogeneous plasma, a lattice of cavities formed within a homogeneous Drude-type metal, e.g., a noble metal (Au, Ag, Cu), would be the most suitable candidate [38]. However, the plasmon bands are extremely lossy due to the intrinsic absorption of noble metals in the optical regime. Instead, one may use an *artificial* plasmonic medium operating in the microwave regime where metals are perfect conductors and losses are minimal. Artificial plasma can be created by a 3D network of thin metallic wires of a few tens of μm in diameter and spaced by a few mm [39]. Such a network is background medium hosting the cavity in Figure 4. The dielectric cavities in Figure 4 are the flat cuboids which are connected via waveguide elements. The waveguide elements are alternately loaded with materials of positive and negative refractive index mimicking the alternation of positive and negative phases. As suggested above, instead of metamaterial-loaded waveguide elements, one can use transmission lines to mimic the phase alteration from cavity to cavity, when operating in the microwave regime [19]. We note that the desired chirality in the host medium can be introduced either by coating the wires with a twisted-nematic liquid crystal with a long pitch or by fabricating a network of metallic helices [40, 41] instead of straight wires.

An alternative EM design which can potentially highlight the topological features studied here has been recently realized in the laboratory [42, 43]. Namely, the dispersion relation of Figure 2 with the Dirac singularity (the curve corresponding to $\eta = 0$) has been reproduced by a 1D periodic array of sinusoidally curved waveguides [44] (see Figure 5) designed so that effective coupling coefficients between adjacent waveguides are alternately positive and negative [42]. The alternative negative and positive coupling is designed by alternating the distances between adjacent waveguides. The sinusoidally curved waveguides are single-mode and are realized by femtosecond direct-writing fabrication [43]. Although the topological nature of the corresponding waveguide modes had not been spotlighted in Refs. [42, 43], an experimental verification of the Dirac singularity in the dispersion relation was achieved by means of 1D conical diffraction [43]. We can therefore hope that a verification of manifold of degenerate zero-dimensional modes reported here is experimentally feasible with the system of Figure 5.

6 Conclusion

We have presented a one-dimensional photonic system, namely an array of coupled cavities in a chiral medium, which is topologically non-trivial and exhibits a manifold of degenerate zero-dimensional modes which are reminiscent of the Majorana states in semiconducting nanowires in touch with a superconducting substrate. This exotic photonic simulator can be realized and probed experimentally system with already fabricated structures such as arrays of sinusoidally curved dielectric waveguides or with cavities coupled with metamaterial elements or transmission lines, embedded in metallic wire networks.

Competing interests
The author declares that he has no competing interests.

References

1. Haldane FDM, Raghu S. Possible realization of directional optical waveguides in photonic crystals with broken time-reversal symmetry. Phys Rev Lett. 2008;100:013904.
2. Wang Z, Chong YD, Joannopoulos JD, Soljačić M. Reflection-free one-way edge modes in a gyromagnetic photonic crystal. Phys Rev Lett. 2008;100:013905.
3. Wang Z, Chong YD, Joannopoulos JD, Soljačić M. Observation of unidirectional backscattering-immune topological electromagnetic states. Nature (London). 2009;461:772.
4. Yu Z, Veronis G, Wang Z, Fan S. One-way electromagnetic waveguide formed at the interface between a plasmonic metal under a static magnetic field and a photonic crystal. Phys Rev Lett. 2008;100:023902.
5. Takeda H, John S. Compact optical one-way waveguide isolators for photonic-band-gap microchips. Phys Rev A. 2008;78:023804.
6. Han D, Lai Y, Zi J, Zhang ZQ, Chan CT. Dirac spectra and edge states in honeycomb plasmonic lattices. Phys Rev Lett. 2009;102:123904.
7. Ao X, Lin Z, Chan CT. One-way edge mode in a magneto-optical honeycomb photonic crystal. Phys Rev B. 2009;80:033105.
8. Onoda M, Ochiai T. Designing spinning Bloch states in 2D photonic crystals for stirring nanoparticles. Phys Rev Lett. 2009;103:033903.
9. Ochiai T, Onoda M. Photonic analog of graphene model and its extension: Dirac cone, symmetry, and edge states. Phys Rev B. 2009;80:155103.
10. Shen R, Shao LB, Wang B, Xing DY. Single Dirac cone with a flat band touching on line-centered-square optical lattices. Phys Rev B. 2010;81:041410.
11. Poo Y, Wu RX, Lin Z, Yang Y, Chan CT. Experimental realization of self-guiding unidirectional electromagnetic edge states. Phys Rev Lett. 2011;106:093903.
12. Yannopapas V. Photonic analog of a spin-polarized system with Rashba spin-orbit coupling. Phys Rev B. 2011;83:113101.
13. Yannopapas V. Gapless surface states in a lattice of coupled cavities: a photonic analog of topological crystalline insulators. Phys Rev B. 2011;84:195126.
14. Chen WJ, Hang ZH, Dong JW, Xiao X, Wang HZ, Chan CT. Observation of backscattering-immune chiral electromagnetic modes without time reversal breaking. Phys Rev Lett. 2011;107:023901.
15. Zhong W, Zhang X. Dirac-cone photonic surface states in three-dimensional photonic crystal slab. Opt Express. 2011;19:13738.
16. Khanikaev AB, Hossein Mousavi S, Tse WK, Kargarian M, MacDonald AH, Shvets G. Photonic topological insulators. Nat Mater. 2012;12:233.
17. Liang GQ, Chong YD. Optical resonator analog of a two-dimensional topological insulator. Phys Rev Lett. 2013;110:203904.
18. Rechtsman MC, Zeuner JM, Plotnik Y, Lumer Y, Podolsky D, Dreisow F, Nolte S, Segev M, Szameit A. Photonic Floquet topological insulators. Nature (London). 2013;496:196.
19. Yannopapas V. Topological photonic bands in two-dimensional networks of metamaterial elements. New J Phys. 2012;14:113017.
20. Kitaev AY. Unpaired Majorana fermions in quantum wires. Phys Usp. 2001;44:131.
21. Kitaev AY. Fault-tolerant quantum computation by anyons. Ann Phys. 2003;303:2.
22. Alicea J, Oreg Y, von Refael G, Oppen F, Fisher MPA. Non-Abelian statistics and topological quantum information processing in 1D wire networks. Nat Phys. 2011;7:412.

23. Pachos JK. Introduction to topological quantum computation. Cambridge: Cambridge University Press; 2012.
24. Yannopapas V. Dirac points, topological edge modes and nonreciprocal transmission in one-dimensional metamaterial-based coupled-cavity arrays. Int J Mod Phys B. 2014;28:1441006.
25. Poddubny A, Miroshnichenko A, Slobozhanyuk A, Kivshar Y. Topological Majorana states in zigzag chains of plasmonic nanoparticles. ACS Photonics. 2014;1:101.
26. Ling CW, Xiao M, Chan CT, Yu SF, Fung KH. Topological edge plasmon modes between diatomic chains of nanoparticles. arXiv:1401.7520.
27. Bardyn CE, Baranov MA, Rico E, İmamoğlu ZP, Diehl S. Majorana modes in driven-dissipative atomic superfluids with a zero Chern number. Phys Rev Lett. 2012;109:130402.
28. Noh C, Rodríguez-Lara BM, Angelakis DG. Proposal for realization of the Majorana equation in a tabletop experiment. Phys Rev A. 2013;87:040102(R).
29. Liew TCH, Savona V. Multimode entanglement in coupled cavity arrays. New J Phys. 2013;15:025015.
30. Purcell EM, Pennypacker CR. Scattering and absorption of light by non-spherical dielectric grains. Astrophys J. 1973;186:705.
31. Dmitriev V. Space-time reversal symmetry properties of eletromagnetic Green's tensors for complex and bianisotropic media. Prog Electromagn Res. 2004;48:145.
32. Iyer AK, Eleftheriades GV. Negative refraction metamaterials: fundamental principles and applications. New York: Wiley; 2005.
33. Caloz C, Itoh T. Electromagnetic metamaterials: transmission line theory and microwave applications. New Jersey: Wiley-IEEE Press; 2006.
34. Sakoda K. Universality of mode symmetries in creating photonic Dirac cones. J Opt Soc Am B. 2012;29:2770.
35. Sakoda K. Proof of the universality of mode symmetries in creating photonic Dirac cones. Opt Express. 2012;20:25181.
36. Laughlin RB. Anomalous quantum Hall effect: an incompressible quantum fluid with fractionally charged excitations. Phys Rev Lett. 1983;50:1395.
37. Cho J, Angelakis DG, Bose S. Fractional quantum Hall state in coupled cavities. Phys Rev Lett. 2008;101:246809.
38. Stefanou N, Modinos A, Yannopapas V. Optical transparency of mesoporous metals. Solid State Commun. 2001;118:69.
39. Pendry JB, Holden AJ, Stewart WJ, Youngs I. Extremely low frequency plasmons in metallic mesostructures. Phys Rev Lett. 1996;76:4773.
40. Wu C, Li H, Wei Z, Yu X, Chan CT. Theory and experimental realization of negative refraction in a metallic helix array. Phys Rev Lett. 2010;105:247401.
41. Wu C, Li H, Yu X, Li F, Chen C, Chan CT. Metallic helix array as a broadband wave plate. Phys Rev Lett. 2011;107:177401.
42. Efremidis NK, Zhang P, Chen Z, Christodoulides DN, Rüter CE, Kip D. Wave propagation in waveguide arrays with alternating positive and negative couplings. Phys Rev A. 2010;81:053817.
43. Zeuner JM, Efremidis NK, Keil R, Dreisow F, Christodoulides DN, Tünnermann A, Nolte S, Szameit A. Optical analogues for massless Dirac particles and conical diffraction in one dimension. Phys Rev Lett. 2012;109:023602.
44. Longhi S, Marangoni M, Lobino M, Ramponi R, Laporta P, Cianci E, Foglietti V. Observation of dynamic localization in periodically curved waveguide arrays. Phys Rev Lett. 2006;96:243901.

6

Characterization of how dissipation and dephasing errors accumulate in quantum computers

Jake Gulliksen, Durga Bhaktavatsala Rao Dasari and Klaus Mølmer*

*Correspondence: moelmer@phys.au.dk
Department of Physics and Astronomy, Aarhus University, Ny Munkegade 120, Aarhus C, DK-8000, Denmark

Abstract
We study the achievements of quantum circuits comprised of several one- and two-qubit gates subject to dissipation and dephasing. Quantum process matrices are determined for the basic one- and two-qubit gate operations and concatenated to yield the process matrix of the combined quantum circuit. Examples are given of process matrices obtained by a Monte Carlo wavefunction analysis of Rydberg blockade gates in neutral atoms. Our analysis is ideally suited to compare different implementations of the same process. In particular, we show that the three-qubit Toffoli gate facilitated by the simultaneous interaction between all atoms may be accomplished with higher fidelity than a concatenation of one- and two-qubit gates.

Keywords: quantum tomography; Monte Carlo wavefunction; Rydberg atoms

1 Introduction

Since the first proposals were made to use quantum effects for computing purposes there has been a strong focus on how errors and imperfections may harm and even prevent successful application of quantum computing. A simple estimate suggests that if each single operation in a computation entails an error with a probability $p > 0$ then the application of k operations will lead to a useful outcome with a probability that decreases exponentially $\sim(1-p)^k$. Error correction codes have provided a way to correct these errors up to a certain probability threshold, thereby allowing scalable, fault-tolerant quantum computing [1, 2].

The errors occurring in a single computational step such as a one- or two-qubit gates are often characterized by a single number, typically related to the overlap between the desired and actual output state, averaged over all input states. There is no guarantee, however, that such a number encapsulates the accumulation of errors in a quantum circuit, where the output state of one operation serves as the input to the next. Errors may build up coherently, so that error probabilities grow quadratically rather than linearly with time, or so that they compensate each other, cf., bang-bang control and composite pulses [3–5]. Thus, a concatenation of two imperfect gates can lead to either unusable results or a correcting mechanism.

Consider the action of a quantum process that takes an input density matrix ρ describing a physical system with Hilbert space dimension D to an output density matrix. Such a process is described as a completely-positive linear map $\mathcal{E} : \rho \rightarrow \mathcal{E}(\rho)$, where $\mathcal{E}(\rho)$ can be

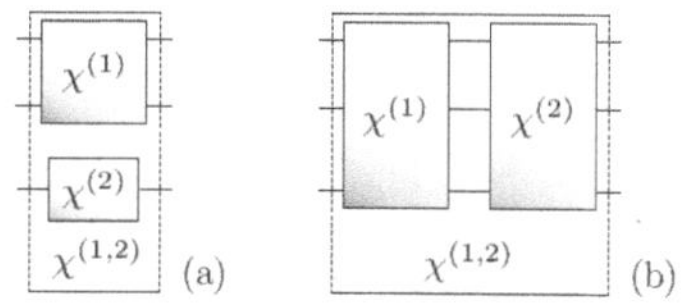

Figure 1 Concatenation of gates carried out in parallel on different qubits (a), and in serial on the same qubits (b) in a quantum computer. Expressions for the resulting process matrices $\chi^{(1,2)}$ are discussed in Section 3.

written [6]

$$\mathcal{E}(\rho) = \sum_{mn} \chi_{mn} E_m \rho E_n^\dagger, \tag{1}$$

by introducing a complete basis of D^2 operators $\{E_n\}$ on the Hilbert space. The coefficients χ_{mn} constitute the process matrix χ.

In quantum computing we aim to implement definite gate operations yielding, ideally, a unitary transformation, $\mathcal{E}(\rho) = U\rho U^\dagger$. Expanding $U = \sum_m \eta_m E_m$, this corresponds to Eq. (1) with $\chi_{mn} = \eta_m \eta_n^*$. The process matrix of an experimental gate implementation differs from this form due to dissipation and decoherence. In this article, we will show that, provided dissipation and decoherence acts locally and is uncorrelated over the quantum computing register χ-matrices calculated once for one- and two-qubit gates can be concatenated, see Figure 1, to characterize circuits built from many of these gates. This assumption may be well justified in our example, which concerns neutral atom quantum computing, where the Rydberg blockade mechanism is used for two-qubit quantum gates [7, 8].

The paper is organized as follows. In Section 2, we review the definition of χ-matrices and how they may be computed with Monte Carlo wave function simulations. In Section 3, we describe how χ-matrices for simple processes on few particles are concatenated to characterize large quantum circuits. In Section 4, we introduce the Rydberg blockade gate scheme for quantum computing with neutral atoms. In Section 5, we concatenate one- and two-qubit gate χ-matrices in a neutral atom system to characterize the circuit performing a Toffoli gate. This we compare to a direct multi-atom Rydberg mediated implementation. In Section 6, we conclude and present an outlook.

2 Process matrix identification

Many techniques now exist to experimentally determine χ. Standard quantum process tomography [6, 9, 10] successfully reproduces χ by measuring all output states via quantum *state* tomography [11, 12]. This has been demonstrated in NMR [13, 14], optical [15, 16], and atomic systems [17]. Alternately, χ may be obtained making use of an ancillary system [18, 19] or avoiding state tomography altogether through the use of suitable 'probe' systems [20–22].

If the system is subject to known dissipation and decoherence mechanisms, the quantum system evolution may be modeled theoretically and the process matrix be calculated by solution of the quantum master equation. A gate operation typically involves application of time dependent laser pulses. Therefore, it is valuable to determine how losses and errors accumulate and contribute to different types of errors in the output. To theoretically characterize a complete quantum circuit is a formidable task and is ultimately at odds with using a physical system to solve computationally hard problems. Still, a theoretical analysis of how errors propagate and accumulate in small systems may guide efforts to pick among different implementations of gates and assess optimal strategies for error correction. Such detailed studies may also serve to confirm the values of experimental parameters [23, 24].

In a recent publication [25], we described how to characterize a quantum controlled-phase gate subject to decay and dephasing. Instead of simulating the evolution of a complete set of D^2 input states we gain access to all elements of χ by evolving a single maximally entangled pure state of the system and an idle ancilla system of the same Hilbert space dimension [18]. The system is propagated stochastically using the Monte Carlo wave function method, which on average reproduces results of a master equation evolution [26–28]. Process characterization using this approach has a number of advantages: First, for large D, an adequate ensemble of wave functions is easier to store and evolve than density matrices. Second, obtaining χ through the output state data from an ensemble of wave functions is less costly, numerically, than from a density matrix [25]. Third, the stochastic evolution consists of a deterministic smooth evolution interrupted by 'quantum jumps.' Since useful quantum gates require excellent fidelity, jumps are rare and a single deterministic 'no-jump' wave function suffices to provide a good estimate and rigorous bound on the process matrices describing the evolution [25].

3 The process matrix for a quantum circuit

Suppose the quantum circuit performing a computational task is composed of N physical units. The Hilbert space of the entire system is then a tensor product of N Hilbert spaces, each of dimension d. An implementation of a quantum process often requires using more than just the qubit states. However, since the physical units only process binary information we shall refer to them as qubits, even if we exploit states from a space larger than dimension 2. On each qubit Hilbert space we assume the complete operator basis $\{e_{n_i}\}$. By merely forming tensor products of the basis operators, we obtain a complete operator basis $\{E_n = e_{n_1} \otimes \cdots \otimes e_{n_N}\}$ for the N qubits, where the single index n represents all values of the set $n_1, \ldots, n_N$. The operator tensor product structure provides a convenient representation of the $D^2 = d^{2N}$ operators E_m ($E_n^\dagger$) in Eq. (1).

If we assume that process matrices χ correctly describe processes acting separately on one and two qubits of the circuit, then the application of several one- and two-qubit operations is exactly represented by an appropriate concatenation of the corresponding process matrices.

3.1 Parallel concatenation

Suppose two subsystems are simultaneously subjected to processes independent of each other. These processes $\mathcal{E}^{(1)}$ and $\mathcal{E}^{(2)}$ may be described by the process matrices $\chi^{(1)}$ and $\chi^{(2)}$ respectively, illustrated as two- and one-qubit gates in Figure 1(a). The combined three-qubit process matrix $\chi^{(1,2)}$ is simply the tensor product of the independent χ matrices. Other systems may be present but idle during the gate operation. They are then represented by the identity operator in the process matrix tensor product.

3.2 Serial concatenation

Most quantum algorithms make use of many computational steps, where the output of every step serves as the input to the subsequent one. In Figure 1(b) we illustrate this situation for two consecutive three-qubit operations $\mathcal{E}^{(1)}$ and $\mathcal{E}^{(2)}$ characterized by $\chi^{(1)}$ and $\chi^{(2)}$ respectively. If the output $\mathcal{E}^{(1)}(\rho)$ of the input density matrix ρ becomes the input of

$\mathcal{E}^{(2)}$, what is the resulting χ matrix? Formally, the output of the sequential application of the operations is given by

$$\mathcal{E}^{(2)}(\mathcal{E}^{(1)}(\rho)) = \sum_{pq} \chi^{(2)}_{pq} E_p \Big(\sum_{mn} \chi^{(1)}_{mn} E_m \rho E_n^\dagger \Big) E_q^\dagger. \tag{2}$$

Since the operators E_r form a complete set, any product $E_p E_m$ can be expanded on these operators, that is, $E_p E_m = \sum_r c^r_{pm} E_r$ and $E_n^\dagger E_q^\dagger = \sum_s (c^s_{qn})^* E_s^\dagger$. Equation (2) then becomes

$$\mathcal{E}^{(2)}(\mathcal{E}^{(1)}(\rho)) = \sum_{rs} \chi^{(1,2)}_{rs} E_r \rho E_s^\dagger, \tag{3}$$

where

$$\chi^{(1,2)}_{rs} = \sum_{mn,pq} c^r_{pm} \chi^{(1)}_{mn} \chi^{(2)}_{pq} (c^s_{qn})^*. \tag{4}$$

Note that although two consecutive processes may act on different subsets of some multi-qubit system both operations may be reformulated to act on the entire system through parallel concatenation.

It now becomes apparent that once the process matrices of all contributing gates in a circuit have been computed conclusively, we limit the cost of finding $\chi^{(1,2)}$ and thus of process matrices for larger quantum circuits. The assessment of how errors accumulate becomes a function of the width and depth of the quantum circuit.

3.3 Example: Toffoli gate

The Toffoli gate, or C_2NOT gate, performs a controlled NOT operation on a target qubit based on the state of two control qubits. The Toffoli gate may be implemented as a sequence of six two-qubit C-NOT gates and nine one-qubit Hadamard and $T = \exp(i\pi\sigma_z/8)$ and $T^\dagger$ phase gates, see Figure 2(a). The gate and its generalization to higher numbers of control qubits (C_k-NOT) have applications as sub-modules in different quantum computing algorithms. Thus, it is relevant to determine the process matrix for its implementation in realistic systems.

In the analysis of the Toffoli gate process matrix we first simulated the propagation of quantum states in the full three-qubit Hilbert space through the sequence of one- and

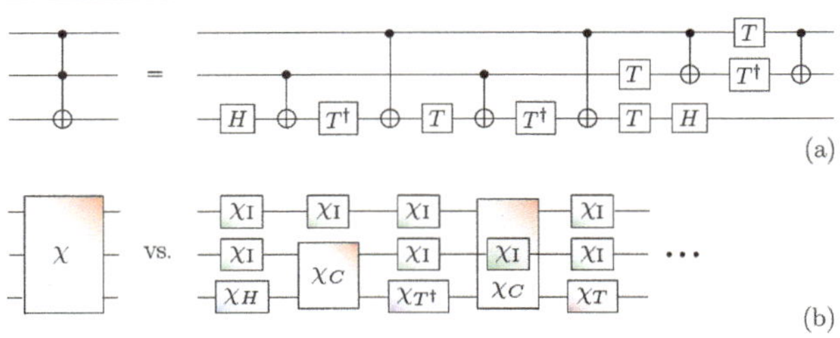

Figure 2 The Toffoli gate: (a) The three-qubit Toffoli gate on the left may be reproduced by a circuit of C-NOT, Hadamard (*H*), $T = \exp(i\pi\sigma_z/8)$ and $T^\dagger$ gates, shown to the right. **(b)** The process matrix χ (left), characterizing the Toffoli gate may be calculated by concatenation of one- and two-qubit process matrices (right). The process matrices χ_C, χ_H, χ_T and $\chi_{T^\dagger}$ characterize the C-NOT, Hadamard, *T* and $T^\dagger$ gates respectively. The 'identity' process matrix χ_I indicates an idle qubit.

two-qubit gates. Such a calculation, e.g. using Monte Carlo wave functions to include dissipation, yields the full circuit process matrix χ_{cir}. Next, we apply the concatenation rules to obtain the circuit's process matrix χ_{cat}. Its repeated use of the same C-NOT χ matrix (cf. Figure 2(b)), which only needs a single calculation on a two-qubit system, attests to the advantage of the latter approach.

4 Rydberg blockade quantum gates

A promising candidate for quantum computing involves neutral atoms held at closely spaced sites in far-off-resonance optical traps. The atoms may be individually addressed with laser fields and excited into high lying Rydberg states that feature strong, long distance dipole and van der Waals forces that can be used to mediate two-qubit interactions [7, 8, 29].

In Rubidium atoms, a convenient choice for the qubit states are the hyperfine ground states $|0\rangle \equiv |5s_{1/2}, F = 1, m_F = 0\rangle$ and $|1\rangle \equiv |5s_{1/2}, F = 2, m_F = 0\rangle$. They can be selectively excited to the Rydberg state $|r\rangle = |97d_{5/2}, m_j = 5/2\rangle$ by a two photon process using a 780-nm (480-nm) laser field, tuned by an amount Δ to the red (blue) of the intermediate $|p\rangle \equiv |5p_{3/2}, F = 3\rangle$ state. The Rabi frequency associated with the red (blue) detuned laser is Ω_R (Ω_B), illustrated in Figure 3(a). An atom that achieves excitation to the Rydberg state shifts the $|r\rangle$ state energy of all other atoms within the so-called blockade radius by an amount $\mathcal{B}$. Thus, one excited atom can prevent the resonant excitation of its neighboring atoms and this is the basis for effective quantum gates between them.

Dephasing of the Rydberg level normally associated with magnetic field noise and atomic motion is modeled by the operator $\hat{L}_{\gamma_d} = \sqrt{\gamma_d}(\mathbb{1} - 2|r\rangle\langle r|)$, where γ_d is the dephasing rate and $\mathbb{1}$ is shorthand for the identity operator. Spontaneous decay from a state $|y\rangle$ to a lower lying state $|z\rangle$ at a rate γ_y is modeled by the jump operator $\hat{L}_{\gamma_y} = \sqrt{\gamma_y}|z\rangle\langle y|$. The effects of both dissipation mechanisms are simulated using the Monte Carlo wave function method [25]. Characteristic parameters are summarized in Table 1.

Adiabatic elimination by the effective operator formalism detailed in Ref. [32] provides a mechanism to decouple the intermediate optically excited state and describe the coherent and incoherent dynamics within the subspace of $|0\rangle, |1\rangle$ and $|r\rangle$. The system is then described by a Hamiltonian coupling a selected qubit state to $|r\rangle$ by an effective Rabi fre-

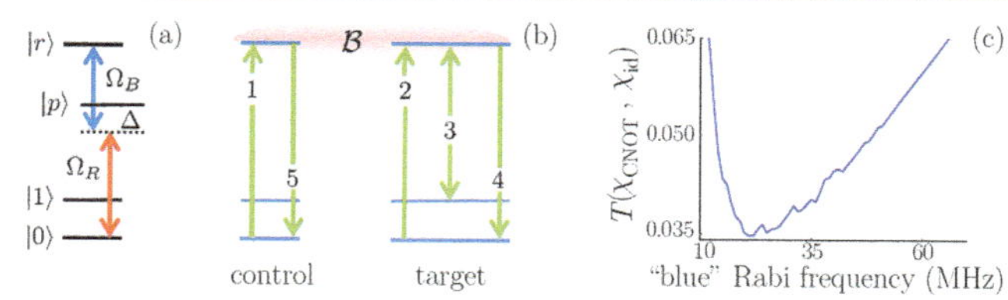

Figure 3 Simulation and characterization of the C-NOT gate: (a) The red (lower) and blue (upper) laser fields drive $|0\rangle$, via $|p\rangle$, into the Rydberg state $|r\rangle$ by two-photon absorption. **(b)** Implementation of the C-NOT gate involves a sequence of two-photon π-pulses: In pulse 1 the control atom makes the transition $|0\rangle$ to $|r\rangle$. The target atom then makes the transitions $|0\rangle \leftrightarrow |r\rangle$ (pulse 2), $|1\rangle \leftrightarrow |r\rangle$ (pulse 3) and finally $|0\rangle \leftrightarrow |r\rangle$ (pulse 4). In pulses 2-4 the target atom's states $|0\rangle$ and $|1\rangle$ are swapped, but only if the control atom is not in $|r\rangle$. Finally, in pulse 5 the control atom is driven from $|r\rangle$ back to $|0\rangle$. **(c)** Trace distance (see text in Section 4.1) between the ideal C-NOT process matrix and the process matrix calculated for the implementation shown in panel (b), with the parameters listed in Table 1. The trace distance is shown as a function of the blue laser Rabi frequency Ω_B.

Table 1 Physical parameters for our simulations based on values discussed in Refs. [30, 31]

Experimental parameter	Symbol	Value
Detuning	$\Delta/2\pi$	2.0 GHz
Red Rabi frequency	$\Omega_R/2\pi$	118 MHz
Blue Rabi frequency	$\Omega_B/2\pi$	10-100 MHz
Rydberg blockade	$\mathcal{B}/2\pi$	20 MHz
Decay rate of $\lvert p\rangle$	$\gamma_p/2\pi$	6.07 MHz
Decay rate of $\lvert r\rangle$	$\gamma_r/2\pi$	0.53 kHz
Dephasing rate of $\lvert r\rangle$	$\gamma_d/2\pi$	1.0 kHz

quency. The formalism also provides an effective form for the operators describing decohering processes [25].

4.1 Rydberg blockade C-NOT gate

In atomic quantum computing proposals, single qubit gates amount to fast, resonant transitions within single atoms and can be made with high precision. Thus, for the purpose of this study we assume that the χ matrices associated with one-qubit gates are identical to the desired ones. The two-qubit C-NOT gate depends on finite interactions between excited state atoms, lengthening gate time and making it prone to dissipation and decoherence.

Figure 3(b) illustrates how a unitary C-NOT gate between two atoms can be implemented by a sequence of five perfect π-pulses. First transferring the control qubit's population from $|0_c\rangle$ to $|r_c\rangle$ (pulse 1), then transferring the target qubit's population between $|0_t\rangle$ and $|1_t\rangle$ via the state $|r_t\rangle$ (pulses 2-4) and finally returning the control qubit's population from $|r_c\rangle$ to $|0_c\rangle$ (pulse 5). If the control qubit initially populates the state $|0_c\rangle$ the Rydberg blockade prevents any transfer during pulses 2-4. Thus, a NOT operation on the target qubit is conditioned on the control qubit initially populating the state $|1_c\rangle$, defining it to be a C-NOT operation.

Monte Carlo wave functions were used to simulate the five π-pulse implementation of the C-NOT gate (Figure 3) with the parameters of Table 1. The performance of the gate was investigated as a function of the blue laser Rabi frequency Ω_B. To provide a simple quantitative measure we applied the trace distance measure $T(\chi_{\text{sim}}, \chi_{\text{id}})$ between the simulated and ideal process matrix, where $T(A,B) \equiv \frac{1}{2}\|A-B\|_{\text{tr}}$ and $\|C\|_{\text{tr}} = \text{Tr}(\sqrt{C^\dagger C})$ is the trace norm. Note that this distance measure is less 'forgiving' than, for example, measures based on the trace overlap [23]. In Figure 3(c) we show the trace distance between a simulated C-NOT gate process matrix and the ideal, unitary process matrix. At low values of Ω_B the gate experiences greater dephasing errors from population in the Rydberg state, due to long gate times. At large Ω_B the blockade mechanism becomes inefficient. Thus, the optimum Rabi frequency lies between these two regimes.

5 The Toffoli gate by Rydberg blockade

We demonstrate the characterization of the Toffoli gate resulting from simulation in Figure 4. The process matrix χ_{cat} of the Toffoli gate in the circuit implementation (Figure 2) may be obtained without further simulation by a concatenation of the single qubit process matrices and the C-NOT process matrix of Section 4.1. Alternatively, we may simulate the circuit implementation in the full three-qubit Hilbert space to obtain χ_{cir}. In the simulation of a Rydberg mediated gate, the characterization of a single qubit has a Hilbert space

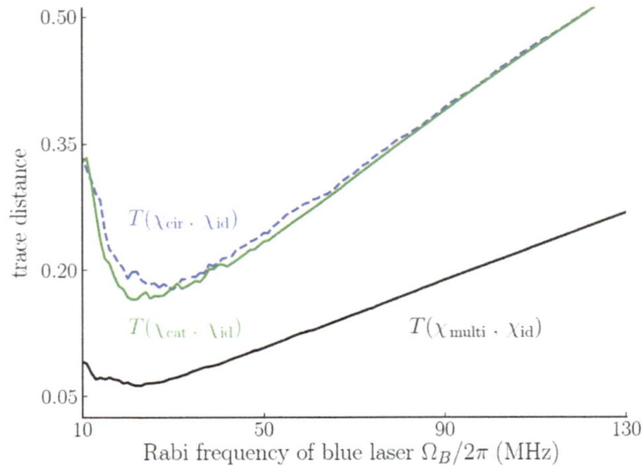

Figure 4 Trace distance between the process matrix χ_{id} for the ideal Toffoli gate and the process matrices for Rydberg interaction implementations subject to dissipation and decoherence. The results are shown as a function of the Rabi frequency Ω_B of the $|p\rangle \rightarrow |r\rangle$ (blue) laser coupling. The dashed curve is obtained by simulating all three qubits as they evolve under the sequence of gates in the Toffoli circuit shown in Figure 2(a). The top solid curve uses concatenation of the one- and two-qubit process matrices to compute the Toffoli circuit process matrix. The bottom (solid) curve results from simulating the multi-qubit implementation shown in Figure 5. In all calculations 500 Monte Carlo trajectories were used with the parameters listed in Table 1.

dimension of $d = 4^2$, which translates into a 4^6 problem for the three qubit circuit characterization. We observe that the Toffoli gate consists of six C-NOT gates and the trace distance to the ideal gate is, indeed, roughly six times the one shown in Figure 3(c).

The top dashed (solid) curve in Figure 4 illustrates trace distance between the full circuit χ_{cir} (concatenated χ_{cat}) process matrix to the ideal process matrix χ_{id}, plotted as a function of Ω_B. Since the decay and decoherence processes apply to the individual atoms, the successive treatment of the evolution of the atoms acted upon by the laser fields and concatenation of the resulting one- and two-qubit process matrices should yield the same results as the solution on the full-register Hilbert space.

Each point in both curves is determined by propagating $n = 500$ wave function trajectories, and the minor discrepancy between the two curves is due to the use of the MCWF method. The final number of propagated wave functions imposes a statistical error $\sim 1/\sqrt{n}$ on both curves. Furthermore, it imposes a small systematical error in the construction of χ_{cat}, because the same simulated C-NOT process matrix is applied several times to form the concatenated process matrix (see Ref. [33] for an analysis of a similar situation). The discrepancy between the curves in Figure 4 depends on Ω_B. For low values of Ω_B, and thus slow gate operation, decay and dephasing dominate the error. More random quantum jumps occur and the resulting scattering of Monte Carlo wave functions explains the relatively large discrepancy between $\text{Tr}(\chi_{cat}, \chi_{id})$ and $\text{Tr}(\chi_{cir}, \chi_{id})$. For high values of Ω_B, imperfect blockade dominates the gate error. This corresponds to a unitary term in the state evolution which is well represented by a single wave function, and the discrepancy between $\text{Tr}(\chi_{cat}, \chi_{id})$ and $\text{Tr}(\chi_{cir}, \chi_{id})$ decreases for large Ω_B.

A Rydberg excited atom blocks excitation of any number of atoms within the Rydberg interaction blockade radius, which may be of order 10 μm. Thus, it is possible to contain an entire qubit register within a single blockade radius, allowing implementation of multi-

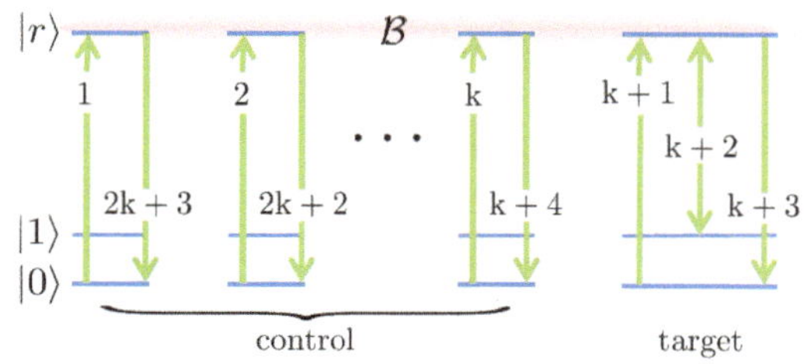

Figure 5 Multi-qubit Rydberg blockade implementation of a C_kNOT gate. Each control atom is sequentially excited from $|0\rangle$ to $|r\rangle$ in k π-pulses. Next, the target atom qubit states are swapped as in Figure 3(b), via the Rydberg state $|r\rangle$ (conditioned on no control atom populating the state $|r\rangle$). The control atoms are then returned to their original state in reverse order. The trace distance between the process matrix using this implementation for $k = 2$ and the ideal Toffoli gate is shown as the lower curve in Figure 4.

qubit gate operations which are faster than the circuit equivalent [34]. One such protocol is the C_kNOT gate operation, illustrated in Figure 5 [35].

For $k = 2$ the gate becomes the Toffoli gate and calculation of the process matrix is only possible by solving the master equation for the complete qubit register. In this paper, simulation of the process, including the decay and decoherence mechanisms detailed above, was carried out using the Monte Carlo method. The trace distance between the process matrix resulting from simulation and the ideal process matrix is shown as the lower, black curve in Figure 4. Remarkably, the multi-qubit implementation, with interactions allowed between all three atoms, performs markedly better than the Toffoli circuit consisting of one- and two-qubit operations. In comparison with the C-NOT gate, the minimal trace distance here is approximately 1.5 times larger. This is consistent with using 7 π-pulses rather than the 5 needed for a single C-NOT gate.

6 Conclusion

In conclusion, we have presented an efficient method to compute the accumulation of errors in quantum circuits comprised of several few-qubit gates. We have shown that a set of concatenation rules on the appropriate few-qubit gate process matrices is enough to reproduce the process matrix of the entire circuit. To demonstrate the method's efficiency at calculating process matrices of large systems we considered the three-qubit Toffoli gate. The Toffoli gate may be implemented as a circuit of one- and two-qubit gates and simulations show that the process matrix obtained via concatenation is in good agreement with the result achieved by propagation through the entire circuit.

Our theory allows comparison of different implementations of gates. In particular, we compared a multi-qubit implementation of the Toffoli gate with its one- and two-qubit circuit implementation. For the parameters chosen, the factor determining gate fidelity was the number of laser π-pulses. More gates lead to a lower fidelity, with a dependence that is almost linear. In this way, our analysis provides the necessary information to choose between different gate implementations. A theory of full error correction may benefit significantly from knowledge of the precise nature of errors incurred, potentially leading to higher thresholds for errors that can be remedied by appropriate error correction. The full process matrix, which remains at our disposal, may be further applied to optimally combine the Toffoli gate with previous and subsequent gate operations along the lines of NMR composite pulses [4].

Our analysis quantifies how erroneous states prepared by one gate are the input states to the subsequent ones and how the resulting accumulation of errors cause the full register quantum state to stray away from the ideal unitary evolution. The calculations assumed a quantum optical master equation with independent Lindblad type relaxation terms. This is well justified for Rydberg blockade quantum computing with optical excitation of neutral atoms, but it also applies for a number of quantum computing proposals with similarly isolated and identifiable qubits. However, if the deleterious interaction between the quantum register and its environment is subject to correlations between qubits and non-Markovian effects, more care must be exercised in the assessment of how errors accumulate, and how they may be corrected [36, 37].

Competing interests
The authors declare that they have no competing interests.

Authors' contributions
All authors contributed to the derivation and analysis as well as to the preparation of the manuscript. JG performed the numerical simulations.

Acknowledgements
The authors gratefully acknowledge discussion with Mark Saffman. This work was supported by the IARPA MQCO program through ARO contract W911NF-10-1-0347 and by the Villum Foundation.

References

1. Calderbank AR, Shor PW. Good quantum error-correcting codes exist. Phys Rev A. 1996;54:1098–105.
2. Steane A. Multiple-particle interference and quantum error correction. Proc R Soc Lond A. 1996;452:2551–77.
3. Viola L, Lloyd S. Dynamical suppression of decoherence in two-state quantum systems. Phys Rev A. 1998;58:2733–44.
4. Levitt MH. Composite pulses. Prog Nucl Magn Reson Spectrosc. 1986;18:61–122.
5. Jones JA. Designing short robust NOT gates for quantum computation. Phys Rev A. 2013;87:052317.
6. Chuang IL, Nielsen MA. Prescription for experimental determination of the dynamics of a quantum black box. J Mod Opt. 1997;44(11-12):2455–67.
7. Jaksch D, Cirac JI, Zoller P, Rolston SL, Côté R, Lukin MD. Fast quantum gates for neutral atoms. Phys Rev Lett. 2000;85:2208–11.
8. Saffman M, Walker TG, Mølmer K. Quantum information with Rydberg atoms. Rev Mod Phys. 2010;82:2313–63.
9. Nielsen MA, Chuang IL. Quantum computation and quantum information. Cambridge: Cambridge University Press; 2010.
10. Poyatos JF, Cirac JI, Zoller P. Complete characterization of a quantum process: the two-bit quantum gate. Phys Rev Lett. 1997;78:390–3.
11. Gross D, Liu Y-K, Flammia ST, Becker S, Eisert J. Quantum state tomography via compressed sensing. Phys Rev Lett. 2010;105:150401.
12. D'Ariano GM, Maccone L, Paini M. Spin tomography. J Opt B. 2003;5:77–84.
13. Childs AM, Chuang IL, Leung DW. Realization of quantum process tomography in NMR. Phys Rev A. 2001;64:012314.
14. Boulant N, Havel TF, Pravia MA, Cory DG. Robust method for estimating the Lindblad operators of a dissipative quantum process from measurements of the density operator at multiple time points. Phys Rev A. 2003;67:042322.
15. Mitchell MW, Ellenor CW, Schneider S, Steinberg AM. Diagnosis, prescription, and prognosis of a Bell-state filter by quantum process tomography. Phys Rev Lett. 2003;91:120402.
16. O'Brien JL, Pryde GJ, Gilchrist A, James DFV, Langford NK, Ralph TC, White AG. Quantum process tomography of a controlled-NOT gate. Phys Rev Lett. 2004;93:080502.
17. Myrskog SH, Fox JK, Mitchell MW, Steinberg AM. Quantum process tomography on vibrational states of atoms in an optical lattice. Phys Rev A. 2005;72:013615.
18. D'Ariano GM, Lo Presti P. Quantum tomography for measuring experimentally the matrix elements of an arbitrary quantum operation. Phys Rev Lett. 2001;86:4195–8.
19. Altepeter JB, Branning D, Jeffrey E, Wei TC, Kwiat PG, Thew RT, O'Brien JL, Nielsen MA, White AG. Ancilla-assisted quantum process tomography. Phys Rev Lett. 2003;90:193601.
20. Mohseni M, Lidar DA. Direct characterization of quantum dynamics. Phys Rev Lett. 2006;97:170501.
21. Mohseni M, Lidar DA. Direct characterization of quantum dynamics: general theory. Phys Rev A. 2007;75:062331.
22. Wang Z-W, Zhang Y-S, Huang Y-F, Ren X-F, Guo G-C. Experimental realization of direct characterization of quantum dynamics. Phys Rev A. 2007;75:044304.
23. Zhang XL, Gill AT, Isenhower L, Walker TG, Saffman M. Fidelity of a Rydberg-blockade quantum gate from simulated quantum process tomography. Phys Rev A. 2012;85:042310.
24. Weinstein YS, Havel TF, Emerson J, Boulant N, Saraceno M, Lloyd S, Cory DG. Quantum process tomography of the quantum Fourier transform. J Chem Phys. 2004;121(13):6117–33.
25. Gulliksen J, Bhaktavatsala Rao DD, Mølmer K. Process characterization with Monte Carlo wave functions. Phys Rev A. 2013;88:052129.

26. Dalibard J, Castin Y, Mølmer K. Wave-function approach to dissipative processes in quantum optics. Phys Rev Lett. 1992;68:580–3.
27. Mølmer K, Castin Y, Dalibard J. Monte Carlo wave-function method in quantum optics. J Opt Soc Am B. 1993;10(3):524–38.
28. Dum R, Zoller P, Ritsch H. Monte Carlo simulation of the atomic master equation for spontaneous emission. Phys Rev A. 1992;45:4879–87.
29. Saffman M, Walker TG. Analysis of a quantum logic device based on dipole-dipole interactions of optically trapped Rydberg atoms. Phys Rev A. 2005;72:022347.
30. Johnson TA, Urban E, Henage T, Isenhower L, Yavuz DD, Walker TG, Saffman M. Rabi oscillations between ground and Rydberg states with dipole-dipole atomic interactions. Phys Rev Lett. 2008;100:113003.
31. Saffman M, Zhang X, Gill A, Isenhower L, Walker T. Rydberg state mediated quantum gates and entanglement of pairs of neutral atoms. J Phys Conf Ser. 2011;264:012023.
32. Reiter F, Sørensen AS. Effective operator formalism for open quantum systems. Phys Rev A. 2012;85:032111.
33. Castin Y, Mølmer K. Monte Carlo wave functions and nonlinear master equations. Phys Rev A. 1996;54:5275–90.
34. Mølmer K, Isenhower L, Saffman M. Efficient Grover search with Rydberg blockade. J Phys B. 2011;44(18):184016.
35. Isenhower L, Saffman M, Mølmer K. Multibit $\mathbf{C}_k$NOT quantum gates via Rydberg blockade. Quantum Inf Process. 2011;10:755–70.
36. Jacobsen SH, Mintert F. Optimal correction of independent and correlated errors. J Phys A, Math Theor. 2014;47:045306.
37. Terhal BM, Burkard G. Fault-tolerant quantum computation for local non-Markovian noise. Phys Rev A. 2005;71:012336.

7

Analog quantum simulation of gravitational waves in a Bose-Einstein condensate

Tupac Bravo*, Carlos Sabín and Ivette Fuentes

*Correspondence:
pmxtb2@nottingham.ac.uk
School of Mathematical Sciences,
University of Nottingham,
Nottingham, NG7 2RD, UK

Abstract

We show how to vary the physical properties of a Bose-Einstein condensate (BEC) in order to mimic an effective gravitational-wave spacetime. In particular, we focus in the simulation of the recently discovered creation of particles by a *real* spacetime distortion in box-type traps. We show that, by modulating the speed of sound in the BEC, the phonons experience the effects of a simulated spacetime ripple with experimentally amenable parameters. These results will inform the experimental programme of gravitational wave astronomy with cold atoms.

Keywords: quantum simulation; gravitational waves; Bose-Einstein condensates

Introduction

Quantum simulators [1] were originally conceived by Feynman as experimentally amenable quantum systems whose dynamics would mimic the behaviour of more inaccessible systems appearing in Nature. Together with the exploration of this visionary insight in countless quantum platforms at an increasingly accelerated rate, last years had witnessed the birth of alternate approaches to quantum simulation. For instance, a quantum simulator can be used to emulate a phenomenon predicted by a well established theory but very hard to test in the laboratory, such as Zitterbewegung [2]. Going a step further, it can also be used to materialise an artificial dynamics that has never been observed in Nature while being theoretically conceivable [3, 4] or even to emulate the action of a mathematical transformation [5].

Einstein's theory of general relativity [6] predicts the existence of gravitational waves [7], namely perturbations of the spacetime generated by accelerated mass distributions. Since sources are typically very far from Earth, the theory predicts that the amplitude of gravitational waves reaching our planet is extremely small and thus, finding experimental evidence of their existence is a difficult task. Indeed the quest for the detection of these spacetime distortions [8] has been one of the biggest enterprises of modern science and the focus of a great amount of work during the last decades, both in theory and experiment. Recently, some of the authors of this manuscript have proposed a novel method of gravitational wave detection based on the generation of particles in a Bose-Einstein condensate produced by the propagation of the gravitational ripple [9]. Detection is carried out through a resonance effect which is possible because the range of frequencies of typical gravitational waves is similar to the one of the Bogoliubov modes of a BEC confined in a box-like potential. Then the gravitational wave is able to hit a particle creation resonance, in a phenomenon resembling the Dynamical Casimir Effect [10, 11]. Due to the frequencies involved, this effect is completely absent in optical cavities. The use of our technique would relax some of the most daunting demands of other programmes for gravitational wave detection such as the use of highly-massive mirrors and *km*-long interferometer arms. However, due to the extremely small amplitude of the gravitational waves when they reach the Earth, their detection is always challenging, since it requires an experimental setup extremely well isolated from possible sources of noise. It would be of great benefit for the experiment if the amplitude of the spacetime ripples were larger.

In this paper, we show how to realise a quantum simulation of the generation of particles by gravitational waves in a BEC. We exploit the fact that the Bogoliubov modes of a trapped BEC satisfy a Klein-Gordon equation on a curved background metric. The metric has two terms [9, 12–14], one corresponding to the real spacetime metric and a second term, corresponding to what we call the analogue gravity metric, which depends on BEC parameters such as velocity flows and energy density. While in [9] we analyse the effect of changes in the real spacetime metric, in this case we consider the manipulation of the analogue gravity [15–18] metric, assuming that the real spacetime is flat. Since in this case the experimentalist is able to manipulate artificially the parameters of the condensate, we are able to simulate spacetime distortions with a much larger amplitude, as if the laboratory were closer to the source of the gravitational ripples. We show that with realistic experimental parameters, a physically meaningful model of gravitational wave can be simulated with current technology.

The paper is organised as follows. First we review the effects of a *real* gravitational wave in the Bogoliubov modes of the BEC. Then we consider the case in which there are non-zero initial velocity flows in the BEC, showing that there is always a reference frame in which we can modulate the speed of sound in such a way that the phonons experience an effect analog to the one produced by the propagation of a real spacetime wave. Finally, for the sake of simplicity we assume that there are no velocity flows in the BEC and we show that the gravitational wave can be simulated with current technology.

Gravitational waves in a BEC

The metric of a gravitational wave spacetime is commonly modelled by a small perturbation $h_{\mu\nu}$ to the flat Minkowski metric $\eta_{\mu\nu}$, i.e. [7],

$$g_{\mu\nu} = \eta_{\mu\nu} + h_{\mu\nu}, \tag{1}$$

where

$$\eta_{\mu\nu} = \begin{pmatrix} -c^2 & 0 & 0 & 0 \\ 0 & 1 & 0 & 0 \\ 0 & 0 & 1 & 0 \\ 0 & 0 & 0 & 1 \end{pmatrix} \tag{2}$$

and c is the speed of light in the vacuum. We consider Minkowski coordinates (t, x, y, z). In the transverse traceless (TT) gauge [7], the perturbation corresponding to a gravitational wave moving in the z-direction can be written as,

$$h_{\mu\nu} = \begin{pmatrix} 0 & 0 & 0 & 0 \\ 0 & h_+(t) & h_\times(t) & 0 \\ 0 & h_\times(t) & -h_+(t) & 0 \\ 0 & 0 & 0 & 0 \end{pmatrix}, \tag{3}$$

where $h_+(t)$, $h_\times(t)$ correspond to time-dependent perturbations in two different polarisations. Later on we will restrict the analysis to 1-dimensional fields, where the line element takes a simple form,

$$ds^2 = -c^2\,dt^2 + \left(1 + h_+(t)\right) dx^2. \tag{4}$$

We are interested in simulating this particular spacetime in a BEC. To this end we use the description of a BEC on a general spacetime metric following references [12, 13]. This description stems from the theory of fluids in a general relativistic background [12] and thus is valid as long as the BEC can be described as a fluid [13]. In the superfluid regime, a BEC is described by a mean field classical background Ψ plus quantum fluctuations $\hat{\Pi}$. These fluctuations, for length scales larger than the so-called healing length, behave like a phononic quantum field on a curved metric. Indeed, in a homogenous condensate, the massless modes of the field obey a Klein-Gordon equation

$$\Box\hat{\Pi} = 0, \tag{5}$$

where the d'Alembertian operator

$$\Box = 1/\sqrt{-\mathfrak{g}}\,\partial_a\left(\sqrt{-\mathfrak{g}}\mathfrak{g}^{ab}\,\partial_b\right) \tag{6}$$

depends on an effective spacetime metric $\mathfrak{g}_{ab}$ - with determinant $\mathfrak{g}$ - given by [12–14]

$$\mathfrak{g}_{ab} = \frac{\rho c}{c_s}\left[g_{ab} + \left(1 - \frac{c_s^2}{c^2}\right) V_a V_b\right]. \tag{7}$$

The effective metric is a function of the real spacetime metric g_{ab} (that in general may be curved) and background mean field properties of the BEC such as the number density n_0, the energy density ρ_0, the pressure p_0 and the speed of sound $c_s = c\sqrt{\partial p/\partial\rho}$. Here p is the total pressure, ρ the total density and V_a is the 4-velocity flow on the BEC. In the absence of background flows $V_a = (c, 0, 0, 0)$ and then,

$$\mathfrak{g}_{ab} = \frac{\rho c}{c_s}\left[g_{ab} + \begin{pmatrix} (c^2 - c_s^2) & 0 & 0 & 0 \\ 0 & 0 & 0 & 0 \\ 0 & 0 & 0 & 0 \\ 0 & 0 & 0 & 0 \end{pmatrix}\right]. \tag{8}$$

In the absence of a gravitational wave, the real spacetime metric is $g_{ab} = \eta_{ab}$, where η_{ab} has been defined in Eq. (2). Therefore, the effective metric of the BEC phononic excitations on the flat spacetime metric is given by,

$$\mathfrak{g}_{ab} = \frac{\rho c}{c_s}\begin{pmatrix} -c_s^2 & 0 & 0 & 0 \\ 0 & 1 & 0 & 0 \\ 0 & 0 & 1 & 0 \\ 0 & 0 & 0 & 1 \end{pmatrix}. \tag{9}$$

Ignoring the conformal factor - which can always be done in 1D or in the case in which is time-independent - we notice that the metric is the flat Minkowski metric with the speed of light being replaced by the speed of sound c_s. By considering a rescaled time coordinate $t' = (c/c_s)t$ we recover the standard Minkowski metric $ds^2 = -c^2\,dt^2 + dx^2$. This means that the phonons live on a spacetime in which, due to the BEC ground state properties, time flows in a different fashion and excitations propagate accordingly.

The real spacetime metric of a gravitational wave is given by Eq. (1) and thus, the effective metric for the phonons is,

$$\mathfrak{g}_{ab} = \frac{\rho c_s}{c}\begin{pmatrix} -c_s^2 & 0 & 0 & 0 \\ 0 & 1+h_+(t) & h_\times(t) & 0 \\ 0 & h_\times(t) & 1-h_+(t) & 0 \\ 0 & 0 & 0 & 1 \end{pmatrix}. \tag{10}$$

For simplicity, we considered a quasi one-dimensional BEC. The line element is conformal to,

$$ds^2 = -c_s^2\,dt^2 + \left(1 + h_+(t)\right)dx^2. \tag{11}$$

In [9], it is shown that the propagation of the gravitational wave generates particles in a BEC confined in a box-like trap. This particle creation is characterised by the Bogoliubov coefficients $\beta_{mn} = -(\hat{\phi}_m, \phi_n^*)$, where $\hat{\phi}_m$, ϕ_n are the m and n mode solutions of Eq. (5) given by the metrics in Eq. (10) and Eq. (9) respectively. In particular, if we model the gravitational wave by a sinusoidal oscillation of frequency Ω that matches the sum of the frequencies of a certain pair of modes m and n, the number of particles grows linearly in time. We refer the reader to [9] for more details.

Quantum simulation

The aim of this section is to show how to get a line element similar to the one in Eq. (11) by manipulating the parameters of the BEC while the real spacetime is assumed to be flat.

Going back to the effective metric Eq. (7) and restricting ourselves to 1D we perform the following coordinate transformation [13]:

$$\chi = \sqrt{c^2t^2 - x^2}, \qquad \zeta = \frac{x}{\sqrt{c^2t^2 - x^2}}. \tag{12}$$

Next, we choose the velocity profile such that in this new coordinate system the BEC is at rest, i.e.,

$$v^\chi = c; \qquad v^\zeta = 0. \tag{13}$$

The velocity profile in the original coordinate system needed in order to do this is therefore

$$v^t = c\sqrt{1+\zeta^2}; \qquad v^x = c\zeta. \tag{14}$$

The line element given by the metric in Eq. (7) then takes the form

$$ds^2 = -\rho(\chi)\frac{c_s(\chi)}{c}\,d\chi^2 + \rho(\chi)\frac{c}{c_s(\chi)}\chi^2\frac{d\zeta^2}{1+\zeta^2} \tag{15}$$

in which we are assuming that the speed of sound c_s and the density ρ might depend on the coordinates. We make another coordinate transformation (this time only in χ), so that

$$c_{s0}^2\,d\tau^2 = \rho(\chi)\frac{c_s(\chi)}{c}\,d\chi^2. \tag{16}$$

Here, we denote by c_{s0} the constant value of the speed of sound in the absence of any manipulation of the BEC parameters. Since we still have the freedom to specify the velocity of sound or the density of the BEC, we choose to let the density be constant, so that $\rho = \rho_0$ and we fix the velocity of sound such that

$$\ell^2(1+h_+(\tau)) \equiv \chi^2\rho_0\frac{c}{c_s(\chi)}, \tag{17}$$

where ℓ is a constant with units of distance. In order to do this, χ as a function of τ must be

$$\chi(\tau) = \sqrt{\frac{2c_{s0}}{\rho_0}\ell\int_0^\tau \sqrt{1+h_+(s)}\,ds}. \tag{18}$$

Modelling the oscillation of the +– polarisation of the gravitational wave in the standard way as

$$h_+(\tau) = A_+\sin(\Omega\tau), \tag{19}$$

expanding the root in the integral in powers of h_+ and keeping only linear terms in A_+, Eq. (18) takes the more friendly form:

$$\begin{aligned}\chi(\tau) &\approx \sqrt{\frac{2\ell c_{s0}}{\rho_0}\int_0^\tau\left(1+\frac{A_+}{2}\sin(\Omega s)\right)ds}\\ &= \sqrt{\frac{2\ell c_{s0}}{\rho_0}\left[\tau+\frac{A_+}{2\Omega}\left(1-\cos(\Omega\tau)\right)\right]},\end{aligned} \tag{20}$$

up to first order in A_+. With this, we can solve Eq. (17) for the velocity of sound in terms of τ - substituting Eq. (20) in Eq. (17) as

$$\begin{aligned}c_s(\tau) &= \frac{cc_{s0}}{\ell}\left[\frac{2\Omega\tau + A_+(1-\cos(\Omega\tau))}{\Omega(1+A_+\sin(\Omega\tau))}\right]\\ &\approx \frac{cc_{s0}}{\ell\Omega}\left(2\Omega\tau + A_+\left[1-\cos(\Omega\tau)-2\Omega\tau\sin(\Omega\tau)\right]\right),\end{aligned} \tag{21}$$

up to first order in A_+. Now that we obtained an explicit expression for the velocity of sound as a function of the coordinate τ, it is natural to ask the significance of the constant parameter ℓ. For this, we see that in the absence of a simulated gravitational wave, Eq. (17) becomes $\ell = \chi_0\sqrt{\rho_0 c/c_{s0}(\chi_0)}$, for a particular value $\chi = \chi_0$. A convenient choice is $\ell = 2c\tau_0$, so that the velocity of sound can be expressed as $c_s(\tau_0) = c_{s0}$ for $\tau = \tau_0$. Hence, ℓ is represented by the hyperbola $c^2t^2 - x^2 = \ell^2 c_{s0}/(\rho_0 c)$ in the Minkowski coordinate system. A final coordinate transformation is made, defining ξ as

$$\zeta = \sinh\left(\frac{\xi}{\ell}\right), \tag{22}$$

such that $d\xi^2 = \ell^2\, d\zeta^2/(1+\zeta^2)$, thus transforming the line element Eq. (15) into

$$ds^2 = -c_{s0}^2\, d\tau^2 + \left(1 + h_+(\tau)\right) d\xi^2, \tag{23}$$

as desired.

The case with no background flows. Experimental implementation

In the last section, we have shown how the speed of sound can be modified to mimic a gravitational wave in a coordinate system (τ, ξ) in which there are no background flows. Now, we will consider for simplicity that this coordinate system is the lab frame (t, x) and relate our results directly with experimental parameters.

As we have already seen above Eq. (9) - and assuming again a 1D spacetime - if $v^t = c$ and $v^x = 0$, then the line element is conformal to:

$$ds^2 = -c_s^2\, dt^2 + dx^2. \tag{24}$$

Now, if we consider that the speed of sound can depend on t:

$$c_s(t) = c_{s0} f(t), \tag{25}$$

the corresponding line element is conformal to:

$$ds^2 = -c_{s0}^2\, dt^2 + \frac{1}{(f(t))^2}\, dx^2. \tag{26}$$

Therefore, if the speed of sound varies in time such that

$$f(t) = \left(1 - \frac{A_+ \sin \Omega t}{2}\right), \tag{27}$$

we find, up to the first order in A_+:

$$ds^2 = -c_{s0}^2\, dt^2 + (1 + A_+ \sin \Omega t)\, dx^2. \tag{28}$$

So the experimental task is to modulate the speed of sound as:

$$c_s(t) = c_{s0}\left(1 - \frac{A_+ \sin \Omega t}{2}\right). \tag{29}$$

In a weakly interacting condensate [19], the speed of sound is

$$c_{s0} = \sqrt{\frac{\rho g}{m}}, \tag{30}$$

where ρ is the density, m the atomic mass and the coupling strength is

$$g = \frac{4\pi \hbar^2 a}{m}, \tag{31}$$

and a is the scattering length. So, finally:

$$c_{s0} = \frac{\hbar}{m}\sqrt{4\pi \rho a}. \tag{32}$$

It is well-known [20] that a can be modulated in time by using the dependence of the scattering length on an external magnetic field around a Feshbach resonance. The aim is thus to achieve:

$$a(t) = a(0)\big(1 - h(t)\big), \tag{33}$$

because then

$$c_s(t) = c_{s0}\left(1 - \frac{h(t)}{2}\right) \tag{34}$$

(up to the first order), where

$$c_{s0} = \frac{\hbar}{m}\sqrt{4\pi \rho a(0)}. \tag{35}$$

To this end, we will exploit the dependence of the scattering length with an external magnetic field

$$a = a_{bg}\left(1 - \frac{\omega}{B - B_0}\right), \tag{36}$$

where a_{bg} is the background scattering length, B_0 is the value of the magnetic field at which the Feshbach resonance takes place, and ω is the width of the Feshbach resonance. Considering a time-dependent magnetic field

$$B(t) = B(0)\big(1 + \delta B(t)\big), \tag{37}$$

and with a little algebra, we can write, to the first order in δB:

$$a(t) = a(0)\left(1 + \frac{B(0)\omega \delta B(t)}{(B(0) - B_0 - \omega)(B(0) - B_0)}\right). \tag{38}$$

So, if we identify:

$$h_+(t) = \frac{B(0)\omega \delta B(t)}{(B(0) - B_0 - \omega)(B(0) - B_0)}, \tag{39}$$

we are simulating a gravitational wave. Writing

$$\delta B(t) = \delta B \sin \Omega t, \tag{40}$$

the amplitude of the simulated wave is:

$$A_+ = \frac{B(0)\omega\delta B}{(B(0) - B_0 - \omega)(B(0) - B_0)} \tag{41}$$

and Ω is the frequency of the simulated wave. Taking experimental values for B_0 and ω, [20] and assuming that we can control the magnetic field in a $0.1G$ scale (so we can take $\delta B \simeq B(0) \simeq 0.1G$), we estimate the amplitude of the simulated wave as

$$A_+ \simeq 10^{-7}. \tag{42}$$

This is much larger that the gravitational waves that we expect to see in the Earth $A_+ \simeq 10^{-20}$, due to the fact that the Earth is far from typical sources of gravitational waves. Therefore it is interesting to think of the physical meaning of the gravitational waves that can be simulated with our techniques.

For instance, the amplitude and frequency of the + polarisation of the gravitational wave generated by a Sun-Earth-like system are:

$$A_+ = \frac{4G^2 mM}{c^4 Rr}; \qquad \Omega = 2\sqrt{\frac{G(m+M)}{r^3}} \tag{43}$$

respectively, where G is Newton's gravitational constant, m and M are the masses of the Earth and the Sun respectively, r the distance between the two bodies and R the distance between the detector and the centre of the mass of the system, which is assumed to be much larger than the wavelength λ of the gravitational wave $R \gg \lambda$. Thus, if we consider the real masses of the Sun and the Earth, a distance between them of $r = 10$ m and $R = 10^7$ m, we find Ω in the kHz range - which is very convenient to generate phonons in the BEC - and the desired value of A_+. In [9] it is predicted that the changes in the covariance matrix of the Bogoliubov modes induced by gravitational waves of typical amplitudes $A_+ < 10^{-20}$ are in principle detectable. Therefore, the changes generated by a simulated wave of much larger amplitude, should be observed in an experiment with current cold-atoms technology.

Conclusions

We have shown how to generate an artificial gravitational wave spacetime for the quantum excitations of a BEC. In the case in which there are no initial background flows, we show that the simulated ripple is obtained through a modulation of the speed of sound in the BEC. In the laboratory, this can be achieved with current technology by exploiting the dependence of the scattering length on the external magnetic field around a Feshbach resonance. With realistic experimental parameters, we find that simulated gravitational waves that can resonate with the Bogoliubov modes of the BEC. The amplitude of these artificial ripples is much larger than the typical amplitude expected for gravitational waves reaching the Earth, due to the fact that the Earth is very far from typical sources.

Thus our simulated waves would mimic the waves generated by sources much closer to the BEC. This feature would enhance the effects of the ripple in the system, facilitating their detection. The experimental test of our predictions would be a proof-of-concept of the generation of particles by gravitational waves and would pave the way for the actual observation of *real* spacetime ripples in a BEC. More generally, our low-cost Earth-based tabletop experiment will inform the whole programme of gravitational wave astronomy.

Competing interests
The authors declare that they have no competing interests.

Authors' contributions
IF conceived the main idea and supervised the project. TB and CS performed the computations. All the authors contributed to the preparation of the manuscript.

Acknowledgements
TB acknowledges funding from CONACYT. IF and CS acknowledge funding from EPSRC (CAF Grant No. EP/G00496X/2 to IF).

References

1. Georgescu IM, Ashhab S, Nori F. Quantum simulation. Rev Mod Phys. 2014;86:153–85.
2. Gerritsma R, Kirchmair G, Zähringer F, Solano E, Blatt R, Roos CF. Quantum simulation of the Dirac equation. Nature. 2010;463:68–71.
3. Casanova J, Sabín C, León J, Egusquiza IL, Gerritsma R, Roos CF, García-Ripoll JJ, Solano E. Quantum simulation of the majorana equation and unphysical operations. Phys Rev X. 2011;1:021018.
4. Sabín C, Casanova J, García-Ripoll JJ, Lamata L, Solano E, León J. Encoding relativistic potential dynamics into free evolution. Phys Rev A. 2012;85:052301.
5. Alvarez-Rodriguez U, Casanova J, Lamata L, Solano E. Quantum simulation of noncausal kinematic transformations. Phys Rev Lett. 2013;111:090503.
6. Rindler W. Relativity: special, general and cosmological. 2nd ed. Oxford: Oxford University Press; 2006.
7. Flanagan EE, Hughes SA. The basics of gravitational wave theory. New J Phys. 2005;7(1):204.
8. Aufmuth P, Danzmann K. Gravitational wave detectors. New J Phys. 2005;7:202.
9. Sabín C, Bruschi DE, Ahmadi M, Fuentes I. Phonon creation by gravitational waves. New J Phys. 2014;16:085003.
10. Moore GT. Quantum theory of the electromagnetic field in a variable length one dimensional cavity. J Math Phys. 1970;11:269.
11. Wilson CM, Johansson G, Pourkabirian A, Simoen M, Johansson JR, Duty T, Nori F, Delsing P. Observation of the dynamical Casimir effect in a superconducting circuit. Nature. 2011;479:376–9.
12. Visser M, Molina-Paris C. Acoustic geometry for general relativistic barotropic irrotational fluid flow. New J Phys. 2010;12:095014.
13. Fagnocchi S, Finazzi S, Liberati S, Kormos M, Trombettoni A. Relativistic Bose-Einstein condensates: a new system for analogue models of gravity. New J Phys. 2010;12:095012.
14. Bruschi DE, Sabín C, White A, Baccetti V, Oi DKL, Fuentes I. Testing the effects of gravity and motion on quantum entanglement in space-based experiments. New J Phys. 2014;16:053041.
15. Barceló C, Liberati S, Visser M. Analogue gravity. Living Rev Relativ. 2011;14:3.
16. Garay LJ, Anglin JR, Cirac JI, Zoller P. Sonic analog of gravitational black holes in Bose-Einstein condensates. Phys Rev Lett. 2000;85:4643.
17. Schley R, Berkovitz A, Rinott S, Shammass I, Blumkin A, Steinhauer J. Planck distribution of phonons in a Bose-Einstein condensate. Phys Rev Lett. 2013;111(5):055301.
18. Steinahauer J. Observation of self-amplifying hawking radiation in an analogue black-hole laser. Nat Phys. 2014;10:864.
19. Pethick CJ, Smith H. Bose Einstein condensation in dilute gases. Cambridge: Cambridge University Press; 2004.
20. Schneider U, Hackermüller L, Ronzheimer JP, Will S, Braun S, Best T, Bloch I, Demler E, Mandt S, Rasch D, Rosch A. Fermionic transport and out-of-equilibrium dynamics in a homogeneous Hubbard model with ultracold atoms. Nat Phys. 2012;8:213–8.

8

Thermalization of strongly interacting bosons after spontaneous emissions in optical lattices

Johannes Schachenmayer[1*], Lode Pollet[2], Matthias Troyer[3] and Andrew J Daley[4,5]

*Correspondence:
Johannes.Schachenmayer@gmail.com
[1] Department of Physics, JILA, NIST, University of Colorado, 440 UCB, Boulder, CO 80309, USA
Full list of author information is available at the end of the article

Abstract

We study the out-of-equilibrium dynamics of bosonic atoms in a 1D optical lattice, after the ground-state is excited by a single spontaneous emission event, i.e. after an absorption and re-emission of a lattice photon. This is an important fundamental source of decoherence for current experiments, and understanding the resulting dynamics and changes in the many-body state is important for controlling heating in quantum simulators. Previously it was found that in the superfluid regime, simple observables relax to values that can be described by a thermal distribution on experimental time-scales, and that this breaks down for strong interactions (in the Mott insulator regime). Here we expand on this result, investigating the relaxation of the momentum distribution as a function of time, and discussing the relationship to eigenstate thermalization. For the strongly interacting limit, we provide an analytical analysis for the behavior of the system, based on an effective low-energy Hamiltonian in which the dynamics can be understood based on correlated doublon-holon pairs.

Keywords: thermalization; optical lattices; spontaneous emissions; Bose-Hubbard model

1 Introduction

An important fundamental question in many-body quantum mechanics is to what extent and under which conditions an isolated system perturbed away from equilibrium will undergo thermalization, in the sense that at long times the system will reach a steady state where simple observables equal the values for a thermal distribution [1–5]. Recently it has been possible to observe integrable dynamics for strongly interacting cold gases confined to move in one dimension [6], and thus reach regimes where systems either do not thermalize in this standard sense [7] or otherwise undergo generalized thermalization [8, 9].

These fundamental questions also have an important impact on the application of cold atoms as quantum simulators [10, 11]. Development in experiments with these systems has reached a stage where it is possible to tailor interesting many-body Hamiltonians, and study the many-body physics of corresponding ground states. However, some of the most interesting physical regimes require the realization of states with small energy gaps, requiring exceptionally low temperatures and entropies, which provide a key challenge for current experiments [12]. This has been particularly true in attempts to observe quantum magnetism in optical lattices within strongly interacting regimes $|U| \gg J$, where U is the on-site interaction energy, and $J/\hbar$ indicates the tunneling amplitude for particles moving between neighboring sites. There, magnetic order in multi-species mixtures is driven by terms of the order of J^2/U, which is typically small for current experiments [12–19].

In this context, it is very important to understand and control the many-body effects of competing heating processes in experiments. Previously, it has often been assumed that all of the energy added to the system will be thermalized, causing an effective increase in temperature. However, in regimes in which the system does not thermalize excitations entirely (or for some types of excitations, at all), the behavior is more complex. A simple example where excitations cannot thermalize on typical experimental timescales is given by spontaneous emission events (incoherent light scattering). In such processes, the main contribution to the increase in the average energy of the system as a function of time comes from atoms being excited to higher Bloch bands of an optical lattice [20–23]. Because the bandgap is usually much larger than the typical energy scales U and J of dynamics in the lowest band, it is not possible for the system to thermalize most of the energy added to the system as a function of time.

At the same time, understanding this process is made more complex by the fact that processes exciting particles to higher Bloch bands are rare compared with processes leaving particles in the lowest band in typical experiments [20, 21]. As a result, the dynamics can actually be dominated by heating in the lowest band of the lattice, the basic effects of which have been studied in several recent articles for bosons [20, 21, 24–28] and for fermions [29–31]. It is then natural to ask, in particular, whether the system can thermalize after spontaneous emission events that leave atoms in the lowest band.

In this article, we study the interplay between spontaneous emissions and thermalization in detail, asking under what conditions key simple quantities such as the quasimomentum distribution and the kinetic energy relax to values given by a thermal distribution. In each case, we ask this question in a practical context by fixing a thermal distribution such that we take a canonical ensemble with the temperature T chosen so that the expectation value of the energy in the thermal distribution matches the expectation value of the energy after a spontaneous emission event. In the case of strong interactions, we might expect that the system reaches integrable limits where the system does not thermalize. On the other hand, one might also expect that as in general, the Bose-Hubbard Hamiltonian is non-integrable, that outside the special cases of strong interactions ($U/J \to \infty$) and non-interacting systems ($U = 0$), thermalization would essentially always occur.

Instead, we find that this generalization is not universally applicable. It is important to note that spontaneous emissions correspond to local quenches for the many-body state, leading to population of low-lying excited states. As we showed previously [24], after a spontaneous emission, depending on the parameter regime and the corresponding different characteristics of the low-energy spectrum: either (i) the system relaxes over short times to thermal values of the quasimomentum distribution and kinetic energy, or (ii) on short timescales, the system relaxes to states that are clearly non-thermal, even if all atoms remain in the lowest Bloch band. The latter regime was found to occur if the interactions are strong enough so that the ground-state of the system corresponds to a Mott insulator state. Below we elaborate on these results especially for the quasimomentum distribution, and discuss their interpretation in terms of eigenstate thermalization.

In the limit where the system is close to an ideal Mott-insulator (a product state with a fixed number of particles per cite) and in the case of unit filing, one can describe the dynamics of the system in an effective model, where one restricts the local Hilbert space to having a maximum of two particles per site. The low-lying excitations can then be described as correlated pairs of holes ("holons") and doubly occupied sites ("doublons") [26, 32]. Here we use this approximation to show that no thermalization is present for spontaneous emission in this limit. We compare our result to exact numerical calculations either using exact diagonalization in small systems, or by making use of t-DMRG techniques [33–36].

This article is organized as follows. We begin by introducing spontaneous emissions and thermalization in these systems in Section 2. Then, in Section 3 we review the explanation of the basic behavior we find in terms of the eigenstate thermalization. In Section 4 we discuss how the behavior can be understood in the strongly interacting limit, making predictions that could be observed for the propagation of excitations when the system does not thermalize in that limit. In Section 5 we then provide a summary and outlook.

2 Thermalization of spontaneous emissions

Spontaneous emissions are a fundamental source of heating in optical dipole potentials [22, 23], and one of the key contributing heating sources in current experiments with cold atoms in optical lattices [20, 21]. As shown in Ref. [21], a multi-band many-body master equation can be derived to describe this process in the case of far detuned lattice light. An adiabatic elimination of the electronic excited states leads to an equation for atoms in their electronic ground states. Taking this, and assuming the typical experimental case in which the lattice spacing a is comparable to the wavelength of scattered photons, $a = \lambda/2$, the dynamics of the many-body density operator ρ follows a master equation of the form ($\hbar \equiv 1$), $\dot{\rho} = -\mathrm{i}[H_{\mathrm{MB}}, \rho] + \mathcal{L}\rho$. Here H_{MB} is a multi-band Bose-Hubbard Hamiltonian and $\mathcal{L}\rho$ includes dissipative Lindblad terms with on-site jump operators $\propto (b_i^{[n]})^\dagger b_i^{[m]}$. Here, $b_i^{[n]}$ is the bosonic annihilation operator at site i, for a particle in band m.

For relatively deep optical lattices, transition rates for inter-band processes coupling neighboring Bloch bands are strongly suppressed by the square of the Lamb-Dicke parameter, $\eta = 2\pi a_T/\lambda$. For a typical experiment with a lattice depth $V_0 \approx 8E_R$ [$E_R = 4\pi^2\hbar^2/(2m\lambda^2)$, where m is the mass of the atom], $\eta^2 \sim 0.1$. In the usual case of a red-detuned optical lattice the dominant processes are thus intra-band processes, which return the atoms to their initial Bloch band. Provided we initially consider the system to be in the lowest band, processes accessing higher Bloch bands are suppressed by a factor of the order η^4 or greater, and the master equation for the 1D model simplifies to:

$$\dot{\rho} = -i[H, \rho] - \frac{\gamma}{2}\sum_i [n_i, [n_i, \rho]], \tag{1}$$

where $n_i = b_i^\dagger b_i$ is the particle number operator for the lowest band, and γ is the scattering rate. The Hamiltonian is the usual single band Bose-Hubbard Hamiltonian

$$H = -J\sum_i \left(b_i^\dagger b_{i+1} + b_{i+1}^\dagger b_i\right) + \frac{U}{2}\sum_i n_i(n_i - 1). \tag{2}$$

In the following we will focus on a system with M sites and $N = M$ particles, so that the filling is $\bar{n} = N/M = 1$. In this case the system undergoes a ground-state quantum phase transition between a superfluid (SF) and a Mott insulating (MI) state at the critical value,

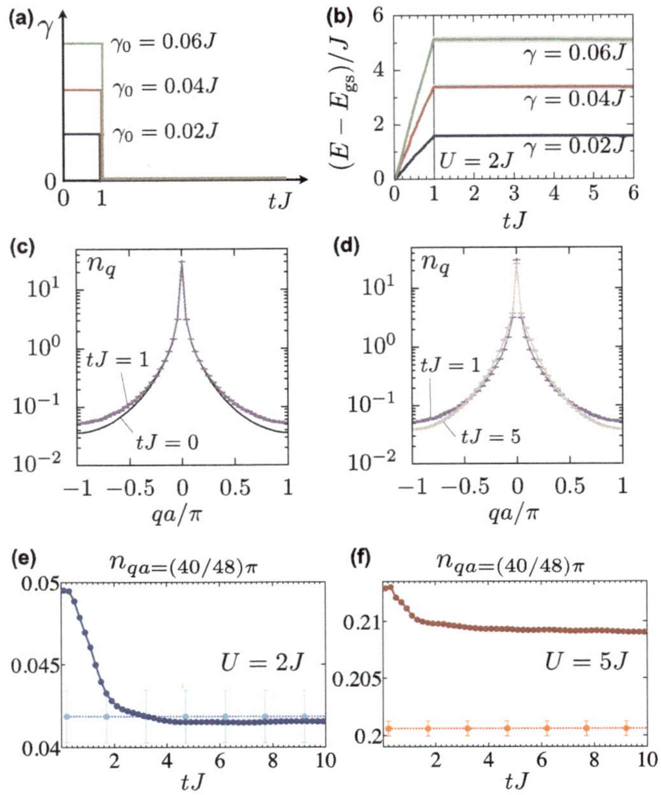

Figure 1 Example of heating and relaxation dynamics. We consider a system with $N = 48$ particles on $M = 48$ sites and $U = 2J$ that is initially in its ground state and subjected to heating with different scattering rates γ_0 for a short time of $tJ = 1$, as sketched in panel **(a)**. **(b)** Depending on γ_0 this leads to an increase in energy. **(c)** Evolution of the quasi-momentum distribution n_q during the heating period ($\gamma_0 = 0.06J$). The tails are lifted due to the localization of particles. **(d)** Relaxation dynamics of n_q after the heating is switched off. The system relaxes to a broadened distribution at $tJ = 5$ (t-DMRG results, bond-dimension, $D = 256, 500$ quantum trajectories). **(e/f)** Relaxation of a large q component of the quasi-momentum distribution after a single spontaneous emission on an arbitrary site (we consider a weighted ensemble average of jumps on the different sites). The dashed horizontal line indicates the equilibrium value (Monte-Carlo calculation). Thermalization only occurs in the superfluid regime **(e)**, and not in the Mott insulating regime **(f)** [t-DMRG results, bond-dimension, $D = 512, 256$ for (e), (f)].

which was estimated to be $U_c \approx 3.25(5)J$ [37, 38]. The underlying physics of the lattice-photon scattering process can be summarized as the environment obtaining information about the position of an atom. Thereby the atoms are localized on the length-scale of a single lattice-site size.

Numerically we can fully simulate the master equation (1) also for large 1D systems by combining t-DMRG techniques with a Monte-Carlo sampling over quantum trajectories [39–42]. There, the density matrix is obtained from a statistical average over many trajectories with randomly applied jump operators [39]. To give an example, in Figure 1 we plot the dynamics for the evolution under equation (1). We start in the ground-state for a system with $U = 2J$ (SF regime, $M = 48$). Then we turn on light scattering for a short time up to $tJ = 1$ and for different values of $\gamma = \gamma_0$ (illustrated in Figure 1(a)). This leads to an increase in energy (Figure 1(b)). Once the heating is switched off, the system then relaxes. This relaxation is for example seen when looking at the evolution of the quasi-momentum distribution

$$n_q = \frac{1}{M}\sum_{n,m} e^{-iq(n-m)}\langle b_n^\dagger b_m\rangle. \tag{3}$$

During the heating period the localization of particles leads to a "lifting" of tails of this distribution, as seen in Figure 1(c). Once the heating is switched off, the high quasi-momentum components become redistributed and the system is relaxing to a broadened distribution, depicted in Figure 1(d). The question whether the system thermalizes or not depends now on the shape of this distribution. In the case that the steady state distribution coincides with a Boltzmann distribution (in the many-body eigenstates) for a temperature corresponding to the same mean energy, we call the system thermalized, otherwise we don't. Obviously, the question of thermalization is connected to the observable under consideration. In Ref. [24] it was found that for example the tails of the quasi-momentum and the kinetic energy $E_{\text{kin}} = \langle H_{\text{kin}}\rangle = -2J\sum_i \langle b_i^\dagger b_{i+1}\rangle + \langle b_{i+1}^\dagger b_i\rangle$ thermalize on experimentally relevant time-scales in the SF regime and that this thermalization breaks down in the MI.

Here, we demonstrate a specific example of this in Figures 1(e),(f). There we show relaxation of a high-q component of the quasimomentum distribution, $q = 40\pi/48$, in a system with $M = 48$ lattice sites. We can clearly see that after a short period of time, this component relaxes to its value in a corresponding thermal distribution provided that we initially start in a superfluid state with $U/J = 2$. In the MI limit, we see that the behavior is similar, in so far as there is a rapid relaxation of this value of quasimomentum on short times. However, this leads to a value that is markedly different to the corresponding thermal value. It is naturally possible that in this regime, there are processes on very long timescales that will eventually lead to thermalization of this quantity. But it is clear that on the MI side of the phase transition, the system exhibits a qualitatively different behavior in terms of how close it is to thermal distributions on the scale of a few tunnelling times. This will significantly impact the ability of the system to thermalize energy added by heating on the timescales of the heating process itself.

We note that the relaxation of the quasimomentum distribution could be directly observed (and compared to a thermal distribution) in a realistic experiment. Here one could use a near-resonant laser beam that is switched on for a particular time and provides the necessary heating through spontaneous emissions. In contrast to our calculations such an experiment would not be limited to relatively low heating rates, i.e. a small number of single spontaneous emissions. Thus, the discrepancies in the tails of the quasimomentum distribution from thermal values can be significantly larger than for the single spontaneous emission event considered in Figures 1(e),(f).

It is important to note that in the regime where the system is relaxing, the thermalization and all of the corresponding dynamics is determined by unitary closed system dynamics alone. In the remainder of this paper we will simplify the question of thermalization to the coherent dynamics after a single spontaneous emission event, i.e. we ask whether expectation values of observables relax to thermal values in a unitary time-dynamics after we apply a single quantum jump. Thus, we consider a unitary time-evolution of an initial state

$$|\psi_l(t = 0^+)\rangle = \frac{n_l|\psi_g\rangle}{\|n_l|\psi_g\rangle\|}, \tag{4}$$

where $|\psi_g\rangle$ denotes the ground-state of the system, and then take a stochastic average over the sites l, which is weighted proportional to the square of the density on site l, $\langle\psi_g|n_l^2|\psi_g\rangle$. The mechanism of this thermalization of closed quantum systems can be explained via eigenstate thermalizaton, a mechanism which we will review in the next section.

3 Eigenstate thermalization

The question whether unitary time-dynamics leads to thermalization or not can be rephrased in terms of the "eigenstate thermalization" [1–3], which we will briefly review here. Consider the situation where the initial state is not an eigenstate of the Hamiltonian (such as it is the case after a spontaneous emission). The state can be expanded into energy eigenstates defined via $H|\alpha\rangle = E_\alpha|\alpha\rangle$, $|\psi\rangle = \sum_\alpha c_\alpha|\alpha\rangle$, and the time-evolution is thus given by

$$|\psi(t)\rangle = \sum_\alpha e^{-\frac{i}{\hbar}E_\alpha t}c_\alpha|\alpha\rangle. \tag{5}$$

We note that the following discussion is restricted to non-degenerate eigenstates, which is the natural situation in large quantum systems [43] and also applies to the situation studied in this paper. Then the time-dependent expectation value of any observable $\hat{O}$ is

$$\langle\hat{O}\rangle_t = \langle\psi(t)|\hat{O}|\psi(t)\rangle = \sum_{\alpha,\beta} e^{-\frac{i}{\hbar}(E_\alpha-E_\beta)t}c_\beta^* c_\alpha\langle\beta|\hat{O}|\alpha\rangle. \tag{6}$$

If the system relaxes into a steady state, this steady state must be identical to the long-time average

$$\overline{\langle\hat{O}\rangle_t} = \lim_{T\to\infty}\frac{1}{T}\int_0^T dt\langle\psi(t)|\hat{O}|\psi(t)\rangle = \sum_\alpha |c_\alpha|^2\langle\alpha|\hat{O}|\alpha\rangle. \tag{7}$$

Thus, the value of the steady state expectation (if such a steady state is developed) only depends on two variables: (i) The expansion coefficients of the initial state, c_α, and (ii) the expectation values of the energy-eigenstates with the particular observable, $\langle\alpha|\hat{O}|\alpha\rangle$. The eigenstate thermalization hypothesis states, that if the eigenenergy expectation values of an observable vary smoothly in the energy window defined by the c_α (and the off-diagonal $\langle\alpha|\hat{O}|\beta\rangle$, $\alpha \neq \beta$ are small), the system relaxes into a state for which the observable can be described with a micro-canonical ensemble [1, 2].

In our situation, spontaneous emission only adds a small amount of energy to the system. Thus, the question whether this energy thermalizes or not will be determined by the *low-energy* spectrum. Thus, we can use an exact numerical diagonalization to obtain the lowest energy eigenstates in systems with $N = M = 10$ sites and particles. As an example, in panels (a)-(c) of Figure 2 we show the matrix elements $\langle\alpha|\hat{O}|\beta\rangle$ for the kinetic energy in the system. As observable we use the kinetic energy $\hat{O} = H_{\text{kin}}$. The eigenvalues are ordered ascending according to their energy. We find that only in the SF case of $U = 2J$, the diagonal elements vary smootly as function, while the off-diagonal elements are zero. This verifies the findings of Ref. [24], and is consistent with the breakdown of the low-energy thermalization when entering the MI regime.

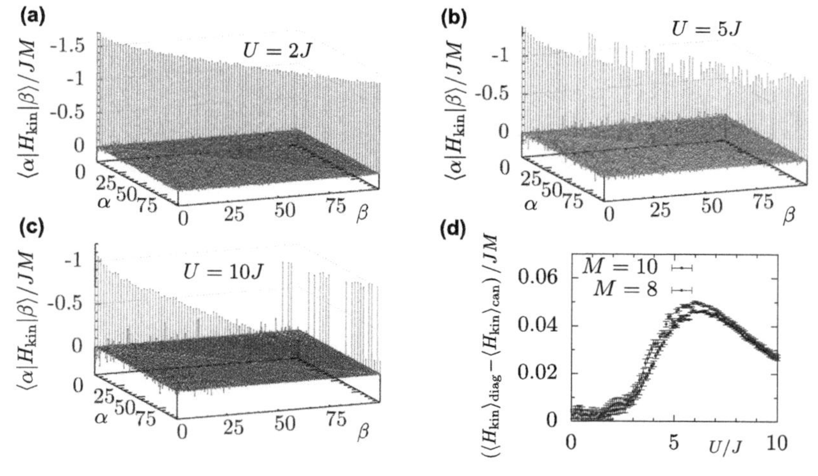

Figure 2 Eigenstate thermalization of spontaneous emissions. (a)-(c) The matrix-elements of the kinetic energy ($N = M = 10$), $\langle\alpha|H_{\rm kin}|\beta\rangle/JM$, in the energy eigenbasis (sorted in ascending order according to their energy). Different panels are for increasing on-site interaction U. Only in the SF regime ($U \lesssim 3.25(5)J$ [37]) the diagonal elements vary smoothly with α, and allow for eigenstate thermalization. **(d)** The difference of the expectations between the diagonal and the canonical ensemble corresponding to the same mean energy after a single spontaneous emission at site $i = M/2$ in a system of $M = 8$ and $M = 10$ sites (errorbars are due to the interpolation to find the canonical ensemble corresponding to the correct energy). Although finite size effects are large the change in behavior at the quantum phase transition is evident.

In addition we plot a direct comparison of the two expectation values in Figure 2(d). We can calculate the expectation value on the one hand from the diagonal ensemble and on the other hand from a canonical ensemble with the same mean-energy. As initial state we choose the state after a single spontaneous emission on site $i = M/2$. Although finite size effects play an important role in this small system, it is already obvious that a discrepancy between the two ensembles arises when entering the MI regime. As was found by t-DMRG calculations in [24], this also holds for larger systems. In the next section we will calculate the low-lying energy eigenstates exactly analytically in the limit of large interactions in the thermodynamic limit.

4 Dynamics in the strongly interacting limit

In this section we will discuss how we can understand the dynamics analytically for the case of strong interactions. In the extreme case of infinite U, at unit filling the particles will be in an idealized Mott insulator state, i.e., $|\psi_{U\to\infty}\rangle = \prod_n |1\rangle_n$. This state is an eigenstate of the jump operator with eigenvalue one, so that light scattering leaving particles in the lowest band does not change the state or introduce any extra energy to the system. Naturally this is a trivial limit that is typically not reached in experiments, and we go beyond this below by considering situations where the amplitude for particles to be found on neighbouring sites is non-zero in the initial state, and light scattering can lead to the formation of pairs consisting of a doubly occupied site and a hole (unoccupied site).

4.1 Hard-core bosons

To lowest order, one can use an approximation in which we assume that the bosons are essentially hard-core bosons (HCBs) [44]. The idea is to restrict the Hilbert space to local states with a maximum of one particle per site. Mathematically, this can be achieved by using the standard Bose-Hubbard model (2) while additionally imposing fermionic anti-commutation relations to the creation/annihilation operators for bosons on the same site, $\{b_i, b_i^\dagger\} = 1$. These particles are not real fermions, since they still commute on different sites, $[b_i, b_j^\dagger] = 1$. However this can be fixed by a Jordan Wigner transformation [45]

$$b_i = \prod_{\alpha=1}^{i-1} e^{i\pi c_\alpha^\dagger c_\alpha} c_i, \tag{8}$$

where the new quasi-particles now are real fermions, $\{c_i, c_j^\dagger\} = \delta_{i,j}$. In this picture the Hamiltonian becomes a Hamiltonian for non-interacting fermions:

$$H = -J \sum_i \left(c_i c_{i+1}^\dagger + c_i^\dagger c_{i+1}\right), \tag{9}$$

and the ground-state takes the product form $|\psi_G\rangle = \prod_n^N \sum_i^N P_{i,n}|i\rangle$ (where $|i\rangle$ denotes the state of a particle at site i). The jump-operators - as is true for all site-local operators - are unaffected by the Jordan-Wigner transformation and so in this transformed notation, $b_i^\dagger b_i \to c_i^\dagger c_i$.

If the system has fewer particles than lattice sites, spontaneous emissions will give rise to heating in the lowest band, as the states are not eigenstates of $c_i^\dagger c_i$, which does not commute with the Hamiltonian. Simple results in this case are discussed in Ref. [39]. However, in the special case of unit filling, which we would like to treat here, the only state described by this simple HCB form is again the trivial state with a single particle on each site, which is an eigenstate of all operators $c_i^\dagger c_i$ and $c_i^\dagger c_{i+1}$. It is then clear that we need to go beyond this treatment in order to understand heating in a Mott Insulator state with finite U.

4.2 Doublon/holon calculation

To study thermalization of spontaneous emission we have to go to the next order in the approximation and allow for states with 0, 1, and 2 particles per site. It turns out that also in this case approximate analytical calculations can be derived [26]. Assuming that the state of the system remains close to the Mott insulator state with $\bar{n} = 1$, then with conservation of total particle number, the lowest-order excitations will be given by combinations of doubly occupied sites (sometimes referred to as population of particles in a second Mott Band), and holes (in the lowest Mott Band). It is now possible to introduce creation/annihilation operators with a vacuum given by the ideal MI state $|\text{vac}\rangle_i = |1\rangle_i$ as

$$\tilde{d}_i^\dagger |\text{vac}\rangle_i = |2\rangle_i, \tag{10}$$

$$\tilde{h}_i^\dagger |\text{vac}\rangle_i = |0\rangle_i. \tag{11}$$

Since there can only be one hole or doublon on each site these quasi-particles must obey the hard-core constraint, which can again be expressed with on-site anti-commutation relations $\{\tilde{d}_i, \tilde{d}_i^\dagger\} = 1 = \{\tilde{h}_i, \tilde{h}_i^\dagger\}$. Note that in principle one also has to add a constraint in order not to have a doublon and holon on the same site. However, based on the assumption that the total number of doubly-occupied sites and holes remains low we will neglect these terms, as was originally done in Ref. [26]. As in the case of the HCB model one can now turn the quasi-particles into proper fermions via the Jordan-Wigner transformation

$$\tilde{d}_i = \prod_{\alpha=1}^{i-1} e^{i\pi d_\alpha^\dagger d_\alpha} d_i, \tag{12}$$

$$\tilde{h}_i = \prod_{\alpha=1}^{i-1} e^{i\pi h_\alpha^\dagger h_\alpha} h_i. \tag{13}$$

In the picture of these fermionic quasi-particles, the Hamiltonian reads

$$H_{dh} = U\sum_i d_i^\dagger d_i - J\Big(2\sum_i d_i d_{i+1}^\dagger + \sum_i h_i h_{i+1}^\dagger + \text{h.c.}\Big)$$
$$- \sqrt{2}J\Big(\sum_i d_i^\dagger h_{i+1}^\dagger - \sum_i h_i d_{i+1}\Big). \tag{14}$$

A discrete Fourier transformation into quasi-momentum space ($x_j = \frac{1}{\sqrt{M}}\sum_q e^{ijqa}x_q$, for the operator x_j, $q \in \{0, 2\pi/M, 2\times 2\pi/M, \ldots, 2\pi\}$) gives the quadratic Hamiltonian (up to a constant)

$$\tilde{H}_{dh} = \sum_q \begin{pmatrix} d_q^\dagger & h_{-q} \end{pmatrix} \begin{pmatrix} E_d(q) & iE_{dh}(q) \\ -iE_{dh}(q) & -E_h(q) \end{pmatrix} \begin{pmatrix} d_q \\ h_{-q}^\dagger \end{pmatrix}. \tag{15}$$

The corresponding dispersion relations are given by $E_d(q) = -4J\cos(q) + U$, $E_h(q) = -2J\cos(q)$, and $E_{dh}(q) = 2J\sqrt{2}\sin(q)$. This Hamiltonian can be diagonalized straightforwardly by a Bogoliubov transformation

$$H = \sum_q \begin{pmatrix} d_q^\dagger & h_{-q} \end{pmatrix} U_B(q)U_B(q)^\dagger \begin{pmatrix} E_d(q) & iE_{dh}(q) \\ -iE_{dh}(q) & -E_h(q) \end{pmatrix} U_B(q)U_B(q)^\dagger \begin{pmatrix} d_q \\ h_{-q}^\dagger \end{pmatrix} \tag{16}$$

$$= \sum_q \begin{pmatrix} d_q^\dagger & h_{-q} \end{pmatrix} U_B(q) \begin{pmatrix} \epsilon_d(q) & 0 \\ 0 & -\epsilon_h(q) \end{pmatrix} U_B(q)^\dagger \begin{pmatrix} d_q \\ h_{-q}^\dagger \end{pmatrix} \tag{17}$$

$$\equiv \sum_q \begin{pmatrix} c_{d,q}^\dagger & c_{h,-q} \end{pmatrix} \begin{pmatrix} \epsilon_d(q) & 0 \\ 0 & -\epsilon_h(q) \end{pmatrix} \begin{pmatrix} c_{d,q} \\ c_{h,-q}^\dagger \end{pmatrix} \tag{18}$$

$$= \sum_q \epsilon_d(q) c_{d,q}^\dagger c_{d,q} + \epsilon_h(q) c_{h,-q}^\dagger c_{h,-q}, \tag{19}$$

where the new fermionic quasi-particles $c_{d/h}$ are linear combinations of the d_q and h_q, i.e. are correlated pairs of doublons and holons. The dispersion relations of the particles are given by (assuming $U > 6J$)

$$\epsilon_{d/h}(q) = \pm J\cos(q) + \frac{1}{2}\sqrt{32J^2\sin^2(q) + \left[U - 6J\cos(q)\right]^2}. \tag{20}$$

The ground state in this approximation is the vacuum $|\text{vac}\rangle$, which is defined by $c_{d/q}|\text{vac}\rangle = 0$ for all q.

4.3 Spontaneous emission in the doublon/holon picture

Using the inverse transformation, $(c^\dagger_{d,q}, c_{h,-q})U^\dagger_B(q) = (d^\dagger_q, h_{-q})$, we can express the local particle number operator for site m, n_m in the Bogoliubov frame. Applying this operator to the ground-state, we find

$$n_m|\mathrm{GS}\rangle = |\mathrm{vac}\rangle + \frac{1}{M}\sum_{q,q'}(u_{q'}v_q - u_q v_{q'})c^\dagger_{d,q'}c^\dagger_{h,-q}e^{ima(q-q')}|\mathrm{vac}\rangle, \tag{21}$$

where we introduced the matrix elements of the transformation as $u_q = U_B(q)[1,1]$, and $v_q = U_B(q)[2,1]$. The time-evolved state after the jump is therefore

$$\sqrt{\mathcal{N}}|\psi(t)\rangle = |\mathrm{vac}\rangle + \frac{1}{M}\sum_{q,q'} e^{ima(q-q')}(u_{q'}v_q - u_q v_{q'})e^{-it[\epsilon_d(q')-\epsilon_h(q)]}c^\dagger_{d,q'}c^\dagger_{h,-q}|\mathrm{vac}\rangle, \tag{22}$$

where $\mathcal{N} \equiv \langle \mathrm{GS}|n_m^2|\mathrm{GS}\rangle$. The initial probablity-distribution of the wavepacket $w(q,q') \equiv u_{q'}v_q - u_q v_{q'}$ is proportional to

$$|w(q,q)'|^2 \propto \sin\left[\arctan\left(\frac{4\sqrt{2}J\sin(q')}{U-6J\cos(q')}\right) - \arctan\left(\frac{4\sqrt{2}J\sin(q)}{U-6J\cos(q)}\right)\right]^2 \tag{23}$$

and examples for $U = 7J$ and $U = 10J$ are depicted in Figure 3. The wavepackets are peaked at opposite momenta and therefore, the jump creates holes and double occupations which move in the opposite direction. For example, for $U/J = 10$, the wavepackets are peaked at $qa = -q'a \approx 0.30\pi$ and we can calculate the group-velocity from (20) as $\frac{d}{dqa}\epsilon_d(q)|_{0.3\pi} \approx 3.75a/J$.

We can test this analytical result with an exact t-DMRG time-evolution calculation after a single jump. We compare the dynamics after a jump in the center of a system with $N = M = 48$ site in Figure 4(a). There we plot the time-evolution of the number of doubly occupied sites as a function of time. The white dashed line in Figure 4(a) indicates the light-cone according to the analytically expected group-velocity and is in excellent agreement with the numerical calculations.

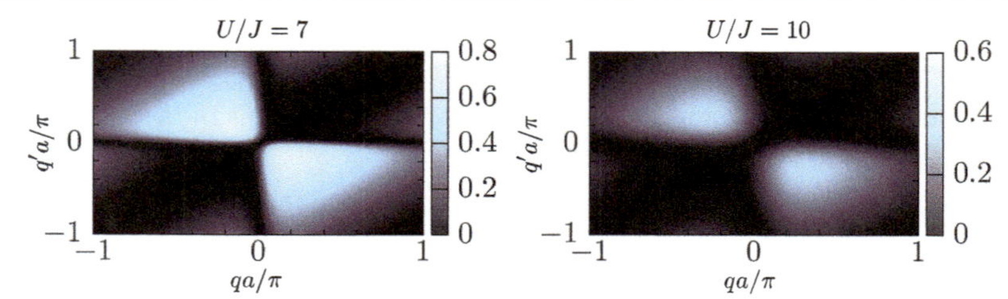

Figure 3 The quasimomentum distribution of the initial state that is created by a spontaneous emission in the doublon-holon model for $U = 7J$ (left panel) and $U = 10J$ (right panel).

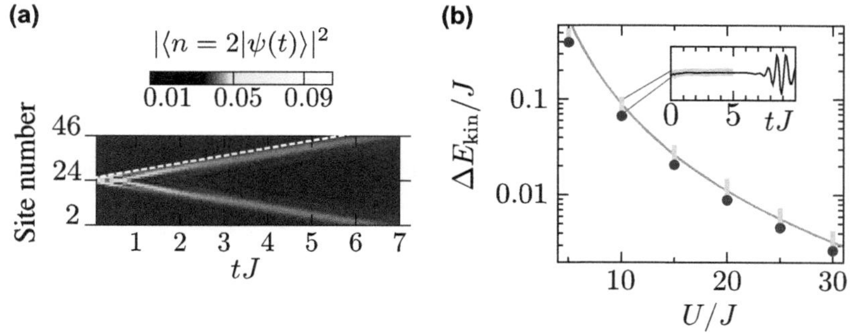

Figure 4 Dynamics after a single spontaneous emission. (a) Time-evolution of the probability of doubly occupied sites after a single spontaneous emission in the center of a 48 site system with $U = 10J$. This observable reveals doublon-holon pair wave-packets, which propagate freely through the system in opposite directions until they reach the boundary (dashed white line: analytical result for the light-cone in the thermodynamic limit). **(b)** Comparison of the corresponding time-evolution of the difference of the kinetic energy to the ground-state. The grey line shows the time-independent analytical result for the thermodynamic limit. The dots show the numerical result right after the jump and the blue line indicates the values that develop before boundary effects become important ($tJ < 5$, see inset). (t-DMRG calculation converged with $D = 256, 512$, $n_m = 6$.)

4.4 Behaviour of the kinetic energy and similar variables

We now consider a specific type of observable, such as the kinetic energy, which in quasi-momentum space and in the Bogoliubov frame can be written as $O = \sum_q O_q$ with

$$O_q \equiv O_{11}(q)c^\dagger_{d,q}c_{d,q} + O_{12}(q)c^\dagger_{d,q}c^\dagger_{h,-q} + O_{21}(q)c_{h,-q}c_{d,q} + O_{22}(q)c_{h,-q}c^\dagger_{h,-q}, \tag{24}$$

with matrix elements O_{ij} for each q. Calculating the expectation value of this observable with the time-evolved state gives

$$\langle\hat{O}\rangle(t) = \frac{1}{N}\sum_q O_{22} + \frac{1}{N}\frac{1}{M^2}\sum_{q,q'}\left(O_{11}(q') - O_{22}(q) + \sum_p O_{22}(p)\right)\left|w(q,q')\right|^2. \tag{25}$$

The important points to note are that all cross-terms corresponding to the O_{12} and O_{21} part of the operator disappear. This is because terms such as $\langle\text{vac}|c_{h,-p}c_{d,p}c^\dagger_{d,q'}c^\dagger_{h,-q}|\text{vac}\rangle = \delta_{p,q'}\delta_{p,q}$ become zero due to the symmetric form of the specific wave-packet that is created in the jump, $w(q,q') \equiv u_{q'}v_q - u_q v_{q'}$, $w(p,p) = 0$. Similarly, Kronecker deltas in the terms of the form $\langle\text{vac}|c_{h,-r}c_{d,r'}c^\dagger_{d,p}c_{d,p}c^\dagger_{d,q'}c^\dagger_{h,-q}|\text{vac}\rangle = \delta_{p,q'}\delta_{r',p}\delta_{r,q}$ lead to a cancellation of time-dependent terms. Thus, we find that any observable of this type is time-independent after the spontaneous emission. This is a remarkable result, given the fact that the state (22) is not an eigenstate of the Hamiltonian. The effect of a jump operator here is that it puts the system into a state whose density matrix is diagonal in the space of single doublon-holon pair excitations, for which case these specific observables don't evolve in time. In particular, the expectation value of observables of the form (24) can be effectively expressed by a density matrix – $\langle O\rangle = \text{tr}(\rho O)$ – of the form

$$\rho = \lambda_{\text{vac}}|\text{vac}\rangle\langle\text{vac}| + \sum_{q,q'}\lambda_{q,q'}\left|q,q'\right\rangle\!\left\langle q,q'\right|, \tag{26}$$

where $|q, q'\rangle \equiv c^{\dagger}_{d,q'} c^{\dagger}_{h,q} |\text{vac}\rangle$. Thus, we conclude that after a single spontaneous emission, observables of the form (24) can immediately be described by the diagonal ensemble, i.e. the effective density matrix for those observables takes a diagonal form in the basis low-lying energy states.

As an example, let us now consider the kinetic energy operator, which can be written as

$$O_q = T_q = \begin{pmatrix} -4J\cos[q] & -2i\sqrt{2}J\sin[q] \\ 2i\sqrt{2}J\sin[q] & 2J\cos[q] \end{pmatrix}. \tag{27}$$

In the thermodynamic limit, we can replace the sum over quasi-momenta with the integral, $(1/M)\sum_q \to \int d(qa)/(2\pi)$, which yields

$$\langle \hat{H}_{\text{kin}} \rangle(t) = -8M\frac{J}{U} - (48M + 88)\left(\frac{J}{U}\right)^3 + \mathcal{O}\left[\left(\frac{J}{U}\right)^5\right]. \tag{28}$$

We compare this analytical result to our time-dependent t-DMRG calculations This is in excellent agreement with our numerical results even for relatively small U/J as shown in Figure 4(b).

4.5 Eigenstate expectation values

We can check the eigenstate expectation values of the kinetic energy analytically. In Ref. [24], it was found that, once one enters the MI regime, these do not coincide with the expectation values for Boltzmann distributions (in the many-body eigenstates) with the corresponding value of total energy anymore (cf. Section 3). Now we can check whether these agree with our analytical calculation. Therefore, using exact diagonalization we compute the 500 lowest eigenstates in a system of $N = M = 10$ sites. These are shown as dots in Figure 5 for the case of $U = 10J$ and $U = 20J$. In addition we add the values we obtain analytically for the ground state (grey dot) and the branch of single doublon-holon pair excitations (grey lines). We find good agreement for large interactions. Since the first branch

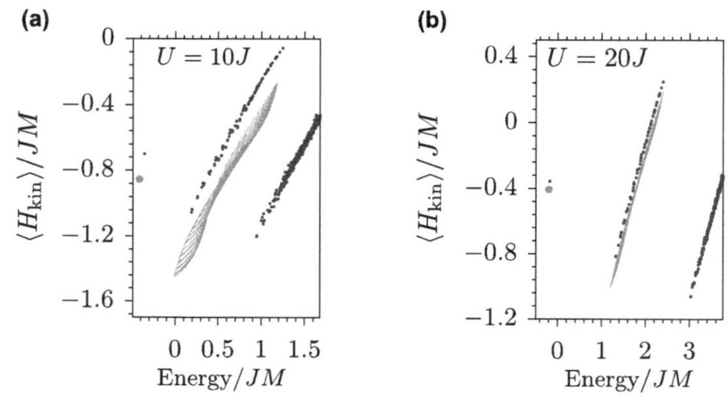

Figure 5 Eigenstate expectation values of the kinetic energy. Violet dots are results from an exact diagonalization calculation in a system with $N = M = 10$ particles/sites. Panel **(a)** is for $U = 10J$, panel **(b)** for $U = 20J$. The grey dot is the result for the ground-state kinetic energy from the analytical doublon-holon calculation in the thermodynamic limit. Grey lines are expectation values for the first branch of single doublon-holon pair excitations. For increasing U the numerical results approach the analytical result as expected (differences arise from contributions with more than two-particles per site and finite size effects).

converges towards the analytical calculation, we conclude that this branch indeed due to the single doublon-holon pair excitations, which is further verified by the fact that it contains $M^2 - M$ points. The second branch must thus correspond to two pair excitations. Already soon after entering the MI $U \sim 5$ the gap and branches start to appear [24]. This indicates that the doublon-holon calculations are indeed a good approximation even for regimes that are not particularly deep in the MI phase.

5 Summary and outlook

In summary, we have investigated the non-equilibrium dynamics of bosons in an optical lattice in the presence of spontaneous emission events. We show clearly parameter regimes in which the system relaxes to thermal distributions for simple quanitites (especially the quasimomentum distribution), and others where it relaxes on short timescales to non-thermal values. We can understand this behavior by applying eigenstate thermalization considerations to small systems for which we can perform exact diagonalization calculations. We also show that we can understand the dynamics in the strongly interacting limit well in terms of propagating doublon-hole pairs, which are analytically tractable.

All of these results are expected to be directly observable in ongoing experiments. In addition to the possibility to measure quasimomentum distributions, the propagation of doublon-holon pairs as investigated here is a key signature for the effects of spontaneous emissions that could be measured in quantum gas microscope experiments with single-site resolution in ways that were previously implemented for quenches within the MI regime [32]. Measurements of this type could be used as a diagnostic tool for heating of many-body states in optical lattices, which could in turn be used to improve the robustness of quantum simulators. In the future, these studies can be continued towards thermalization for fermions in optical lattices undergoing spontaneous emissions [29–31], and the incorporation of partial thermalization from excitations to higher Bloch bands.

Competing interests

The authors declare that they have no competing interests.

Authors' contributions

All authors contributed equally to the writing of this paper. All authors read and approved the final manuscript.

Author details

[1]Department of Physics, JILA, NIST, University of Colorado, 440 UCB, Boulder, CO 80309, USA. [2]Department of Physics, Arnold Sommerfeld Center for Theoretical Physics and Center for NanoScience, University of Munich, Munich, 80333, Germany. [3]Theoretische Physik, ETH Zurich, Zurich, 8093, Switzerland. [4]Department of Physics and SUPA, University of Strathclyde, Glasgow, G4 0NG, UK. [5]Department of Physics and Astronomy, University of Pittsburgh, Pittsburgh, PA 15260, USA.

Acknowledgements

We would like to thank I. Bloch, W. Ketterle, S. Langer, H. Pichler, U. Schneider, and P. Zoller for discussions. This work was supported in part by AFOSR grant FA9550-13-1-0093, by a grant from the US Army Research Office with funding from the DARPA OLE program, and by the Aspen Center for Physics support under NSF grant 1066293. JS was financially supported by NSF grants JILA-NSF-PFC-1125844 and NSF-PIF- 1211914. MT was supported by ERC Advanced Grant SIMCOFE and by the Swiss National Science Foundation through the National Competence Center in Research QSIT. Computational resources were provided by the Center for Simulation and Modeling at the University of Pittsburgh.

References

1. Deutsch JM: *Phys. Rev. A* 1991, **43**(4):2046.
2. Srednicki M: *Phys. Rev. E* 1994, **50**:888.
3. Rigol M, Dunjko V, Olshanii M: *Nature* 2008, **452**(7189):854.
4. Cazalilla MA, Citro R, Giamarchi T, Orignac E, Rigol M: *Rev. Mod. Phys.* 2011, **83**:1405.
5. Rigol M, Srednicki M: *Phys. Rev. Lett.* 2012, **108**:110601.

6. Kinoshita T, Wenger T, Weiss DS: *Nature* 2006, **440**(7086):900.
7. Rigol M: *Phys. Rev. Lett.* 2009, **103**:100403.
8. Cassidy AC, Clark CW, Rigol M: *Phys. Rev. Lett.* 2011, **106**:140405.
9. Rigol M, Fitzpatrick M: *Phys. Rev. A* 2011, **84**:033640.
10. Bloch I, Dalibard J, Nascimbene S: *Nat. Phys.* 2012, **8**(4):267.
11. Cirac JI, Zoller P: *Nat. Phys.* 2012, **8**(4):264.
12. McKay DC, DeMarco B: *Rep. Prog. Phys.* 2011, **74**(5):054401.
13. Jördens R, Tarruell L, Greif D, Uehlinger T, Strohmaier N, Moritz H, Esslinger T, De Leo L, Kollath C, Georges A, Scarola V, Pollet L, Burovski E, Kozik E, Troyer M: *Phys. Rev. Lett.* 2010, **104**(18):180401.
14. Hofstetter W, Cirac JI, Zoller P, Demler E, Lukin MD: *Phys. Rev. Lett.* 2002, **89**(22):220407.
15. Mathy CJM, Huse DA, Hulet RG: *Phys. Rev. A* 2012, **86**:023606.
16. Jördens R, Strohmaier N, Günter K, Moritz H, Esslinger T: *Nature (London)* 2008, **455**(7210):204.
17. Paiva T, Scalettar R, Randeria M, Trivedi N: *Phys. Rev. Lett.* 2010, **104**:066406.
18. De Leo L, Bernier JS, Kollath C, Georges A, Scarola VW: *Phys. Rev. A* 2011, **83**:023606.
19. Fuchs S, Gull E, Pollet L, Burovski E, Kozik E, Pruschke T, Troyer M: *Phys. Rev. Lett.* 2011, **106**:030401.
20. Gerbier F, Castin Y: *Phys. Rev. A* 2010, **82**:013615.
21. Pichler H, Daley AJ, Zoller P: *Phys. Rev. A* 2010, **82**(6):063605.
22. Gordon JP, Ashkin A: *Phys. Rev. A* 1980, **21**:1606.
23. Dalibard J, Cohen-Tannoudji C: *J. Phys. B, At. Mol. Opt. Phys.* 1985, **18**(8):1661.
24. Schachenmayer J, Pollet L, Troyer M, Daley AJ: *Phys. Rev. A* 2014, **89**(1):011601.
25. Poletti D, Bernier JS, Georges A, Kollath C: *Phys. Rev. Lett.* 2012, **109**:045302.
26. Barmettler P, Poletti D, Cheneau M, Kollath C: *Phys. Rev. A* 2012, **85**(5):053625.
27. Poletti D, Barmettler P, Georges A, Kollath C: *Phys. Rev. Lett.* 2013, **111**:195301.
28. Sorg S, Vidmar L, Pollet L, Heidrich-Meisner F: arXiv:1405.5404; 2014.
29. Bernier JS, Barmettler P, Poletti D, Kollath C: *Phys. Rev. A* 2013, **87**:063608.
30. Sarkar S, Langer S, Schachenmayer J, Daley AJ: *Phys. Rev. A* 2014, **90**(2):023618.
31. Bernier JS, Poletti D, Kollath C: arXiv:1407.1098; 2014.
32. Cheneau M, Barmettler P, Poletti D, Endres M, Schauß P, Fukuhara T, Gross C, Bloch I, Kollath C, Kuhr S: *Nature* 2012, **481**(7382):484.
33. Vidal G: *Phys. Rev. Lett.* 2004, **93**(4):040502.
34. Daley AJ, Kollath C, Schollwöck U, Vidal G: *J. Stat. Mech. Theory Exp.* 2004, **2004**:P04005.
35. White SR, Feiguin AE: *Phys. Rev. Lett.* 2004, **93**(7):076401.
36. Verstraete F, Murg V, Cirac JI: *Adv. Phys.* 2008, **57**:143.
37. Schönmeier-Kromer J, Pollet L: *Phys. Rev. A* 2014, **89**:023605.
38. Kühner TD, White SR, Monien H: *Phys. Rev. B* 2000, **61**:12474.
39. Daley AJ: *Adv. Phys.* 2014, **63**(2):77.
40. Mølmer K, Castin Y, Dalibard J: *J. Opt. Soc. Am. B* 1993, **10**(3):524.
41. Gardiner CW, Zoller P: *Quantum Noise*. Berlin: Springer; 2005.
42. Carmichael HJ: *An Open Systems Approach to Quantum Optics*. Berlin: Springer; 1993.
43. Linden N, Popescu S, Short AJ, Winter A: *Phys. Rev. E* 2009, **79**(6):061103.
44. Rigol M, Muramatsu A: *Phys. Rev. A* 2004, **70**(3):031603.
45. Jordan P, Wigner E: *Z. Phys.* 1928, **47**:631.

9

Qubit decoherence due to detector switching

Ioana Serban[1,2] and Frank K Wilhelm[2,3*]

*Correspondence: fwm@lusi.uni-sb.de
[2]IQC and Department of Physics and Astronomy, University of Waterloo, Waterloo, Ontario, Canada
[3]Theoretical Physics, Saarland University, Saarbrücken, Germany
Full list of author information is available at the end of the article

Abstract

We discuss the noise occurring during a classical dissipative switching process as it would be detected by a quantum bit used as a detector for noise. In particular we study the switching-induced decoherence *during* escape events. We present a simple method to obtain analytical results for the qubit dephasing and bit-flip, which goes beyond the simple Born-Markov notion of the qubit as a spectrometer for noise but still allows us to correlate its behavior with the noise source. These results also provide insight into the qubit measurement process involving a switching type of detector, showing under which conditions switching detectors can be operated fast and with low error. In particular, the state in the end recovers from temporary bit flip errors due to an intrinsic approximate time reversal symmetry, whereas the switching process mostly produces low frequency pure dephasing noise.

Keywords: switching process; noise; qubit

1 Introduction

Noise-activated switching out of a metastable state is a common phenomenon in a wide range of physical systems, including Josephson junctions, nanomechanical devices, and chemical reactions [1, 2]. Starting with Kramers seminal work [3], such processes have been studied close to equilibrium [4], as well as in driven systems [5]. The activated escape paths have been studied theoretically and observed experimentally [6, 7]. Due to the multiple time scales involved, some of them very fast, observing these fluctuations remains challenging.

Recently, noise-activated switching has gained attention due to its role in qubit detection. Examples of switching detectors include the superconducting quantum interference device (SQUID) [8–10], where switching occurs between the superconducting and dissipative state and is driven by quantum fluctuations, and the Josephson bifurcation amplifier (JBA) [11–14], which has been employed in the delicate task of detecting a qubit state in a minimally invasive fashion [15]. In this case, the detector can switch between different, weakly dissipative, dynamical states based on quantum activation [16] or dynamical tunneling [17]. Using an appropriate choice of a reference frame, switching between such dynamical states can also be described as escape from a *static* metastable well [16, 17]. Switching detectors are currently receiving renewed attention for their potential as microwave photon counters [18–20].

Switching is a highly nonlinear phenomenon, driven by large, non-equilibrium environmental fluctuations. Consequently, the noise produced during a switching process has long correlation times and can be very large. Its impact on a qubit is thus far outside the

Markov paradigm and detection is far from the weak measurement scenario. Some understanding for switching-type detectors has been provided by numerical studies [21, 22] and in a simplified two-state detector version in Refs. [23, 24]. General properties of binary outcome detectors have been discussed in [25]. Our approach is complementary in taking a rather concrete model of the detector.

In this paper we propose a simple analytical method to investigate qubit decoherence due to a switching process that can be part of a detector. Qubits are excellent spectrometers for quantum noise, most simply based on studying their relaxation rate [26], with a bandwidth that can be extended when dephasing and echo data are also taken into account [27]. Here, we model the noise source as a classical, overdamped particle trapped in a metastable well. This model is a strong simplification of the various detectors described above, yet capturing essential elements of the underlying switching process. These elements include that switching is a rare event and is driven by fluctuations, either thermal or quantum. Here, the escape of the particle is driven by large, rare fluctuations in the environment. We investigate the qubit dephasing and bit flips induced by the switching during the escape event. We adopt the simple model of quantum measurement that dephasing is a necessary ingredient as the real-time representation of wavefunction collapse, whereas bit-flips are unwanted falsifications of the measurement outcome. Note that these bit-flips occur if the qubit Hamiltonian and its coupling to the detector do not commute. They can be seen as an interplay of the coherent evolution of the qubit during the measurement process and the noise that is inevitable during detection. Note that an actual detector does not have to switch - in the 'down' state, the detector potential is deformed to make switching highly unlikely, leading to another class of trajectories that stay confined during the time of the experiment. Note that the trajectories studies here are entirely classical, quantum physics only comes in by the qubit exposed to them, making our approach different from the *quantum* trajectories method [28].

2 Method

The overdamped classical particle performs Brownian motion according to

$$\dot{x} = K(x) + f(t), \tag{1}$$

where K is the deterministic force experienced by the particle resulting from the metastable potential and f is white Gaussian noise with a probability density functional given by [29]

$$P[f(t)] = \exp\left(-\int_0^{t_f} \frac{f^2(t)}{2D}\,dt\right), \tag{2}$$

where we assume [30] the noise intensity D to be small compared to the barrier height ΔU, see Figure 1. The probability density functional is not normalized *a priori*, subsequent expressions have to be normalized by hand. For a noise driven trajectory it can be obtained by substituting Eq. (1)

$$P[x(t)] = \exp\left(-\frac{S[x(t)]}{D}\right), \qquad S[x(t)] = \frac{1}{2}\int_0^{t_f} dt(\dot{x} - K(x))^2. \tag{3}$$

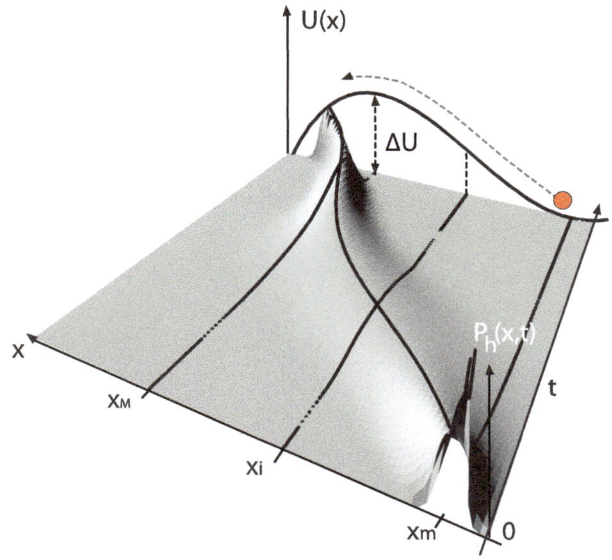

Figure 1 Prehistory probability density *P*(*x*, *t*) for the potential *U*(*x*), where *U*′(*x*) = −*K*(*x*), and optimal trajectory $x_{opt}(t)$. Here $x_{m,M,i}$ are the positions of the minimum, maximum and inflection points of the potential.

For the study of qubit decoherence during a switching event one will need expectation values for a generic observable O of the type

$$O(t_0) = \langle \exp(\lambda\phi[x(t), s(t, t_0)]) \rangle_{sw}, \tag{4}$$

$$\phi[x(t), s(t, t_0)] = \int_0^{t_f} x(t) s(t, t_0)\, dt, \tag{5}$$

were $s(t, t_0)$ is a time dependent generating field for computing dephasing and bit-flip as will be detailed later. The coupling constant λ has been singled out to parameterize the coupling to the qubit and the time t_0 indicates when the observable is measured. Effectively, it constitutes a modulation of $x(t)$ and λ. This approach is fairly standard in stochastic path integrals and the choice of $s(t, t_0)$ suitable for our problem will be clarified later. This formulation allows for the present formalism to include detectors such as the JBA, where, in a rotating frame, the coupling between the qubit and detector becomes time dependent and periodic. We are interested in the qubit decoherence *during* switching. Thus, the averaging $\langle\rangle_{sw}$ is performed only over switching trajectories of the detector, which satisfy the boundary conditions $x(0) = x_m$ and $x(t_f) = x_f$, with x_m inside and x_f outside the metastable well. By choosing $s(t, t_0) = 0$ at $t_f > t > t_0$, the average becomes post-conditioned by a switching event having taken place at the final time t_f, i.e., experimental verification would need to include recording the time of the switch and post-selecting for different times. This condition makes sure that we only study realizations of the measurement when switching occurs. In order to not over-constrain the trajectories we have verified the dependence of our result on the precise choice of x_f and found minuscule changes only.

Since the exact switching trajectory between the initial and final point remain random, we average over all possible paths using the weight $e^{-S[x]/D}$

$$O(t_0) = \int_{(x_m,0)}^{(x_f,t_f)} \mathcal{D}x(t) \exp\left(\lambda\phi\left[x(t), s(t,t_0)\right] - \frac{S[x(t)]}{D} \right) \times P(x_m, 0|x_f, t_f)^{-1}, \tag{6}$$

where the total switching probability

$$P(x_m, 0|x_f, t_f) = \int_{(x_m,0)}^{(x_f,t_f)} \mathcal{D}x(t) \exp\left(-\frac{S[x(t)]}{D} \right) \tag{7}$$

serves as a normalization.

Switching over a high barrier is a rare event and thus requires very specific noise realizations. Thus, the switching trajectories form a narrow tube in phase space centered around an optimal trajectory [31, 32] which minimizes S, and for the present case satisfies

$$\ddot{x}_{\text{opt}} = K(x_{\text{opt}})K'(x_{\text{opt}}), \qquad x_{\text{opt}}(0) = x_m, \qquad x_{\text{opt}}(t_f) = x_f. \tag{8}$$

This optimal trajectory is driven by an optimal realization of the environmental noise. Thus $S[x(t)] = S[x_{\text{opt}}(t)] + S_2[x(t) - x_{\text{opt}}(t)]$ and we perform a saddlepoint approximation around this optimal solution

$$S_2\left[x(t)\right] \approx \frac{1}{2}\int_0^{t_f} dt\left(\dot{x}(t)^2 - \Lambda(t)^2 x(t)^2\right), \tag{9}$$

where $\Lambda(t)^2 = -(K'(x)^2 + K(x)K''(x))|_{x=x_{\text{opt}}(t)}$. Divergences due to the emergence of a slow mode on the barrier top [30, 33] are avoided by the appropriate choice of the initial (kinetic) energy, $0 < \dot{x}(0)^2/2 \ll \Delta U$ in order to satisfy the boundary conditions (8). Thus, the switching event takes place, with non-vanishing probability, within a *finite* time t_f. Within the saddle point approximation (9), the path integral (6) becomes Gaussian and can be evaluated analytically [34]

$$O(t_0) = \exp\left(\lambda\phi\left[x_{\text{opt}}(t) + \frac{x_0(t)}{2}, s(t,t_0) \right] \right), \tag{10}$$

where x_0 is the solution of

$$\ddot{x}_0 + \Lambda^2 x_0 + D\lambda s(t,t_0) = 0, \qquad x_0(0) = x_0(t_f) = 0. \tag{11}$$

The two linearly independent solutions of the homogeneous part of Eq. (11) are

$$x_1(t) = \dot{x}_{\text{opt}}(t), \qquad x_2(t) = \dot{x}_{\text{opt}}(t)\int_0^t \frac{dt'}{x_{\text{opt}}^2(t')}, \tag{12}$$

and the full $x_0(t) = x_1(t)c_1(t) + x_2(t)c_2(t)$ can be determined by variation of parameters.

We consider the case of a potential given by $U(x)/\Omega = x/2 - x^3/6$, which can, for example, approximate a Josephson junction biased at half the critical current. Note that this approximation does not account for retrapping of the Josephson junction in the supercurrent state after n full rotations of the phase, hence it applies when the bias is large enough to precvent such retracing or phase diffusion. $\Omega = K'(x_{\mathrm{m}})$ plays the role of a characteristic frequency in our effective system but physically corresponds to the damping coefficient.

3 Results

Pure dephasing: We are now going to discuss the impact of these distributions on dephasing and bit-flip. The stochastic process and distribution of trajectories is the same in both cases. Dephasing typically probes the low-frequency part of the noise spectrum. We assume a qubit coupling Hamiltonian of the form

$$\hat{H} = \hbar\omega\hat{\sigma}_z + \eta_d\hat{\sigma}_z x(t), \tag{13}$$

where $x(t)$ is the coordinate of the classical particle. The only effect of the environment on the qubit in this case is the irreversible decay of the phase coherence $C(t_0) = O(t_0)$, which can be obtained by replacing in Eq. (4) $\lambda = \lambda_d = \eta_d/(i\hbar)$ and the generating function

$$s(t,t_0) = \begin{cases} 1, & t < t_0, \\ 0, & t > t_0, \end{cases} \tag{14}$$

which has been chosen to fix the initial condition of the particle. This choice can be physically understood to collect all low-frequency fluctuations in the past of the evaluation of the observable at time t.

Figure 2 shows that the escape process is driven by strong noise (c), with a maximum intensity at the time where the most probable trajectory (d) reaches the inflection point of the barrier. In the vicinity of the same point, a sharp drop in qubit coherence (a) is observed. After the inflection point the motion is slowed down, and becomes diffusive close to the barrier top. During this stage, the qubit coherence remains at an almost constant value. We observe that the optimal noise becomes stronger for shorter values of t_f. However, the strongest drop in qubit coherence was observed for the longest t_f. In this case the

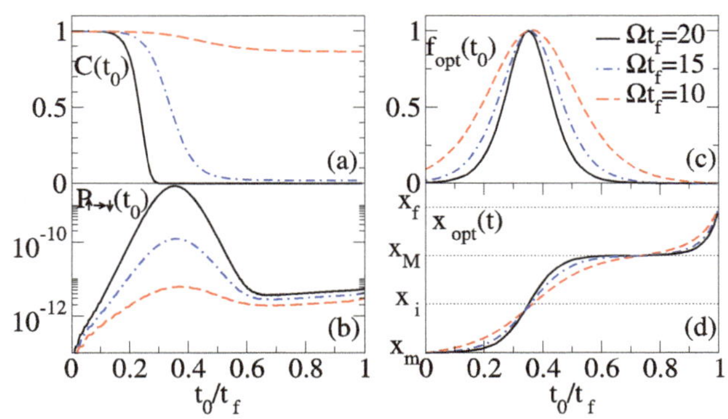

Figure 2 Qubit coherence during a switching event ($t_0 < t_f$) (a), probability for an induced bit flip error as a function of t_0 (b), the corresponding most probable noise trajectory $f_{\mathrm{opt}}(t_0)$ and (c) most probable switching trajectory $x_{\mathrm{opt}}(t_0)$ (d) for various values of t_f. The particle starts from the minimum of the well with a speed adjusted to the switching condition and a diffusion coefficient $D = 0.01$.

optimal trajectory spends more time close to the barrier top, where the motion is diffusive and broadly distributed, driven by low amplitude noise.

Bit flips: We consider a qubit-environment coupling which allows for bit flips

$$\hat{H} = \hbar\omega\hat{\sigma}_z + \eta_f x(t)\hat{\sigma}_x. \tag{15}$$

Bit-flip typically probes the noise spectrum around the qubit frequency. Under this generic scenario, the interplay of unitary qubit evolution and the measurement not being instantaneous leads to bit-flips compromising the post-measurement state. We concentrate on the case of that unwanted evolution being slow, i.e., weak qubit-detector coupling. The probability of noise induced transitions during a switching event at $t_0 < t_f$ is given by the propagator in the interaction picture (henceforth labelled by a subscript I)

$$P_{\uparrow\to\downarrow}(t_0) = \left|\left\langle\downarrow\left|\hat{U}_I(t_0)\right|\uparrow\right\rangle\right|^2 P(x_m, 0|x_f, t_f)^{-1}, \tag{16}$$

where in the limit of short time and weak coupling

$$\begin{aligned}\hat{U}_I(t_0) &= \mathcal{T}\exp\left(\int_0^{t_0} dt \frac{H_I(t)}{i\hbar}\right) \approx 1 + \int_0^{t_0} dt \frac{H_I(t)}{i\hbar},\\ \hat{H}_I(t) &= \eta_f \hat{U}_0^\dagger(t)\hat{\sigma}_x\hat{U}_0(t)x(t),\end{aligned} \tag{17}$$

and $\hat{U}_0$ describes the free qubit evolution. We follow the standard steps of relaxation calculations by time-dependent perturbation theory as outlines in [26, 35] but do *not* perform the Markov approximation which would be inappropriate given the strong temporal correlations of the noise. Still, the transition probability can be expressed through the autocorrelation function of ϕ which in turn can be generated out of O by formal differentiation of Eq. (4). We obtain

$$\begin{aligned}P_{\uparrow\to\downarrow}(t_0) = &\lim_{\lambda_f\to 0}\frac{\partial^2}{\partial\lambda_f^2}\frac{\eta_f^2}{\hbar^2}\big\langle\exp\left(\lambda_f\phi\left[x(t), s(t,t_0)\mathcal{R}(t)\right]\right)\\ &+ \exp\left(\lambda_f\phi\left[x(t), s(t,t_0)\mathcal{I}(t)\right]\right)\big\rangle_{sw},\\ \mathcal{R}(t) + i\mathcal{I}(t) = &\left\langle\downarrow\left|\hat{U}_0^\dagger(t)\hat{\sigma}_x\hat{U}_0(t)\right|\uparrow\right\rangle.\end{aligned} \tag{18}$$

In Figure 2(b) we observe, again, a sharp feature in $P_{\uparrow\to\downarrow}(t_0)$ at the point in time where the most probable trajectory (d) reaches the steepest point on the potential barrier, and the driving noise (c) reaches it maximum strength. Despite the optimal noise being strongest for short switching time t_f, the peak in $P_{\uparrow\to\downarrow}(t_0)$ is stronger for longer t_f. Another notable feature is the quasi-reversion of the bit flip error which occurs at this point. This feature cannot be explained by the single, deterministic trajectory x_{opt} alone, which causes only the steady increase of $P_{\uparrow\to\downarrow}(t_0)$.

4 Discussion

The results presented above can be understood from the distribution of switching trajectories. We calculate the probability $P_h(x, t)$ for the classical particle to occupy the position

x at time t during a switching event, i.e., a prehistory density distribution [30]

$$P_h(x,t) = \frac{P(x_\mathrm{m},0|x,t)P(x,t|x_\mathrm{f},t_\mathrm{f})}{P(x_\mathrm{m},0|x_\mathrm{f},t_\mathrm{f})}. \tag{19}$$

Within the approximation (9), the probability for a transition between any pair of points (x_1,t_1) and (x_2,t_2), with $t_{1,2} < t_\mathrm{f}$ reads

$$P(x_1,t_1|x_2,t_2) = \int_{(\delta x_1,t_1)}^{(\delta x_2,t_2)} \mathcal{D}x(t) \exp\left(-\frac{S[x_\mathrm{opt}(t)]_1^2 + S_2[x(t)]_1^2}{D}\right), \tag{20}$$

where $S[x(t)]_1^2$ implies that the time integral is taken between t_1 and t_2 and $\delta x_{1,2} = x_{1,2} - x_\mathrm{opt}(t_{1,2})$. One can show that

$$P(x_1,t_1|x_2,t_2) = \exp\left(-\frac{S[x_\mathrm{opt}(t)]_1^2 + S_2[x_b(t)]_1^2}{D}\right)F(t_1|t_2), \tag{21}$$

where $\ddot{x}_b + \Lambda^2(t)x_b(t) = 0$ and $x_b(t_{1,2}) = \delta x_{1,2}$ and

$$F(t_1|t_2) = \int_{(0,t_1)}^{(0,t_2)} \mathcal{D}x(t) \exp\left(-\frac{S_2[x(t)]}{D}\right)$$
$$= \left(2\pi D\dot{x}_\mathrm{opt}(t_1)\dot{x}_\mathrm{opt}(t_2)\int_{t_1}^{t_2}\frac{dt}{\dot{x}_\mathrm{opt}^2(t)}\right)^{-1/2}. \tag{22}$$

We obtain a Gaussian distribution, centered around $x_\mathrm{opt}(t)$

$$P_h(x,t) = \frac{1}{\sqrt{\pi w(t)}} \exp\left(-\frac{(x - x_\mathrm{opt}(t))^2}{w(t)}\right), \tag{23}$$

where $w(t) = F(0|t_\mathrm{f})^2/(F(0|t)^2F(t|t_\mathrm{f})^2)$.

Figures 1 and 3 reveal a narrow tube of trajectories close to the bottom of the well, followed by a strong widening of the distribution in the process of climbing up the potential

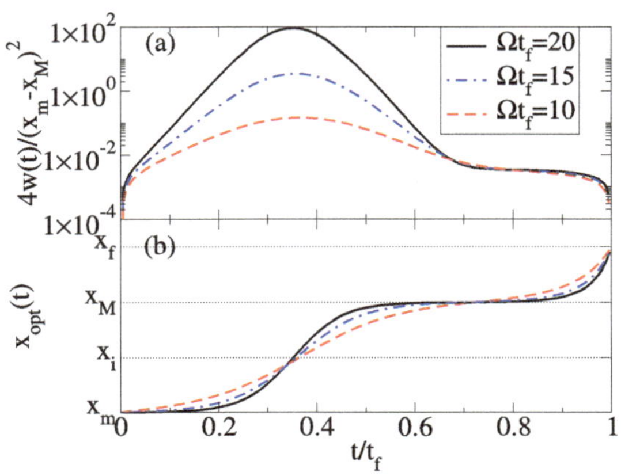

Figure 3 Width of the prehistory probability distribution $P_h(x,t)$ (a) and optimal trajectory (b) for a particle starting from the minimum at rest, parameters as in previous figure.

barrier. This event is driven by a sharp noise pulse. On the barrier top we see again a fairly localized density distribution, again driven by low-amplitude noise. The tube narrows even more on the outer side of the barrier. These results are in agreement with the findings of Ref. [6], for a different system. It should be emphasized that this narrow tube is due to the fact that the necessary realization of environmental noise is a rare event in sync with the connection to a fixed final time - if the latter were not imposed, we would find a mixture of those tubes.

We found, see Figure 2 that the qubit suffers the strongest decoherence at the point in time when the optimal trajectory reaches the steepest point on the barrier wall. This is true for both bit flips and dephasing. The magnitude of both effects depends strongly on the total time necessary for the switching event, such that longer t_f leads to enhanced coherence loss, and higher bit flip rate. The observed effect can be explained by the strong widening of the prehistory distribution $P_h(x,t)$ at the same point in time, see Figure 3.

The peak in $P_{\uparrow\rightarrow\downarrow}$ originates in the strong widening of the trajectory tube. Thus, large excursions around x_{opt}, see Figure 1, are very probable during this time. However, since each of these switching trajectories must return to the narrow tube of trajectories on the other side of the widening, i.e. in the region close to the barrier top, they all show an approximate time reversal symmetry. In other words, part of the change to the qubit state - the one originating from the average of the trajectory, is deterministic and is in parts undoing itself. Thus any induced bit flip errors will be partially reversed when the particle reaches the barrier top.

In conclusion, we have studied the interaction between a switching particle that could be part of a detector with a qubit. We have shown how the stochastic dynamics of the detector are correlated with the qubit. A switching detector, such as the one modeled here, presents several qualities which makes it desirable as a nonlinear qubit detector. The strong decoherence suffered by the qubit as the classical particle climbs up the potential barrier affects strongly the coherence, leading to a fast measurement. The equally strong bit flip errors acquired during the process are reversed by the quasi time-reversal symmetry of most trajectories.

We observe that a fast switching event causes less decoherence, despite stronger noise being required for surmounting the barrier. Figure 3 reveals that longer switching time t_f allows more freedom in the choice of the particular time the particle climbs up the potential barrier and such incoherent behavior causes decoherence between different realizations of the experiment.

We note that the widening of the trajectory tube at the inflection point of the potential appears throughout the literature on Brownian motion as a general feature of the tube of escape trajectories out of metastable states, but the connection to quantum measurement has not been made yet. Having identified it as the major cause for the features observed in qubit decoherence, we expect these features to be common to various potentials. Therefore we expect that our results have applicability to existing experimental setups, e.g. the JBA and the DC-SQUID.

Competing interests

We declare to have no competing interests, financial or otherwise.

Authors' contributions

IS performed the calculations, made the figures, and drafted paper and discussions. FKW conceived the project, guided the methodology and finalized paper and discussion. All authors read and approved the final manuscript.

Author details
[1]Instituut Lorentz, Leiden University, Leiden, The Netherlands. [2]IQC and Department of Physics and Astronomy, University of Waterloo, Waterloo, Ontario, Canada. [3]Theoretical Physics, Saarland University, Saarbrücken, Germany.

Acknowledgements
We are grateful to MI Dykman for pointing out the prehistory probability distribution approach and for many useful suggestions. We acknowledge useful discussions with J Gambetta, WA Coish and TC Wei as well as LCG Govia for his careful reading of the manuscript. This work has been supported by NSERC Discovery Grants and Quantum Works. Supported by the Army Research Office under contract W911NF-14-1-0080.

References

1. Aldridge JS, Cleland AN. Noise-enabled precision measurements of a Duffing nanomechanical resonator. Phys Rev Lett. 2005;94:156403.
2. Wales D. Energy landscapes: applications to clusters, biomolecules and glasses. Cambridge: Cambridge University Press; 2003.
3. Kramers H. Brownian motion in a field of force and the diffusion model of chemical reactions. Physica. 1940;7:284.
4. Hanggi P, Talkner P, Borkovec M. Reaction rate theory: fifty years after Kramers. Rev Mod Phys. 1990;62:252.
5. Lehmann J, Reimann P, Hänggi P. Surmounting oscillating barriers. Phys Rev Lett. 2000;84:1639.
6. Hales J, Zhukov A, Roy R, Dykman MI. Dynamics of activated escape, and its observation in a semiconductor laser. Phys Rev Lett. 2000;85:78.
7. Chan HB, Dykman MI, Stambaugh C. Paths of fluctuation induced switching. Phys Rev Lett. 2008;100:130602.
8. van der Wal C, ter Haar A, Wilhelm F, Schouten R, Harmans C, Orlando T, Lloyd S, Mooij J. Quantum superposition of macroscopic persistent-current states. Science. 2000;290:773.
9. Vion D, Aassime A, Cottet A, Joyez P, Pothier H, Urbina C, Esteve D, Devoret M. Manipulating the quantum state of an electrical circuit. Science. 2002;296:866.
10. Martinis JM, Nam S, Aumentado J, Urbina C. Rabi oscillations in a large Josephson-junction qubit. Phys Rev Lett. 2002;89:117901.
11. Siddiqi I, Vijay R, Pierre F, Wilson CM, Metcalfe M, Rigetti C, Frunzio L, Devoret MH. RF-driven Josephson bifurcation amplifier for quantum measurement. Phys Rev Lett. 2004;93:207002.
12. Siddiqi I, Vijay R, Pierre F, Wilson CM, Frunzio L, Metcalfe M, Rigetti C, Schoelkopf RJ, Devoret MH, Vion D, Esteve D. Direct observation of dynamical bifurcation between two driven oscillation states of a Josephson junction. Phys Rev Lett. 2005;94:027005.
13. Metcalfe M, Boaknin E, Manucharyan V, Vijay R, Siddiqi I, Rigetti C, Frunzio L, Schoelkopf RJ, Devoret MH. Measuring the decoherence of a quantronium qubit with the cavity bifurcation amplifier. Phys Rev B. 2007;76:174516.
14. Schmitt V, Zhou X, Juliusson K, Blais A, Bertet P, Vion D, Esteve D. Multiplexed Readout of transmon qubits with Josephson bifurcation amplifiers. Phys Rev A. 2008;90:062333.
15. Lupascu A, Saito S, Picot T, de Groot PC, Harmans CJPM, Mooij JE. Quantum non-demolition measurement of a superconducting two-level system. Nat Phys. 2007;3:119.
16. Marthaler M, Dykman M. Switching via quantum activation: a parametrically modulated oscillator. Phys Rev A. 2006;73:042108.
17. Serban I, Wilhelm FK. Dynamical tunneling in macroscopic systems. Phys Rev Lett. 2007;99:137001.
18. Chen Y-F, Hover D, Sendelbach S, Maurer L, Merkel ST, Pritchett EJ, Wilhelm FK, McDermott R. Microwave photon counter based on Josephson junctions. Phys Rev Lett. 2011;107:217401.
19. Poudel A, McDermott R, Vavilov MG. Quantum efficiency of a microwave photon detector based on a current-biased Josephson junction. Phys Rev B. 2012;86:174506.
20. Goiva LCG, Pritchett EJ, Xu C, Plourde BLT, Vavilov MG, Wilhelm FK, McDermott R. High-fidelity qubit measurement with a microwave-photon counter. Phys Rev A. 2014;90:062307.
21. Nakano H, Saito S, Semba K, Takayanagi H. Quantum time-evolution in qubit readout process with a Josephson bifurcation amplifier. Phys Rev Lett. 2009;102:257003.
22. Nakano H, Tanaka H, Saito S, Semba K, Takayanagi H, Ueda M. A theoretical analysis of flux-qubit measurements with a dc-SQUID (unpublished). arXiv:cond-mat/0406622.
23. Ashhab S, You JQ, Nori F. Weak and strong measurement of a qubit using a switching-based detector. Phys Rev A. 2009;79:032317.
24. Ashhab S, You JQ, Nori F. The information about the state of a qubit gained by a weakly coupled detector. New J Phys. 2009;11:083017.
25. Korotkov AN. Weak and strong measurement of a qubit using a switching-based detector. Phys Rev B. 2008;78:174512.
26. Schoelkopf RJ, Clerk AA, Girvin SM, Lehnert KW, Devoret MH. Qubits as spectrometers of quantum noise. In: Nazarov YV, Blanter YM, editors. Quantum noise. Dordrecht: Kluwer Academic; 2006.
27. Bylander J, Gustavsson S, Yan F, Yoshihara F, Harrabi K, Fitch G, Cory DG, Nakamura Y, Tsai J-S, Oliver WD. Noise spectroscopy through dynamical decoupling with a superconducting flux qubit. Nat Phys. 2011;7:565.
28. Wiseman HM, Milburn GJ. Quantum measurement and control. Cambridge: Cambridge University Press; 2014.
29. Feynman R, Hibbs A. Quantum mechanics and path integrals. New York: McGraw-Hill; 1965.
30. Dykman MI, McClintock PVE, Smelyanski VN, Stein ND, Stocks NG. Optimal paths and the pre-history problem for large fluctuations in noise-driven systems. Phys Rev Lett. 1992;68:2718.
31. Onsager L, Machlup S. Fluctuations and irreversible processes. Phys Rev. 1953;91:1505.
32. Dykman M, Krivoglaz M, Soskin S. Transition probabilities and spectral density of fluctuations of noise-driven bistable systems. In: Moss F, McClintock PVE, editors. Noise in nonlinear dynamical systems. Cambridge: Cambridge University Press; 1989. p. 347.
33. Luchinsky D, McClintock P, Dykman M. Analogue studies of nonlinear systems. Rep Prog Phys. 1998;61:889.

34. Kleinert H. Path integrals in quantum mechanics, statistics, polymer physics, and financial markets. Singapore: World Scientific; 2006.
35. Wilhelm FK, Hartmann U, Storcz MJ, Geller MR. Quantum computing with superconductors II: decoherence in manipulating quantum coherence. In: Flatte ME, Tifrea I, editors. Solid state systems. Dordrecht: Springer; 2006. p. 195.

10

Phase diagram of a QED-cavity array coupled via a N-type level scheme

Jiasen Jin[1*], Rosario Fazio[1,2] and Davide Rossini[1]

*Correspondence: jiasen.jin@sns.it
[1]NEST, Scuola Normale Superiore and Istituto di Nanoscienze, CNR, Piazza dei Cavalieri, 7, Pisa, 56126, Italy
Full list of author information is available at the end of the article

Abstract

We study the zero-temperature phase diagram of a one-dimensional array of QED cavities where, besides the single-photon hopping, an additional coupling between neighboring cavities is mediated by an N-type four-level system. By varying the relative strength of the various couplings, the array is shown to exhibit a variety of quantum phases including a polaritonic Mott insulator, a density-wave and a superfluid phase. Our results have been obtained by means of numerical density-matrix renormalization group calculations. The phase diagram was obtained by analyzing the energy gaps for the polaritons, as well as through a study of two-point correlation functions.

PACS Codes: 42.50.Pq; 85.25.Cp; 64.70.Tg

Keywords: cavity quantum electrodynamics; strongly correlated polariton systems; quantum phase transitions

1 Introduction

The recent impressive advances in the field of quantum simulators allowed to probe the many-body physics of strongly correlated systems at the level of the single quantum object. At present cold atoms trapped in optical lattices can be considered among the most promising examples of quantum simulators. By means of ultracold atomic and molecular gases, it is nowadays possible to reach a degree of control and accuracy in engineering the dynamics of many-body systems that were unimaginable in previously. As a consequence, the coherent quantum dynamics emerging from carefully tailored microscopic Hamiltonians can now be tested experimentally [1]. It has been possible, just to recall one example, to implement the Bose-Hubbard (BH) model [2, 3] and to detect its zero-temperature superfluid (SF) to Mott insulator (MI) quantum phase transition [4]. Other models involving spinor gases, Fermi systems, Bose-Fermi mixtures, or dipolar gases have been also devised and realized, providing an even richer phase diagram (see for example the review [5]). We mention the stabilization of density-wave (DW) phases for bosons, as well as more peculiar topological or supersolid orderings, which can arise in the presence of finite-range interactions [6].

More recently a novel kind of many-body quantum simulator has been introduced, based on the idea to use single photons as quantum objects. Since photons hardly interact in open space, the most natural way to significantly increase their interactions is to trap them into an optical QED cavity, and couple the field with atoms/molecules inside it in order to create an optical nonlinearity. If the nonlinearity is sufficiently large, the so called photon blockade sets in [7, 8], namely, the presence of a single photon inside a cavity prevents a second one to enter it. In the rotating-wave approximation, the simplest light-matter interaction scheme of this type can be accurately described by the Jaynes-Cummings (JC) model. By arranging an array of cavities coupled through the photon hopping, such to generate a competition between the hopping and the on-site nonlinearities, one can devise a setup that is well described by the so called Jaynes-Cummings-Hubbard (JCH) model [9–11].

In many respects, if one ignores dissipation, the physics emerging from the JCH Hamiltonian resembles, at low-energies, that of an effective BH model. Probably the main difference between the two systems is that, instead of having neutral bosons as building blocks of the model, in the JCH Hamiltonian one has to think in terms of polaritons, *i.e.*, combined photonic/atomic excitations. Many different works already addressed the JCH equilibrium phase diagram with analytical, as well as numerical methods, leading to a fairly complete theoretical understanding of the nature and the location of the emerging phases and phase transitions in terms of the parameters governing the system (the field has been recently reviewed in, *e.g.*, Refs. [12–15]).

Additional interest in cavity arrays comes from the fact that these systems can be naturally considered as open-system quantum simulators. Some related features have been recently explored [16–24]. In the following we will not touch on this and consider only the 'equilibrium' phase diagram.

This intense activity has been very recently boosted by the first experiments on QED cavity arrays [25–27]. As of today, the most concrete possibility to realize controllable and scalable quantum simulators with cavity arrays involves circuit-QED cavities [28–30].

So far the coupling between cavities has been mostly considered through photon hopping. Only few works started addressing more general schemes, where the cavity coupling can be induced also by means of non-linear elements [23, 24, 31, 32]. Such configurations include cross-Kerr interactions and/or correlated hopping terms, which lead to generalizations of the JCH model in a way similar to the extended BH (EBH) Hamiltonian for atoms with large dipole momentum loaded in optical lattices [33]. The underlying physical model is believed to possess a much richer structure, with the emergence of exotic phases of correlated polaritons. It is particularly interesting to address these schemes in one-dimensional (1D) systems, where interactions become crucial to stabilize exotic phases of matter [33–37]. These notably include a series of nontrivial density-wave (DW) states, which can arise in the strong coupling regime [38], as well as supersolidity and phase-separation effects [39, 40]. Extension to consider also counter-rotating terms in the ultra-strong coupling regime, thus leading to the so called Rabi-Hubbard model [41], have been investigated [42]. However we are not aware of numerical investigations of coupled cavity models beyond the JCH and Rabi-Hubbard model.

In all such situations, non-perturbative, either numerical or analytical calculations are necessary. Here the density-matrix renormalization group (DMRG) algorithm [43, 44] has been employed to work out the quantitative zero-temperature phase diagram of the JCH model [45–47]. This is a particularly efficient method for the statics of 1D many-body problems. Its key strategy consists in constructing a portion of the system (called block) and then recursively enlarge it. At each step, the basis of the corresponding Hamiltonian is truncated to a given value m, so that one can manage the Hamiltonian in an effective

Hilbert space of fixed dimensions, as the physical system grows. This truncation is performed by retaining the eigenstates corresponding to the m highest eigenvalues of the reduced density matrix of the block.

The aim of this paper is to quantitatively study a generalization of the JCH Hamiltonian, aimed at taking into account an effective nearest-neighbor nonlinearity between cavities mediated by an N-type four-level system as discussed for two cavities in Ref. [48]. The presence of this coupling leads to an effective cross-Kerr nonlinearity. An analysis at the mean-field level of a dissipative open EBH as an effective model for nonlinearly coupled cavities has been performed, unveiling the emergence of novel photon crystal and supersolid phases [23, 24]. Here we do not resort to the effective EBH model and analyze the full model as introduced in [48]. Using the DMRG algorithm, we work out the 1D ground-state phase diagram. We show that a physics similar to the EBH model appears, with a rich phase diagram including gapless SF, as well as MI and DW phases of polaritons. We postpone the analysis of the interplay of driving and dissipation to a future work.

The paper is organized as follows. In the next two sections we introduce the model of coupled cavities of our interest (Section 2) and the quantities we are going to address, namely the energy gaps, and the staggered number-number correlations (Section 3). In Section 4 we discuss the zero-temperature equilibrium phase diagram, focusing on the MI/SF boundary and on the boundary separating the DW from the other phases. Finally, in Section 5 we draw our conclusions.

2 The model

Let us consider a 1D array of QED cavities, where photons can hop between neighboring cavities. Moreover two adjacent resonators are also nonlinearly coupled to each other via a N-type four-level system, as shown in Figure 1(a). For the sake of clarity in our description, we shall divide the 1D array into coupled effective sites composed of a cavity and an atom. The four levels are denoted by $\{|i\rangle\}_{i=1,\ldots,4}$, and are depicted in Figure 1(b). An external laser with frequency Ω resonantly drives the transition $|3\rangle \leftrightarrow |2\rangle$. The transition $|1\rangle \leftrightarrow |3\rangle$ is resonantly coupled to the cavity mode of the same site with strength g_1, while the transition $|2\rangle \leftrightarrow |4\rangle$ couples to the cavity mode of its right nearest-neighbor site with strength g_2, and a detuning Δ.

The use of such N-type atom for generating large Kerr nonlinearity has been extensively studied in the literature [7, 8, 49], however the vast majority of the scenarios only focused on a single-mode cavity. Our work is inspired by the idea of Ref. [48], where the cross-Kerr

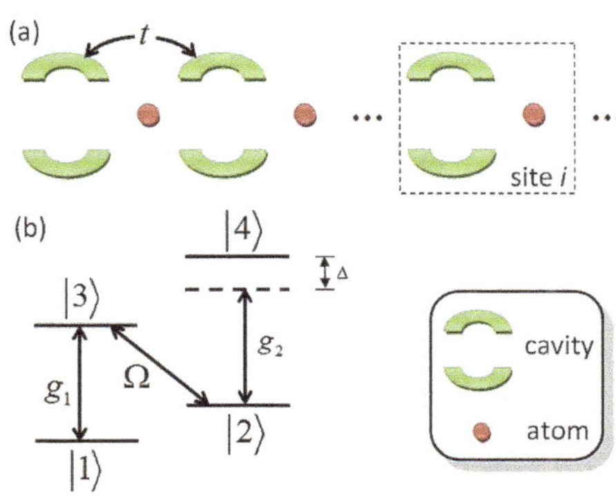

Figure 1 Scheme for nonlinearly coupled QED cavities. (a) An array of QED cavities nonlinearly coupled by N-type atoms. The photon hopping between nearest-neighbor cavities has a strength t. Each effective site is composed of a cavity and an atom (dashed box). **(b)** Level structure of the N-type atoms. The transition $|1\rangle \leftrightarrow |3\rangle$ is resonantly coupled to the cavity mode of its own site with strength g_1, while $|2\rangle \leftrightarrow |4\rangle$ is coupled to the cavity mode of its right nearest-neighbor site with strength g_2, and has a detuning Δ. The transition $|2\rangle \leftrightarrow |3\rangle$ is in resonance with an external laser field of strength Ω.

nonlinearity is generated between two different and neighboring cavities, in circuit-QED systems. In practice, we use the unbalanced couplings of atomic transition $|1\rangle \leftrightarrow |3\rangle$ with left cavity mode, and $|2\rangle \leftrightarrow |4\rangle$ with right cavity mode respectively, in order to generate the local (g_1) and nonlocal (g_2) nonlinearities of our many-body system. This kind of four-level artificial molecule can be realized using two Josephson transmon qubits coupled by a superconducting quantum interference device.

Using the interaction picture and in the rotating-wave approximation, the system Hamiltonian reads

$$\mathcal{H} = \sum_i \left[\Delta\sigma_i^{44} + \left(\Omega\sigma_i^{23} + g_1\sigma_i^{13}a_i^\dagger + \text{H.c.}\right) + \left(-ta_i a_{i+1}^\dagger + g_2\sigma_i^{24}a_{i+1}^\dagger + \text{H.c.}\right)\right], \tag{1}$$

where $\sigma^{mn} = |m\rangle\langle n|$ $(m, n = 1, 2, 3, 4)$, and a $(a^\dagger)$ is the annihilation (creation) operator of the cavity mode. The subscripts denote the site position along the 1D chain. The first three terms in the r.h.s. of Eq. (1) describe the local terms and the nonlinearities on each site. Inside the latter brackets, the first term is the photon hopping, while the second term describes the coupling of the atom to its right neighboring cavity, which generates an effective nonlocal cross-Kerr nonlinearity between the two cavities.

Hereafter we concentrate on the 1D model in Eq. (1) at zero temperature, specifically addressing the case without dissipation with DMRG. Let us also fix the Hamiltonian quantities in units of Ω, set $\hbar = 1$, and work with open boundary conditions. We recall that, in the presence of dissipation, the problem becomes much more difficult to be handled numerically.[a]

For the system we are considering here, in the strong coupling regime atoms and photons cannot be considered as two separate entities. It is thus natural to investigate the phase diagram in terms of combined atomic/photonic modes, named polaritons. The polaritonic number operator on each site i, representing the number of local excitations, is defined as $n_i^{\text{pol}} = 2\sigma_i^{44} + \sigma_i^{33} + \sigma_i^{22} + a_i^\dagger a_i$. For the closed system described by the Hamiltonian (1), the total number $N^{\text{pol}} = \sum_i n_i^{\text{pol}}$ of such polaritons is a conserved quantity. In the following we work in the canonical ensemble for polaritons, and focus on the integer filling situation.

3 Energy gaps and correlation functions

The different nature of the various phases is sensitive to a number of properties which we are going to focus on. Here we are going to study quantities that resemble those characterizing the various phases of the EBH model [36].

First of all, the ground-state energy gap is an important indicator which characterizes the presence or absence of criticality in the model. In particular, in the critical SF phase, the *charge gap* vanishes in the thermodynamic limit. On the other side, in the insulating MI and DW phases, such gap remains finite. In order to make connection with a similar notation in the EBH model, below we introduce the so called charge and neutral gaps referring respectively to the gaps corresponding to adding one extra particle ('charge' sector) or remaining with the same number of particles ('neutral' sector). We stress however that in the present model the excitation carry no real charge. This has to be understood only as a convention.

The charge gap is defined as

$$\Delta E_c = \Delta E^+ - \Delta E^-, \tag{2}$$

where, in the canonical ensemble, ΔE^{+} (ΔE^{-}) denotes the extra energy needed to add (remove) one particle, *i.e.* one polariton, in the system. In the specific, focusing on the unit filling, $\Delta E^{+} = E_{L+1} - E_L$ and $\Delta E^{-} = E_L - E_{L-1}$, where E_L is the ground-state energy per site of an L-sites cavity-array with exactly L excitations, and E_{L+1} (E_{L-1}) is the corresponding energy per site with one excitation more (less). It is therefore possible to extrapolate ΔE_c by running three different DMRG simulations with fixed number of polaritons $N^{\mathrm{pol}} = L-1$, $L, L+1$ [52, 53].

While the charge gap is able to detect particle-hole excitations, in some circumstances it is possible that the dominant low-energy excitations are of a different type. Their presence can be detected only by the so called *neutral gap* at a fixed number of particles,

$$\Delta E_n = E_L^1 - E_L, \tag{3}$$

where, again working in the canonical ensemble, E_L^1 denotes the first excited energy per site of an L-site system with L excitations.

In the following, we also focus on the analysis of the staggered diagonal order for the polaritons, in order to distinguish the DW from the other phases. We do this by investigating the two-point correlation function

$$C_{\mathrm{DW}}(r) = (-1)^r \langle \delta n_i^{\mathrm{pol}} \delta n_{i+r}^{\mathrm{pol}} \rangle, \tag{4}$$

where $\delta n_i^{\mathrm{pol}} = n_i^{\mathrm{pol}} - \bar{n}$ denotes the polariton fluctuation from the average filling $\bar{n}$. The order parameter identifying the DW phase is thus given by: $\mathcal{O}_{\mathrm{DW}} \equiv \lim_{r\to\infty} C_{\mathrm{DW}}(r)$. A finite value of $\mathcal{O}_{\mathrm{DW}}$ indicates a tendency to establish, in the thermodynamic limit, a staggered occupation of the polaritons. On the other side, in the MI as well as the SF phases, $C_{\mathrm{DW}}(r)$ vanishes exponentially with increasing distance r.

4 Phase diagram

The zero-temperature phase diagram of model (1), at unit polariton filling $\bar{n} = 1$ and in the $g_2 - t$ plane, is summarized in Figure 2. We observe that three different phases can be stabilized. Their boundaries have been obtained by means of a finite-size scaling of the numerical data, for systems up to $L = 300$ sites. In our simulations we imposed a cutoff

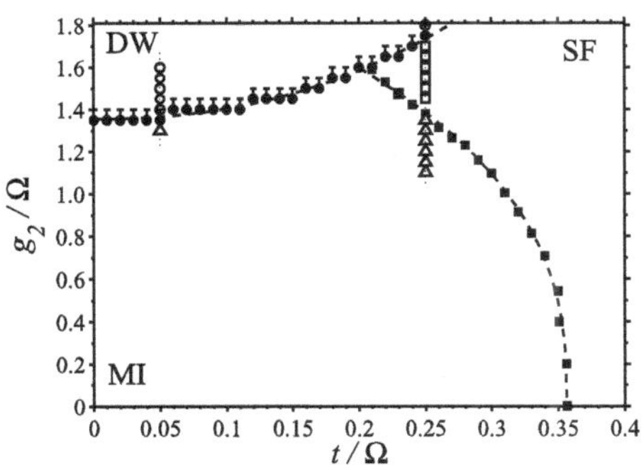

Figure 2 Phase diagram. Ground-state phase diagram for a 1D system of coupled cavities described by Hamiltonian (1) in the $g_2 - t$ parameter space. Here and in the subsequent figures we choose $g_1/\Omega = 0.8$, $\Delta/\Omega = -2$. The various symbols correspond to points belonging to different phases (triangles = MI, circles = DW, squares = SF). Filled points lie close to a phase transition, and are used to draw the interpolating curves (dashed lines). The DW-to-MI/SF boundary (filled circles) has been obtained by analyzing the density-wave order parameter, while the MI-SF boundary (filled squares) through the charge gap. The two vertical dotted lines denoted two cuts in the phase diagram which will be analyzed in details below. The error bars in the points characterizing the DW-to-MI/SF boundary take into account the discretization of the g_2 values that we adopted in our numerical simulations (see the text).

photon number in each cavity, such that $n_i^{\text{phot}} \leq 3$. We also truncated the effective Hilbert space dimension to a value $m = 80$ in all the simulations, except for those shown in Figure 6 for the neutral energy gap (see the discussion in Section 4.3). We checked that, by increasing m and the local fock-space truncation over the photon number, the results concerning the charge gap and the DW order parameter do not change on the scales shown here.

For small photon hopping ($t/\Omega \lesssim 0.2$), by increasing the nonlocal nonlinearity g_2 the system exhibits a direct transition from the MI to the DW phase. On the other hand, for $t/\Omega \gtrsim 0.2$, the MI-to-DW transition is mediated by an extended region appearing at intermediate g_2 values, where the system stabilizes into a gapless SF. In the following we are going to elucidate our finite-size scaling procedure and how we were able to distinguish between the different phases.

4.1 Boundary between MI and SF phases

In the limit of small g_2 and t values, the dominant presence of the on-site interactions stabilize the system into a MI phase with exactly one polariton per cavity ($\bar{n} = 1$), and where the charge energy gap has a finite value. As long as the hopping strength t is progressively increased (and for fixed g_1, g_2), the system eventually enters a SF phase, with a vanishing gap. The filled squares of Figure 2 denoting the MI/SF boundaries have been obtained by means of a finite-size scaling of the charge gap. We performed simulations up to $L = 100$ sites and analyzed whether the gap closes or remains finite in the thermodynamic limit $L \to \infty$.

In Figure 3, left panel, we highlight the size-dependence of ΔE_c as a function of $1/L$ for two points in the phase space close to the MI/SF transition (see points along the dotted line in Figure 2). We expect to see a quadratic dependence $\Delta E_c \sim L^{-2}$ (dashed line) at large L [52, 53], however a linear extrapolation (solid line) is already a good approximation to the scaling, and we can use it to determine ΔE_c in the thermodynamics limit. Indeed, we observe that the difference between quadratic and linear extrapolation is tiny ($\lesssim 10^{-3}$) and does not produce any distinguishable modification on the scale of Figure 2. In the specific case of Figure 3, we fixed $t/\Omega = 0.25$ and chose two different values of g_2/Ω corresponding

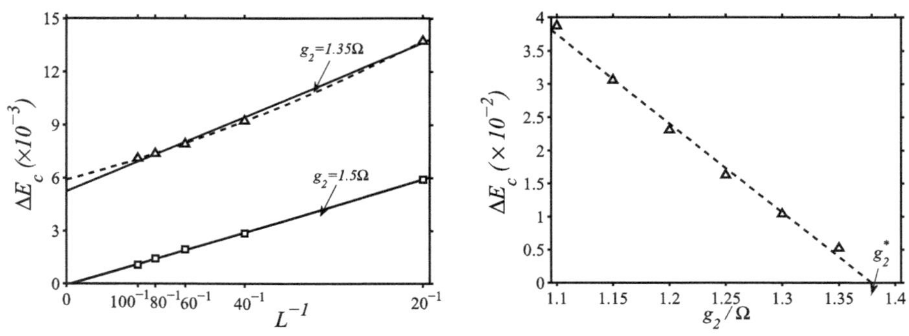

Figure 3 Analysis of the MI-SF boundary. Left panel: system-size dependence of the charge gap ΔE_c per site, for fixed $t/\Omega = 0.25$ and two values of g_2 in the MI ($g_2/\Omega = 1.35$) and in the SF ($g_2/\Omega = 1.5$) phase. Symbols denote the DMRG results. Solid and dashed lines are linear and quadratic fitting curves, respectively. The difference between the extrapolated values in the two fits $\Delta E_c^{\infty} = \lim_{L\to\infty} \Delta E_c^{(L)}$ is negligible. Right panel: determination of the critical g_2^* value for the quantum phase transition. The triangles denote the charge gap per site ΔE_c^{∞} at the thermodynamic limit, as extrapolated in the left panel. The dashed line is a best linear fit of the data *vs.* g_2. The critical point is obtained when ΔE_c vanishes. For $t/\Omega = 0.25$, we get $g_2^*/\Omega \approx 1.379$.

to configurations in the gapped MI ($g_2/\Omega = 1.35$, triangles) and in the gapless SF phase ($g_2/\Omega = 1.5$, squares). The MI is signaled by an extrapolated finite value of $\lim_{L\to\infty} \Delta E_c > 0$, while in the SF this is zero.

In order to locate the critical g_2 for a given value of t (filled squares in Figure 2) we perform a linear extrapolation of the charge gaps in the vicinity of the critical value of g_2. An example of such procedure is shown in the right panel of Figure 3, where we plot ΔE_c as a function of g_2, when this is close to the phase transition (in the specific, here we set $t/\Omega = 0.25$). After a linear extrapolation, we get a critical g_2 value corresponding to $g_2^*/\Omega \approx 1.379$. An analogous procedure is repeated for all the filled squares shown in Figure 2, thus identifying the MI/SF boundary.

4.2 Boundary of the DW phase

The DW phase is characterized by a finite order parameter $\mathcal{O}_{\rm DW}$. Let us therefore look at the two-point staggered correlator in Eq. (4). Since in DMRG simulations we are employing open boundary conditions, to minimize the border effects we analyze the correlations in such a way that the two points are taken symmetrically with respect to the center of the system.[b] The left panel of Figure 4 shows how differently such polariton correlations behave when the system goes from the MI to DW phase, for a fixed system size.

To be more accurate, in the right panel we performed a finite-size scaling and showed that the staggered correlation $C_{\rm DW}(r)$ approaches the zero value exponentially with L, in the MI phase (a similar behavior occurs in the SF region). On the other hand, in the DW such correlator asymptotically converges to a finite value. In the specific, here we fix $t/\Omega = 0.05$ and show that for $g_2/\Omega = 1.3, 1.35$ the DW order is exponentially suppressed with L, while for $g_2/\Omega = 1.4$ it remains finite. The $\mathcal{O}_{\rm DW}$ order parameter reached for $L \to \infty$ is displayed in the inset as a function of g_2.

In order to determine the DW boundary in the phase diagram of Figure 2, we adopted the following protocol. For a fixed value of t/Ω, we start increasing g_2 from zero up to a finite value, with a fixed increment $\delta g_2 = 0.05\Omega$, and to compute the DW order parameter

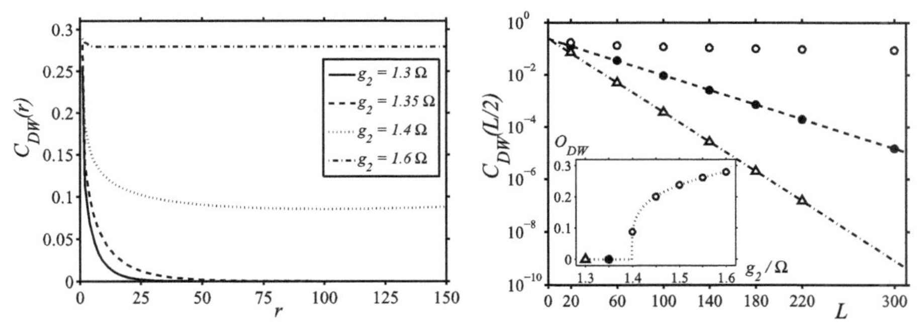

Figure 4 Determination of the DW boundaries. The two-point correlation function $C_{\rm DW}(r)$ for the polariton number and its asymptotic value near the MI-DW quantum phase transition. Here we fix $t/\Omega = 0.05$ and vary g_2/Ω. Left panel: behavior at fixed system size $L = 300$, as a function of the distance r and for different values of $g_2/\Omega = 1.3$, 1.35, 1.4, and 1.6. To minimize boundary effects, we chose the two points $(i, i + r)$ symmetrically with respect to the center of the array. Right panel: finite-size scaling close to the transition. Empty circles, filled circles, and triangles respectively are for $g_2/\Omega = 1.4$ (DW), 1.35 (near the critical point), and 1.3 (MI phase). In the MI phase, $C_{\rm DW}$ vanishes exponentially with L, according to the fits: $C_{\rm DW}^{g_2=1.35} \approx 0.241 \times e^{-0.032L}$ and $C_{\rm DW}^{g_2=1.3} \approx 0.253 \times e^{-0.065L}$. In the DW phase, $C_{\rm DW}$ converges to a finite value. The inset shows such obtained asymptotic value $\mathcal{O}_{\rm DW}$, as a function of g_2.

Figure 5 DW phase order parameter as a function of detuning. For a positive detuning, the system will never present as a DW phase, meaning that $C_{DW}(L/2) = 0$. Taking a negative value of the detuning, we observe that $C_{DW}(L/2)$ increases with $|\Delta|$. The parameters in this figure are $t/\Omega = 0.05$, $g_1/\Omega = 0.8$, $g_2/\Omega = 1.6$ and the system size is $L = 100$.

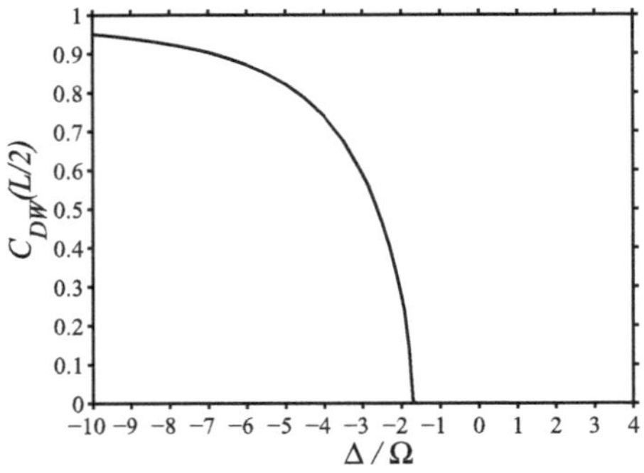

for all such values of g_2. The boundary of DW phase in the $g_2 - t$ plane (filled circles in Figure 2), for any fixed t, is located by the $g_2^*(t)$ that gives the first non-vanishing order parameter $\mathcal{O}_{DW}$.

Here we stress that, because of the arrangement of our 1D array [see Figure 1(a)] and of the asymmetric coupling between the atom and its right/left cavity, the antiferromagnetic symmetry of the system is spontaneously broken. In particular, the state $|4\rangle$ of the L-th atom will be never occupied, since the transition $|2\rangle \leftrightarrow |4\rangle$ does not couple to any cavity mode [see Eq. (1)]. As a consequence, in our simulations we do not need any symmetry-breaking potential. We can observe that the expectation value for the onsite number of polaritons explicitly exhibits a staggered behavior, in that the occupation of the $(2n-1)$-th site is always higher than that of the $(2n)$-th site (for any integer value of n). Finally we notice that such staggering persists at finite size, also for the set of parameters corresponding to the MI phase, although it is extremely tiny and decreases with L. This effect eventually disappears in the thermodynamic limit.

The extension of the DW phase depends on the cavity detuning Δ. In particular, the robustness of the order parameter increases with increasing the modulus of the detuning (see Figure 5). Quite remarkably, we note that a positive Δ will never stabilize an antiferromagnetic DW ordering.

4.3 Neutral gap

The analysis leading to the phase diagram in Figure 2 has been corroborated by a study of the neutral gap, which vanishes both in proximity of the phase transitions and in the entire superfluid region. Differently for the charge gap, it is able to detect the presence of excitations other than particle-hole, and thus locates the boundaries of insulating regions (as the DW) beyond the MI.

The data displayed in Figure 6 show the behavior of ΔE_n as a function of g_2, for a fixed value of t/Ω. In particular we analyzed a vertical cut in the phase diagram of Figure 2 (see the rightmost vertical dotted line in that figure), where the system can be in three different phases according to the value of g_2. With increasing g_2, it goes from the MI phase (nonzero ΔE_n, for $t/\Omega \lesssim 1.45$) to the SF phase (zero ΔE_n, for $1.45 \lesssim t/\Omega \lesssim 1.8$), and then to the DW phase (nonzero ΔE_n, for $t/\Omega \gtrsim 1.8$).

While we cannot see a clear signature of a finite gap for $g_2 = 1.8\Omega$, the scaling with the size displayed in the left panel of Figure 6 seems to suggest the scenario depicted above. It is however important to stress that the DMRG simulations needed to compute the neutral gap have to target the two lowest-lying eigenstates in a single run. Thus they generally

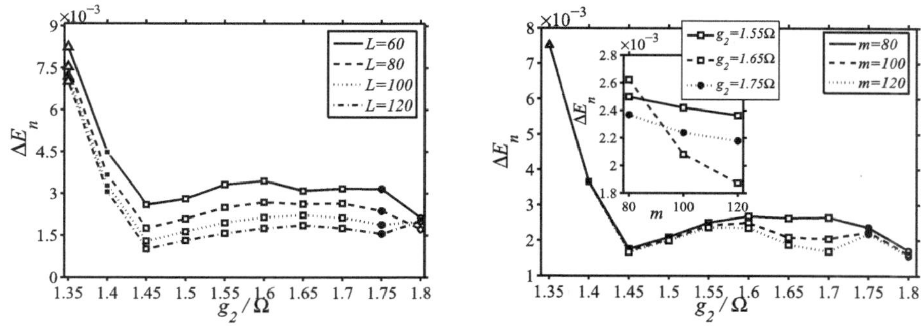

Figure 6 Analysis of the neutral gap. Neutral energy gap as a function of g_2 for $t/\Omega = 0.25$, *i.e.*, along the vertical cut depicted in Figure 2. In the left panel, the different curves are for various system sizes according to the legend, and for a fixed number of kept states $m = 80$ in the DMRG algorithm. The right panel evidences the convergence of the data, at fixed $L = 80$, by increasing m (see also the inset, where we show the behavior of ΔE_n as a function of m, for three different values of g_2).

require a larger dimension m of the effective Hilbert space, as compared to all the other ground-state calculations discussed before. The analysis of the neutral gap requires a careful convergence test of the results with m, which we provide in the right panel of Figure 6. We observe that the non monotonic features that are visible in the region $1.45 \lesssim t/\Omega \lesssim 1.8$ have to be probably ascribed to the inaccuracy of the method at small m values. This signals the presence of the gapless SF phase there, in agreement with the results provided by the charge gap (MI/SF boundary) and for the DW order parameter (SF/DW boundary).

5 Summary

Using the density-matrix renormalization group with open boundary conditions, we studied the equilibrium phase diagram of a system of coupled QED cavities in one dimension. We provided results beyond the standard model of couplings through photon hopping, and also considered nearest-neighbor cross-Kerr nonlinearities. Our analysis is based on a finite-size scaling of the ground-state charge and neutral gaps, as well as of the density-wave order parameter, for systems up to 300 sites. We showed that, beyond the conventional Mott insulator and superfluid phases, the presence of a nearest-neighbor nonlinear coupling can also stabilize a density-wave ordering of polaritons.

Competing interests

The authors declare that they have no competing interests.

Authors' contributions

All the authors participated in the design of the research, analysis of the results, and writing of the paper. The DMRG code used to run all the simulations of this research has been developed and written by DR and coworkers [44] (see also www.dmrg.it). The DMRG simulations were performed by JJ.

Author details

[1]NEST, Scuola Normale Superiore and Istituto di Nanoscienze, CNR, Piazza dei Cavalieri, 7, Pisa, 56126, Italy. [2]Center for Quantum Technologies, National University of Singapore, Singapore, 117543, Singapore.

Acknowledgements

We would like to acknowledge our previous collaboration with M. Hartmann and M. Leib which was inspiring for the present work. This work was supported by Italian MIUR via FIRB Project RBFR12NLNA and PRIN Project 2010LLKJBX, by EU through IP-SIQS, and by National Natural Science Foundation of China under Grant No. 11175033 and No. 11305021.

Endnotes

[a] It is however possible to address the effect of dissipation with a DMRG approach in a 1D chain, when this is described by a master equation within the Lindblad formalism. In the language of tensor networks, one has to

generalize the matrix-product-state ansatz to a matrix-product-density-operator ansatz for mixed states, as originally proposed in Refs. [50, 51]. The computational complexity is greater than for static computations, and is eventually related to the amount of entanglement in the steady state.

b The two points of $\langle \delta n_i^{\mathrm{pol}} \delta n_j^{\mathrm{pol}} \rangle$, with $|i-j| = r$, have been chosen such that $i = (L-r+1)/2, j = (L+r+1)/2$ for odd r, and $i = (L-r)/2, j = (L+r)/2$ for even r (*e.g.* for $L = 100$ sites, $r = 1$ corresponds to $i = 50, j = 51$; $r = 2$ corresponds to $i = 49, j = 51$; $r = 3$ to $i = 49, j = 52$, and so on).

References

1. Bloch I, Dalibard J, Zwerger W. Many-body physics with ultracold gases. Rev Mod Phys. 2008;80:885.
2. Fisher MPA, Weichman PB, Grinstein G, Fisher DS. Boson localization and the superfluid-insulator transition. Phys Rev B. 1989;40:546.
3. Jaksch D, Bruder C, Cirac JI, Gardiner CW, Zoller P. Cold bosonic atoms in optical lattices. Phys Rev Lett. 1998;81:3108.
4. Greiner M, Mandel O, Esslinger T, Hänsch TW, Bloch I. Quantum phase transition from a superfluid to a Mott insulator in a gas of ultracold atoms. Nature. 2002;415:39.
5. Lewenstein M, Sanpera A, Ahufinger V, Damski B, Sen(De) A, Sen U. Ultracold atomic gases in optical lattices: mimicking condensed matter physics and beyond. Adv Phys. 2007;56:243.
6. Lahaye T, Menotti C, Santos L, Lewenstein M, Pfau T. The physics of dipolar bosonic quantum gases. Rep Prog Phys. 2009;72:126401.
7. Schmidt H, Imamoğlu A. Giant Kerr nonlinearities obtained by electromagnetically induced transparency. Opt Lett. 1996;21:1936.
8. Imamoğlu A, Schmidt H, Woods G, Deutsch M. Strongly interacting photons in a nonlinear cavity. Phys Rev Lett. 1997;79:1467.
9. Hartmann MJ, Brandão FGSL, Plenio MB. Strongly interacting polaritons in coupled arrays of cavities. Nat Phys. 2006;2:849.
10. Greentree AD, Tahan C, Cole JH, Hollenberg LCL. Quantum phase transitions of light. Nat Phys. 2006;2:856.
11. Angelakis DG, Santos MF, Bose S. Photon-blockade-induced Mott transitions and XY spin models in coupled cavity arrays. Phys Rev A. 2007;76:031805(R).
12. Hartmann MJ, Brandão FGSL, Plenio MB. Quantum many-body phenomena in coupled cavity arrays. Laser Photonics Rev. 2008;2:527.
13. Tomadin A, Fazio R. Many-body phenomena in QED-cavity arrays. J Opt Soc Am. 2010;27:A130.
14. Houck AA, Türeci HE, Koch J. On-chip quantum simulation with superconducting circuits. Nat Phys. 2012;8:292.
15. Schmidt S, Koch J. Circuit QED lattices: towards quantum simulation with superconducting circuits. Ann Phys. 2013;525:395.
16. Carusotto I, Gerace D, De Liberato S, Ciuti C, Imamoğlu A. Fermionized photons in an array of driven dissipative nonlinear cavities. Phys Rev Lett. 2009;103:033601.
17. Tomadin A, Giovannetti V, Fazio R, Gerace D, Carusotto I, Türeci HE, Imamoğlu A. Signatures of the superfluid-insulator phase transition in laser-driven dissipative nonlinear cavity arrays. Phys Rev A. 2010;81:061801(R).
18. Hartmann MJ. Polariton crystallization in driven arrays of lossy nonlinear resonators. Phys Rev Lett. 2010;104:113601.
19. Nunnenkamp A, Koch J, Girvin SM. Synthetic gauge fields and homodyne transmission in Jaynes-Cummings lattices. New J Phys. 2011;13:095008.
20. Nissen F, Schmidt S, Biondi M, Blatter G, Türeci HE, Keeling J. Nonequilibrium dynamics of coupled qubit-cavity arrays. Phys Rev Lett. 2012;108:233603.
21. Grujic T, Clark SR, Angelakis DG, Jaksch D. Non-equilibrium many-body effects in driven nonlinear resonator arrays. New J Phys. 2012;14:103025.
22. Grujic T, Clark SR, Jaksch D, Angelakis DG. Repulsively induced photon superbunching in driven resonator arrays. Phys Rev A. 2013;87:053846.
23. Jin J, Rossini D, Fazio R, Leib M, Hartmann MJ. Photon solid phases in driven arrays of nonlinearly coupled cavities. Phys Rev Lett. 2013;110:163605.
24. Jin J, Rossini D, Leib M, Hartmann MJ, Fazio R. Steady-state phase diagram of a driven QED-cavity array with cross-Kerr nonlinearities. Phys Rev A. 2014;90:023827.
25. Underwood DL, Shanks WE, Koch J, Houck AA. Low-disorder microwave cavity lattices for quantum simulation with photons. Phys Rev A. 2012;86:023837.
26. Abbarchi M, Amo A, Sala VG, Solnyshkov DD, Flayac H, Ferrier L, Sagnes I, Galopin E, Lemaître A, Malpuech G, Bloch J. Macroscopic quantum self-trapping and Josephson oscillations of exciton polaritons. Nat Phys. 2013;9:275.
27. Toyoda K, Matsuno Y, Noguchi A, Haze S, Urabe S. Experimental realization of a quantum phase transition of polaritonic excitations. Phys Rev Lett. 2013;111:160501.
28. Lucero E, Barends R, Chen Y, Kelly J, Mariantoni M, Megrant A, O'Malley P, Sank D, Vainsencher A, Wenner J, White T, Yin Y, Cleland AN, Martinis JM. Computing prime factors with a Josephson phase qubit quantum processor. Nat Phys. 2012;8:719.
29. Steffen L, Salathe Y, Oppliger M, Kurpiers P, Baur M, Lang C, Eichler C, Puebla-Hellmann G, Fedorov A, Wallraff A. Deterministic quantum teleportation with feed-forward in a solid state system. Nature. 2013;500:319.
30. Chen Y, Roushan P, Sank D, Neill C, Lucero E, Mariantoni M, Barends R, Chiaro B, Kelly J, Megrant A, Mutus JY, O'Malley PJJ, Vainsencher A, Wenner J, White TC, Yin Y, Cleland AN, Martinis JM. Emulating weak localization using a solid-state quantum circuit. Nat Commun. 2014;5:5184.
31. Zueco D, Mazo JJ, Solano E, García Ripoll JJ. Microwave photonics with Josephson junction arrays: negative refraction index and entanglement through disorder. Phys Rev B. 2012;86:024503.
32. Peropadre B, Zueco D, Wulschner F, Deppe F, Marx A, Gross R, García Ripoll JJ. Tunable coupling engineering between superconducting resonators: from sidebands to effective gauge fields. Phys Rev B. 2013;87:134504.
33. Sowiński T, Dutta O, Hauke P, Tagliacozzo L, Lewenstein M. Dipolar molecules in optical lattices. Phys Rev Lett. 2012;108:115301.

34. Dalla Torre EG, Berg E, Altman E. Hidden order in 1D Bose insulators. Phys Rev Lett. 2006;97:260401.
35. Berg E, Dalla Torre EG, Giamarchi T, Altman E. Rise and fall of hidden string order of lattice bosons. Phys Rev B. 2008;77:245119.
36. Rossini D, Fazio R. Phase diagram of the extended Bose-Hubbard model. New J Phys. 2012;14:065012.
37. Deng X, Citro R, Orignac E, Minguzzi A, Santos L. Polar bosons in one-dimensional disordered optical lattices. Phys Rev B. 2013;87:195101.
38. Wikberg E, Larson J, Bergholtz EJ, Karlhede A. Fractional domain walls from on-site softening in dipolar bosons. Phys Rev A. 2012;85:033607.
39. Batrouni GG, Scalettar RT, Rousseau VG, Grémaud B. Competing supersolid and Haldane insulator phases in the extended one-dimensional bosonic Hubbard model. Phys Rev Lett. 2013;110:265303.
40. Batrouni GG, Rousseau VG, Scalettar RT, Grémaud B. Competing phases, phase separation, and coexistence in the extended one-dimensional bosonic Hubbard model. Phys Rev B. 2014;90:205123.
41. Schiró M, Bordyuh M, Öztop B, Türeci HE. Phase transition of light in cavity QED lattices. Phys Rev Lett. 2012;109:053601.
42. Kumar B, Jalal S. Quantum Ising dynamics and Majorana-like edge modes in the Rabi lattice model. Phys Rev A. 2013;88:011802(R).
43. Schollwöck U. The density-matrix renormalization group. Rev Mod Phys. 2005;77:259.
44. De Chiara G, Rizzi M, Rossini D, Montangero S. Density matrix renormalization group for dummies. J Comput Theor Nanosci. 2008;5:1277.
45. Rossini D, Fazio R. Mott-insulating and glassy phases of polaritons in 1D arrays of coupled cavities. Phys Rev Lett. 2007;99:186401.
46. Rossini D, Santoro GE, Fazio R. Photon and polariton fluctuations in arrays of QED-cavities. Europhys Lett. 2008;83:47011.
47. D'Souza AG, Sanders BC, Feder DL. Fermionized photons in the ground state of one-dimensional coupled cavities. Phys Rev A. 2013;88:063801.
48. Hu Y, Ge GQ, Chen S, Yang XF, Chen YL. Cross-Kerr-effect induced by coupled Josephson qubits in circuit quantum electrodynamics. Phys Rev A. 2011;84:012329.
49. Rebić S, Twamley J, Milburn GJ. Giant Kerr nonlinearities in circuit quantum electrodynamics. Phys Rev Lett. 2009;103:150503.
50. Verstraete F, García-Ripoll JJ, Cirac JI. Matrix product density operators: simulation of finite-temperature and dissipative systems. Phys Rev Lett. 2004;93:207204.
51. Zwolak M, Vidal G. Mixed-state dynamics in one-dimensional quantum lattice systems: a time-dependent superoperator renormalization algorithm. Phys Rev Lett. 2004;93:207205.
52. Kühner TD, Monien H. Phases of the one-dimensional Bose-Hubbard model. Phys Rev B. 1998;58:R14741.
53. Kühner TD, White SR, Monien H. One-dimensional Bose-Hubbard model with nearest-neighbor interaction. Phys Rev B. 2000;61:12474.

11

Minimal ancilla mediated quantum computation

Timothy J Proctor[1*] and Viv Kendon[1,2]

*Correspondence: py08tjp@leeds.ac.uk
[1] School of Physics and Astronomy, University of Leeds, E C Stoner Building, Leeds, LS2 9JT, UK
Full list of author information is available at the end of the article

Abstract

Schemes of universal quantum computation in which the interactions between the computational elements, in a computational register, are mediated by some ancillary system are of interest due to their relevance to the physical implementation of a quantum computer. Furthermore, reducing the level of control required over both the ancillary and register systems has the potential to simplify any experimental implementation. In this paper we consider how to minimise the control needed to implement universal quantum computation in an ancilla-mediated fashion. Considering computational schemes which require no measurements and hence evolve by unitary dynamics for the global system, we show that when employing an ancilla qubit there are certain fixed-time ancilla-register interactions which, along with ancilla initialisation in the computational basis, are universal for quantum computation with no additional control of either the ancilla or the register. We develop two distinct models based on locally inequivalent interactions and we then discuss the relationship between these unitary models and the measurement-based ancilla-mediated models known as ancilla-driven quantum computation.

Keywords: ancilla; universal gates; minimal control; quantum computation; quantum bus; ancilla-driven; ancilla-controlled

1 Introduction

The original theoretical setting for quantum computation is the gate model [1] in which a global unitary that acts on a register of qubits, which computes the solution to some problem, is decomposed into a sequence of fundamental gates that are applied to the register. As in classical computation, it is desirable for these fundamental gates to be members of some *finite* and *universal* gate set, from which any global unitary can be composed up to arbitrary accuracy. There has been extensive research on such universal sets, and a significant example is the set composed of any entangling gate in conjunction with any universal set of single-qubit unitaries [2, 3]. Furthermore, almost any two-qubit entangling gate is universal on its own provided that it can be applied to arbitrary pairs of qubits [4, 5]. These results are of significant theoretical importance for the understanding of quantum computation.

However, the physical implementation of these models requires direct interactions between arbitrary pairs of register qubits and, often, direct application of single-qubit rotations and measurements. This is a huge practical challenge and most experimentally implemented or proposed schemes mediate the required multi-qubit interactions using some ancillary system. An example of such an ancilla-mediated scheme is the original Cirac-Zoller ion-trap gate [6], where the ancilla in this case is the collective quantized motion of the ions. Further examples include superconducting qubits coupled to nitrogen-vacancy (NV) centres [7–10] or transmission line resonators [11, 12], spin qubits coupled via ancillary photonic qubits [13, 14] and the coupling of a Cooper-pair box to a micro-mechanical resonator [15]. It is therefore of both practical and theoretical interest to study the effect of incorporating the ancillary system into the computational model. Indeed, those gates that have been shown to be universal in a direct implementation of the circuit model cannot in general be utilised to implement quantum computations entirely mediated via an ancilla. We will refer to schemes in which all the multi-qubit interactions are mediated via some ancillary system as *ancilla-mediated quantum computation* (AMQC).

An extensively researched model of AMQC is *quantum bus* (qubus) computation [16–21]. This model employs a field-mode ancilla to mediate two-qubit gates on pairs of register qubits with the interaction between the ancilla and a register qubit being a controlled displacement of the field-mode. Recently, we have developed an analogous model which employs a d-dimensional *qudit* ancilla [22] with a displacement operator defined in the discrete phase space of the qudit [23, 24]. These models have been shown to require a lower number of operations to implement certain gate sequences than a direct implementation of the circuit model [19, 22, 25]. However, neither of these models can implement a universal gate set on the register using *only* this ancilla-register interaction and so, although no interactions between register qubits are required, some further direct access is needed to the register qubits to implement some basis-changing single-qubit unitary [22, 26].

In order to implement useful quantum computations, it will be necessary to have register qubits with as long a coherence time as possible. However, if it is necessary to access each register qubit to implement multiple forms of control this will potentially introduce many sources of decoherence. Limiting the forms of access required to the register qubits may help to isolate the register more effectively, and so, motivated by this, the measurement-based *ancilla-driven quantum computation* (ADQC) [27–29] and more recently the globally unitary *ancilla-controlled quantum computation* (ACQC) [30] schemes were developed. In both of these models the access to the register qubits is limited to one *fixed-time* interaction between a single register qubit (at a time) and an ancilla qubit, where the qubits are not necessarily of the same physical type. The additional direct access to the register that was required in the qubus architecture, and analogous qudit model, is replaced by local unitary operations on the ancilla and, in the case of ADQC, ancilla measurements.

Although both the ancilla-driven and ancilla-controlled models require the minimum possible access to the computational register, they replace the local control of the register with local control of the ancillary system and so still require more than one fixed quantum gate to implement universal computation. Halil-Shah and Oi [31] have recently shown that the measurement-based ancilla-driven model can be adapted so that no local control, beyond ancilla preparation in a fixed state, is required of either the ancilla or the register. In this model, the computation can be achieved using a fixed interaction and ancilla measurement in the computational basis alone. However, this requires a stochastic repeat-until-success style gate scheme [32], whereby one has to wait until a random walk through the set of unitaries is within the required precision of the desired unitary. In this paper we will show that it is possible to develop deterministic models that require only a single fixed ancilla-register interaction and ancilla preparation in the computational basis with no ancilla measurements necessary. Such schemes require a minimal level of control of both the ancillary and register systems whilst allowing for universal quantum computation. Hence, we will refer to such models as *minimal control* models of ancilla-mediated quantum computation, and we will often drop the reference to ancilla-mediation for brevity. In Section 3 we introduce two such schemes based on locally inequivalent interactions. The first of these models requires multiple ancillas to implement entangling two-qubit gates on the register and so, although it requires minimal control, it has an overhead in terms of ancilla use. We then develop an alternative minimal control model which does not have this ancilla overhead and requires only three ancilla-register interactions per two-qubit entangling gate, the minimum possible in any unitary scheme [33]. We briefly discuss the physical implementation of these models before concluding in Section 4. We begin in Section 2 with some essential definitions.

2 Definitions

We denote the Pauli operators acting on the jth qubit by X_j, Y_j and Z_j and take $|0\rangle$ and $|1\rangle$ to be the positive and negative eigenstates of the Pauli Z operator respectively. Using standard definitions, we take the Hadamard gate H to be

$$H := \frac{1}{\sqrt{2}}\big(|0\rangle\langle 0| + |0\rangle\langle 1| + |1\rangle\langle 0| - |1\rangle\langle 1|\big), \tag{1}$$

and the single-qubit phase gate to be

$$R(\theta) := |0\rangle\langle 0| + e^{i\theta}|1\rangle\langle 1|. \tag{2}$$

We furthermore define $T := R(\pi/4)$ and the two-qubit SWAP gate

$$\text{SWAP} := |00\rangle\langle 00| + |01\rangle\langle 10| + |10\rangle\langle 01| + |11\rangle\langle 11|. \tag{3}$$

Except for those gates defined above and the identity operator $\mathbb{I}$, where standard notation is used, all single-qubit gates will be denoted by lower case roman letters. We define a general controlled gate, with a control qubit j and a target qubit k, by

$$C_k^j(u,v) := |0\rangle\langle 0|_j \otimes u_k + |1\rangle\langle 1|_j \otimes v_k, \tag{4}$$

where $u, v \in U(2)$. The subscripts j and k will be dropped from the notation when no ambiguity will arise and we let $Cu := C(\mathbb{I}, u)$ and $SCu := \text{SWAP} \cdot Cu$. Two operators $U, V \in U(4)$ are called *locally equivalent* [34] with respect to a decomposition into qubit subsystems j and k if

$$U = u_j \otimes v_k \cdot V \cdot p_j \otimes q_k, \tag{5}$$

for some $u, v, p, q \in U(2)$.

3 Minimal control ancilla-mediated quantum computation

We now present two schemes of ancilla-mediated quantum computation that require only a single fixed-time ancilla-register interaction and ancilla preparation in the computational basis and hence are minimal control models.

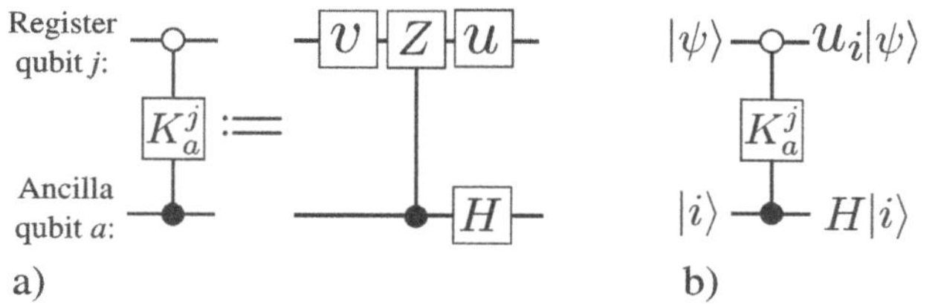

Figure 1 The general K_a^j ancilla-register interaction. (a) Decomposition into local and non-local parts. **(b)** In this minimal control model the elements of a universal set for single-qubit unitaries, $\{u_0, u_1\}$, are applied to the register by the K_a^j interaction and initialising the ancilla in the state $|i\rangle$ where $i = 0, 1$.

3.1 A first minimal control model

We introduce our first model by giving a general form for an ancilla-register interaction which under certain conditions can implement a universal gate set on a register of qubits within the constraints of minimal control. We give an explicit construction for the application of a universal gate set on the register before comparing this model to the measurement-based scheme of Halil-Shah and Oi [31] and giving a simple example of an interaction that obeys the required constraints.

3.1.1 A general interaction

We consider a general fixed ancilla-register interaction of the form

$$K_a^j := u_j \otimes H_a \cdot CZ \cdot v_j \otimes \mathbb{I}_a, \tag{6}$$

where $u, v \in U(2)$. This interaction is shown in Figure 1a and it is locally equivalent to CZ. We define $u_0 := uv$ and $u_1 := uZv$ and note that we may also write the interaction in the form $K_a^j = \mathbb{I} \otimes H \cdot C_j^a(u_0, u_1)$. We will show that for any K such that $\{u_0, u_1\}$ is a universal single-qubit gate set we may implement a minimal control model. We do this by showing how we may implement a universal gate set on the register qubits. It follows directly from the definition of the ancilla-register interaction K_a^j that

$$K_a^j |\psi\rangle_j |i\rangle_a = u_i |\psi\rangle_j \otimes H|i\rangle_a, \tag{7}$$

where $i = 0, 1$. Hence, we can deterministically apply the elements of $\{u_0, u_1\}$ on any register qubit, which we assume is a universal set for $SU(2)$, and so we may simulate any gate in $SU(2)$ up to arbitrary accuracy using only K and the initialisation of ancilla in the computational basis. This gate method is depicted in the circuit diagram of Figure 1b.

We now show how to implement a maximally entangling gate between two register qubits, j and k, using only K and ancillas prepared in the computational basis. A straightforward explicit calculation, utilising the identities $vu_0^\dagger u = \mathbb{I}$, $HZH = X$, $XX = ZZ = \mathbb{I}$ and $XZXZ = -\mathbb{I}$, shows that

$$K_a^k K_a^j \cdot u_{0_j}^\dagger \otimes u_{0_k}^\dagger \otimes \mathbb{I} \cdot K_a^k K_a^j = M_k^j \otimes \mathbb{I}_a, \tag{8}$$

where the induced entangling gate on the register qubits j and k is

$$M_k^j = u_j \otimes u_k \cdot CZ \cdot v_j \otimes v_k. \tag{9}$$

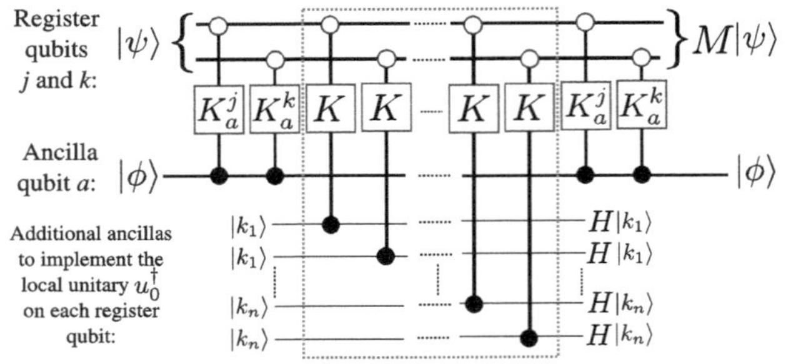

Figure 2 Two-qubit gates in the first minimal control model with interactions of the form K_a^j. This gate sequence has no overall effect on the 'entangling' ancilla qubit but implements the entangling unitary M, given by Eq. (9) and which is locally equivalent to CZ, on the register qubits. $2n$ additional ancillas with appropriate initial states in the computational basis are required to implement the unitary $u_0^\dagger$ (up to arbitrary accuracy) on each register qubit and these extra interactions are shown in the blue dashed box. The number of these additional ancillas depends on the form of u_0 and u_1.

Although Eq. (8) is an ancilla-mediated sequence which implements an entangling two-qubit gate on the register, as written it requires local unitaries on the register qubits and it is not decomposed into only K gates. However, we may decompose the $u_0^\dagger$ gate on each register qubit into further K gates. This is because u_0 and u_1 are a universal set for $SU(2)$ and hence there is a choice of $k_1, \ldots, k_n = 0, 1$ such that

$$\left(\prod_{i=1}^{n} K_{a_i}^j\right)|\psi\rangle_j|k_1\rangle_{a_1}\cdots|k_n\rangle_{a_n} = \tilde{u}_0^\dagger|\psi\rangle_j H|k_i\rangle_{a_1}\cdots H|k_n\rangle_{a_n}, \tag{10}$$

with $\tilde{u}_0^\dagger = u_{k_n}\cdots u_{k_1}$ approximating $u_0^\dagger$ up to arbitrary accuracy with finite n. In certain cases $u_0^\dagger$ may be implemented exactly. Hence we may implement an entangling gate which is locally equivalent to CZ between pairs of register qubits using only K gates and ancillas initialised in the computational basis.[a] The circuit diagram for this two-qubit gate on the register is given in Figure 2. Therefore, under the assumption that u_0 and u_1 are universal for $SU(2)$, we have shown that K, along with ancillas prepared in the computational basis, can implement a minimal control model of ancilla-mediated quantum computation.

3.1.2 Discussion and comparison with a measurement-based model

We compare this minimal unitary model to the recently proposed measurement-based minimal scheme of Halil-Shah and Oi [31]. In their model the computation is achieved using only a fixed interaction, ancilla preparation in a fixed state, and measurement in the computational basis. The fixed initial ancilla state is compensated for by the measurement which projects the ancilla onto states in the computational basis. Only one and two ancilla-register interactions are required to implement the single and two-qubit gates respectively. However this model results in a probabilistic repeat-until-success style gate scheme [32], whereby one has to wait until a random walk through the set of unitaries is within the required precision of the desired unitary. Although the model presented here is entirely deterministic, there is an overhead, that in general may be large, to implement the two-qubit gates. We note that this overhead does not appear if an ancilla measurement is performed after the first two interactions in Figure 2, for an appropriately initialised ancilla and measurement basis, and in this case both models are of a similar form.

We note that if vu is diagonal in the computational basis the additional ancillas are not required, however in this case u_0 and u_1 will commute and so cannot form a universal set for $SU(2)$. In general, the required additional ancillas for the two-qubit gates may create an impractical overhead. However, we now demonstrate that there exists an appropriate form for K that is universal and has a low overhead for additional ancillas for each entangling gate.

3.1.3 A specific interaction

We can parametrise a general unitary operator $p \in U(2)$ by the matrix in the computational basis

$$p(\eta, \phi, \psi, \theta) = e^{i\eta} \begin{pmatrix} e^{i\phi} \cos\theta & e^{-i\psi} \sin\theta \\ e^{i\psi} \sin\theta & -e^{-i\phi} \cos\theta \end{pmatrix}. \tag{11}$$

A specific suitable choice for the ancilla-register interaction K is given by taking $u = p(\eta, \zeta, \zeta, \frac{\pi}{8})$ and $v = p(\frac{\pi}{8} - \eta, -\zeta - \frac{\pi}{8}, \zeta - \frac{\pi}{8}, \frac{\pi}{8})$. It is straightforward to show that this gives $u_0 = T$ and $u_1 = HT$. We have that $T^7 = T^\dagger$ and so $u_1 u_0^7 = H$. It then follows that u_0 and u_1 form a universal set for $SU(2)$ as H and T are a universal set for single-qubit unitaries [35]. It is necessary to implement $u_0^\dagger$ on each register qubit to implement the sequence of Eq. (8) and Figure 2. We have that $u_0^\dagger = u_0^7$ and so the sequence of Eq. (8) and Figure 2 can be implemented using 14 ancillas prepared in the state $|0\rangle$ and one 'entangling' ancilla, that mediates the gate, prepared in any state.

3.2 A second minimal control model

We now present an alternative minimal control model which does not require additional ancillas. As before, we will introduce the model with a general interaction which with certain restrictions can be used to implement a minimal control model and then give a specific example of a simple suitable interaction.

3.2.1 A general interaction

Take an ancilla-register interaction of the form

$$L_a^j := \mathbb{I}_j \otimes u_a \cdot SCR(\theta) \cdot R(\theta_r)_j \otimes R(\theta_a)_a. \tag{12}$$

This is decomposed into local and non-local parts in Figure 3a. We note that this can also be expressed as $L_a^j = SC_j^a(uR(\theta_r), uR(\theta + \theta_r)) \cdot \mathbb{I}_j \otimes R(\theta_a)_a$. We show that an interaction of this form, along with ancillas prepared in the computational basis, can implement universal quantum computation on the register if θ is such that $CR(\theta)$ is entangling (all non-trivial θ) and $\{v_0, v_1\}$ is a universal set for $SU(2)$ where $v_i := R(\theta^i + \theta_a)uR(\theta^i + \theta_r)$. As before, we do this by showing how we may implement a two-qubit entangling gate and a universal set for single-qubit unitaries on the register. We note that it is possible to set θ_r and θ_a to zero and obtain a universal interaction and these local rotations are included to increase the generality of the interaction.

We may implement an entangling two-qubit gate between register qubits j and k using an ancilla initialised in the state $|0\rangle$ by interacting the ancilla sequentially with qubits j

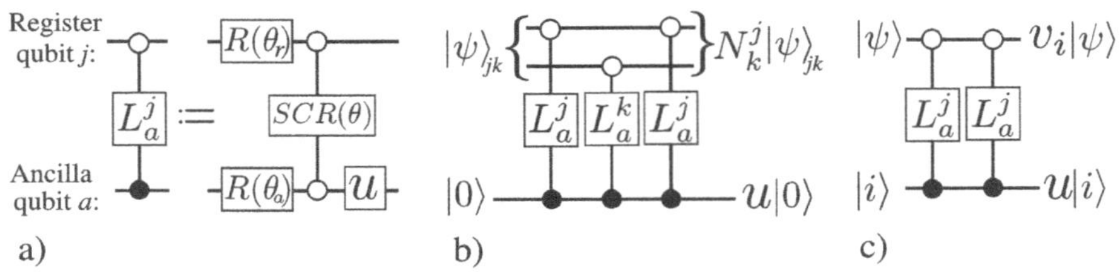

Figure 3 The fixed ancilla-register interaction L_a^j. (a) Decomposition into local and non-local parts. **(b)** The two-qubit entangling gate N_k^j is implemented on the register qubits j and k via the sequence of Eq. (13) with an ancilla initialised in the state $|0\rangle$, where the form of N_k^j is given in Eq. (14). This two-qubit gate is not symmetric with respect to the exchange of j and k. **(c)** The single-qubit gate $v_i = R(\theta^i + \theta_a)uR(\theta^i + \theta_r)$ is applied to a register qubit j by two applications of L with an ancilla initialised in the state $|i\rangle$, $i = 0, 1$.

and k before completing the gate with a second interaction with the j qubit. This is the interaction sequence

$$L_a^j L_a^k L_a^j |\psi\rangle_{jk} |0\rangle_a = N_k^j |\psi\rangle_{jk} \otimes u|0\rangle_a, \tag{13}$$

where N_k^j is entangling for non-trivial θ and is given by

$$N_k^j = R(\theta_a)u_j \otimes \mathbb{I}_k \cdot SCR(\theta) \cdot R(\theta_a)uR(\theta_r)_j \otimes R(\theta_r)_k. \tag{14}$$

This is represented in the circuit diagram of Figure 3b and can be shown with a simple explicit calculation. We may decompose any single-qubit gate on a register qubit into only ancilla-register interactions L and ancilla state-preparation in the computational basis. This is because

$$L_a^j L_a^j |\psi\rangle_j |i\rangle_a = v_i |\psi\rangle_j \otimes u|i\rangle_a, \tag{15}$$

where $i = 0, 1$ and we assume that $\{v_0, v_1\}$ is a universal set for $SU(2)$. This is represented in the circuit diagram of Figure 3c. Hence, as we have shown how to implement a two-qubit entangling gate and a universal set for $SU(2)$ on the register then this is a minimal control model of ancilla-mediated quantum computation.

3.2.2 Discussion and comparison with other models

This model requires three interactions for each entangling two-qubit gate on the register which, although greater than the two needed with the aid of ancilla measurement in ADQC [28] and the minimal extension of Halil-Shah and Oi [31], is the minimum possible in any measurement-free scheme [33]. Furthermore, in contrast to the first minimal control model, there is no requirement for multiple ancilla qubits to implement the entangling gates. Finally, we note that the two-qubit gates on the register are implemented in an identical fashion to those in the ACQC model [30] and L obeys the required conditions to be universal for that model.

3.2.3 A specific interaction

A simple example of a specific form for the interaction L_a^j such that v_0 and v_1 form a universal set for $SU(2)$ (and hence L_a^j may implement this minimal control model) is given by taking $u = H$, $\theta = \frac{\pi}{4}$ and $\theta_r = \theta_a = 0$. We then have that $L_a^j = \mathbb{I}_j \otimes H_a \cdot SCT$ and hence $v_0 = H$ and $v_1 = THT$. A proof of the universality of $\{H, THT\}$ for $SU(2)$ is given in the Appendix. The entangling gate induced on a pair of register qubits from the sequence of Eq. (13) is then $N = H \otimes \mathbb{I} \cdot SCT \cdot H \otimes \mathbb{I}$ and this can easily simulate CNOT := CX as in this case $(N_k^j)^4 = C_j^k X$.

3.3 The physical implementation of minimal control models

The models introduced herein are motivated by the physical challenges of implementing the multiple forms of control required in most universal models of quantum computation and hence it is interesting to briefly consider systems that may be appropriate for a physical realisation. We initially concentrate on a simple Hamiltonian for implementing the second minimal control model presented in Section 3.2. We set $\hbar = 1$ and consider the two-qubit interaction Hamiltonian

$$\mathbf{H}(\theta) = \pi(X \otimes X + Y \otimes Y) + (\pi - \theta)Z \otimes Z, \tag{16}$$

which, applied for a time $t = 1/4$, implements (up to an irrelevant global phase) the unitary operator

$$U(\theta) = e^{-i\mathbf{H}(\theta)/4} = SCR(\theta) \cdot R(-\theta/2) \otimes R(-\theta/2). \tag{17}$$

If we consider the second minimal control model and take the fixed ancilla-register gate L_a^j to be the unitary implemented by applying $\mathbf{H}(\theta)$ to the ancilla and register qubit for a time $t = 1/4$, i.e. $U(\theta)$, followed by a fixed ancilla rotation of the form $R(\theta/2)HR(\theta/2)$, we have that $L_a^j = \mathbb{I}_j \otimes R(\theta/2)HR(\theta/2)_a \cdot U(\theta)$. Hence, this L_a^j gives $v_0 = H$ and $v_1 = R(\theta)HR(\theta)$ which we have shown to be a universal set for $SU(2)$ when $\theta = \pi/4$ and so this form for L_a^j is appropriate for implementing the second minimal control model. With this simple interaction Hamiltonian, $\mathbf{H}(\pi/4)$, local control of the ancilla is required. However, we see that this is a *fixed* gate on the ancilla after every ancilla-register interaction via $\mathbf{H}(\pi/4)$ and hence this can be a fixed element in an experimental setup or incorporated into the natural evolution of the ancilla between interactions.[b] For example, if the ancillary qubit is photonic the local operation can be performed by fixed linear optics [36] after each ancilla-register interaction. Indeed, the use of ancillary photons to mediate gates has been demonstrated in many experimental setups, for example with atomic [37, 38] or spin [13, 14] qubits.

Interactions with the form $\mathbf{H}(\theta)$ arise naturally in spin systems, with one example of an implementation of $\mathbf{H}(\pi/4)$ given by the coupling between quantum dot resonant exchange qubits [39]. A particularly relevant physical system to ancilla-mediated models is the coupling of nuclear spins via ancillary electronic spins in nitrogen-vacancy (NV) defects in diamond [40–42] and in such setups it may be possible to engineer the Hamiltonian $\mathbf{H}(\pi/4)$ [43]. Although in some physical realisations, such as the photonic case discussed above, the fixed local operation on the ancillary qubit after each interaction is convenient or natural, in others it may be problematic and negate the benefits of the models introduced herein. However, it is also possible to find Hamiltonians that directly implement suitable interactions for either of the models proposed in Sections 3. In certain systems the Hamiltonian is highly tuneable, with an example being those involving superconducting qubits

[44], and due to the long coherence times of ensembles of electron spins in NV centers [45] a particularly promising physical system for ancilla-mediated quantum computing is arrays of spin ensembles in diamond (each spin ensemble is an effective qubit) coupled by ancillary flux qubits [7–9]. Indeed, coherent coupling in such a system as been demonstrated [10]. It would be interesting to consider which physical systems have Hamiltonians that are naturally suited to generating appropriate interactions for the models introduced herein and we leave a more detailed study of this for future work.

4 Conclusions

We have presented two unitary models of ancilla-mediated quantum computation that require only minimal control of both the ancillary and register systems. The only control necessary in these models to implement universal quantum computation on a register of qubits is a single fixed-time ancilla-register interaction between one ancilla qubit and one register qubit (at a time) and ancilla preparation in the computational basis. The first of these models is based on maximally entangling interactions that are locally equivalent to CZ and requires multiple ancilla qubits to mediate two-qubit entangling gates on the register. This model is similar in many respects to the minimal measurement-based ancilla-mediated model of Halil-Shah and Oi [31], in which the requirement for ancilla preparation is replaced with the need for ancilla measurements in the computational basis, but is deterministic rather than stochastic. The second of these models removes the need for multiple ancillas to mediate each entangling gate by employing interactions that utilise the SWAP gate in a similar manner to the model known as ancilla-controlled quantum computation [30]. As in the ancilla-controlled model, only three ancilla-register interactions are required to implement a two-qubit entangling gate on the register, which is the minimum possible in any scheme that does not include measurements [33], and two for a single-qubit gate. Again, due to the global unitarity of the model the computation is deterministic and is based on a finite gate set composed of one two-qubit entangling gate and two single-qubit gates that form a universal set for single-qubit unitaries. We conjecture that these models require the minimal possible level of control for a unitary ancilla-mediated scheme.

Appendix

Here we prove that $v_0 = H$ and $v_1 = THT$ are a universal set for $SU(2)$. Using similar notation to Boykin *et al.* [35], we denote the nth roots of the X and Z operators by $X^{\frac{1}{n}}$ and $Z^{\frac{1}{n}}$. Any $u \in SU(2)$ can be written as

$$u = \exp(i\varphi\hat{n}\cdot\vec{\sigma}), \tag{18}$$

where $\vec{\sigma} = (X, Y, Z)$ is the vector of Pauli operators, $\hat{n} = (n_x, n_y, n_z)$ is some unit vector in $\mathbb{R}^3$, $\hat{n}\cdot\vec{\sigma} = n_xX + n_yY + n_zZ$ and $\varphi \in \mathbb{R}$ is some rotation angle. We have that

$$\exp(i\varphi\hat{n}\cdot\vec{\sigma}) = \cos\varphi\mathbb{I} + i\sin\varphi(\hat{n}\cdot\vec{\sigma}). \tag{19}$$

Up to irrelevant global phases, which we ignore from now on, $Z = \exp(i\frac{\pi}{2}Z)$ and $X = \exp(i\frac{\pi}{2}X)$ and hence $Z^{\frac{1}{n}} = \exp(i\frac{\pi}{2n}Z)$ and $X^{\frac{1}{n}} = \exp(i\frac{\pi}{2n}X)$. Using these, and the identity

$HZH = X$, we have that $X^{\frac{1}{n}} = HZ^{\frac{1}{n}}H$. It is straightforward to confirm that $T = Z^{-\frac{1}{4}}$ and so $v_+ := v_0 v_1 = X^{-\frac{1}{4}} Z^{-\frac{1}{4}}$ and $v_- := v_1 v_0 = Z^{-\frac{1}{4}} X^{-\frac{1}{4}}$. From a simple explicit calculation, we have that

$$v_\pm = \cos^2\frac{\pi}{8} - i\sin^2\frac{\pi}{8}\left(\cot\frac{\pi}{8}(Z+X) \mp Y\right). \tag{20}$$

Therefore, for both v_+ and v_- we have that $\cos\varphi = \cos^2\frac{\pi}{8}$ and hence φ is an irrational multiple of π [35] and $\hat{n}_\pm = n_\pm/\|n_\pm\|$ where $n_\pm = -(\cot\frac{\pi}{8}, \mp 1, \cot\frac{\pi}{8})$. As φ is an irrational multiple of π we can approximate to arbitrary accuracy any rotation around the $n_\pm$ axis by m applications of $v_\pm$, with m a finite integer. As these axes of rotation are not parallel then any arbitrary rotation can be decomposed into rotations around these axes [46]. This then proves that v_+ and v_- and hence v_0 and v_1 are a universal set for $SU(2)$.

Competing interests
The authors declare that they have no competing interests.

Authors' contributions
The main ideas in this paper were proposed and developed by TJP. All authors read and approved the final manuscript.

Author details
[1]School of Physics and Astronomy, University of Leeds, E C Stoner Building, Leeds, LS2 9JT, UK. [2]*Present address:* Department of Physics, Durham University, South Road, Durham, DH1 3LE, UK.

Acknowledgements
TJP was supported by a University of Leeds Research Scholarship.

Endnotes

[a] The two-qubit gate in this first minimal control model can be seen to employ essentially the same geometric methods as the qudit ancilla model we introduced in Ref. [22] and hence also the qubus model. In the qudit ancilla model the ancilla interacts with the register via displacements of the ancilla (with the displacements defined in the discrete lattice phase space of a qudit) controlled by a register qubit. In this model with a qudit of dimension 2, i.e. a qubit ancilla, controlled displacements are the $C_a^j X$ and $C_a^j Z$ gates. The two-qubit gates between register qubits j and k are then mediated by a sequence of the form $C_a^k X C_a^j Z C_a^k X C_a^j Z = C_k^j Z$. This requires two different interactions between the ancilla and register. The model presented here essential uses this gate method but removes the need for two different interactions by including a Hadamard gate on the ancilla in the interaction definition as $HZH = X$. It is the additional local gates u_l and u_r (needed to make the gate universal without additional single-qubit gates) that then results in the need for additional ancillas to mediate the two-qubit gates in the first minimal control model. Finally, this gate method can be considered to be geometric as *XZXZ* can be considered to create a closed loop in the discrete lattice phase space of a qubit - the details of this are in Ref. [22].

[b] Note that this is different to ADQC and ACQC in which the required rotations on the ancilla depend on the gate that is to be implemented (and previous measurement outcomes in the case of ADQC).

References

1. Feynman R: **Quantum mechanical computers.** *Opt. News* 1985, **11**(1):11-20.
2. Brylinski J-L, Brylinski R: *Universal Quantum Gates*, pp. 101-116. Boca Raton: Chapman & Hall/CRC Press; 2002.
3. Bremner MJ, Dawson CM, Dodd JL, Gilchrist A, Harrow AW, Mortimer D, Nielsen MA, Osborne TJ: **Practical scheme for quantum computation with any two-qubit entangling gate.** *Phys. Rev. Lett.* 2002, **89**(24):247902.
4. Lloyd S: **Almost any quantum logic gate is universal.** *Phys. Rev. Lett.* 1995, **75**(2):346.
5. Barenco A: **A universal two-bit gate for quantum computation.** *Proc. R. Soc. Lond. Ser. A, Math. Phys. Sci.* 1995, **449**(1937):679-683.
6. Cirac JI, Zoller P: **Quantum computations with cold trapped ions.** *Phys. Rev. Lett.* **74**, 4091 (1995)
7. Marcos D, Wubs M, Taylor JM, Aguado R, Lukin MD, Sørensen AS: **Coupling nitrogen-vacancy centers in diamond to superconducting flux qubits.** *Phys. Rev. Lett.* 2010, **105**(21):210501.
8. Lü X-Y, Xiang Z-L, Cui W, You JQ, Nori F: **Quantum memory using a hybrid circuit with flux qubits and nitrogen-vacancy centers.** *Phys. Rev. A* 2013, **88**(1):012329.
9. Qiu Y, Xiong W, Tian L, You JQ: **Coupling spin ensembles via superconducting flux qubits.** *Phys. Rev. A* 2014, **89**(4):042321.
10. Zhu X, Saito S, Kemp A, Kakuyanagi K, Karimoto S-i, Nakano H, Munro WJ, Tokura Y, Everitt MS, Nemoto K, et al: **Coherent coupling of a superconducting flux qubit to an electron spin ensemble in diamond.** *Nature* 2011, **478**(7368):221-224.
11. Wang Y-D, Kemp A, Semba K: **Coupling superconducting flux qubits at optimal point via dynamic decoupling with the quantum bus.** *Phys. Rev. B* 2009, **79**(2):024502.

12. Wallraff A, Schuster DI, Blais A, Frunzio L, Huang R-S, Majer J, Kumar S, Girvin SM, Schoelkopf RJ: **Strong coupling of a single photon to a superconducting qubit using circuit quantum electrodynamics.** *Nature* 2004, **431**(7005):162-167.
13. Carter SG, Sweeney TM, Kim M, Kim CS, Solenov D, Economou SE, Reinecke TL, Yang L, Bracker AS, Gammon D: **Quantum control of a spin qubit coupled to a photonic crystal cavity.** *Nat. Photonics* 2013, **7**(4):329-334.
14. Luxmoore I, Wasley N, Ramsay A, Thijssen A, Oulton R, Hugues M, Kasture S, Achanta V, Fox A, Skolnick M: **Interfacing spins in an InGaAs quantum dot to a semiconductor waveguide circuit using emitted photons.** *Phys. Rev. Lett.* 2013, **110**(3):037402.
15. Armour AD, Blencowe MP, Schwab KC: **Entanglement and decoherence of a micromechanical resonator via coupling to a cooper-pair box.** *Phys. Rev. Lett.* 2002, **88**(14):148301.
16. Milburn GJ: **Simulating nonlinear spin models in an ion trap.** arXiv:quant-ph/9908037; 1999.
17. Wang X, Zanardi P: **Simulation of many-body interactions by conditional geometric phases.** *Phys. Rev. A* 2002, **65**(3):032327.
18. Spiller TP, Nemoto K, Braunstein SL, Munro WJ, van Loock P, Milburn GJ: **Quantum computation by communication.** *New J. Phys.* 2006, **8**(2):30.
19. Brown KL, De S, Kendon VM, Munro WJ: **Ancilla-based quantum simulation.** *New J. Phys.* 2011, **13**(9):095007.
20. Louis SGR, Nemoto K, Munro WJ, Spiller TP: **The efficiencies of generating cluster states with weak nonlinearities.** *New J. Phys.* 2007, **9**(6):193.
21. Munro WJ, Nemoto K, Spiller TP, Barrett SD, Kok P, Beausoleil RG: **Efficient optical quantum information processing.** *J. Opt. B, Quantum Semiclass. Opt.* 2005, **7**(7):135.
22. Proctor TJ, Dooley S, Kendon V: **Quantum computation mediated by ancillary qudits and spin coherent states.** *Phys. Rev. A*, to appear. *arXiv preprint* arXiv:1402.6674v3; 2014.
23. Wootters WK: **A Wigner-function formulation of finite-state quantum mechanics.** *Ann. Phys.* 1987, **176**(1):1-21.
24. Vourdas A: **Quantum systems with finite Hilbert space.** *Rep. Prog. Phys.* 2004, **67**(3):267.
25. Unpublished work in progress.
26. Brown KL: **Using the qubus for quantum computing.** *PhD thesis*. University of Leeds; 2011
27. Kashefi E, Oi DKL, Browne D, Anders J, Andersson E: **Twisted graph states for ancilla-driven universal quantum computation.** *Electron. Notes Theor. Comput. Sci.* 2009, **249**:307-331
28. Anders J, Oi DKL, Kashefi E, Browne DE, Andersson E: **Ancilla-driven universal quantum computation.** *Phys. Rev. A* 2010, **82**(2):020301.
29. Halil Shah K, Oi DKL: **Ancilla driven quantum computation with arbitrary entangling strength.** In: *Theory of Quantum Computation, Communication, and Cryptography, 8th Conference, TQC 2013*, 2013. *LIPIcs-Leibniz International Proceedings in Informatics*, vol. 23.
30. Proctor TJ, Andersson E, Kendon V: **Universal quantum computation by the unitary control of ancilla qubits and using a fixed ancilla-register interaction.** *Phys. Rev. A* 2013, **88**(4):042330.
31. Halil-Shah K, Oi DKL: **A minimum control ancilla driven quantum computation scheme with repeat-until-success style gate generation.** *arXiv preprint* arXiv:1401.8004; 2014.
32. Paetznick A, Svore KM: **Repeat-until-success: non-deterministic decomposition of single-qubit unitaries.** *Quantum Inf. Comput.* 2014, **14**(15-16):1277-1301.
33. Lamata L, León J, Pérez-García D, Salgado D, Solano E: **Sequential implementation of global quantum operations.** *Phys. Rev. Lett.* 2008, **101**(18):180506.
34. Makhlin Y: **Nonlocal properties of two-qubit gates and mixed states, and the optimization of quantum computations.** *Quantum Inf. Process.* 2002, **1**(4):243-252.
35. Boykin PO, Mor T, Pulver M, Roychowdhury V, Vatan F: **A new universal and fault-tolerant quantum basis.** *Inf. Process. Lett.* 2000, **75**(3):101-107.
36. Kok P, Munro WJ, Nemoto K, Ralph TC, Dowling JP, Milburn GJ: **Linear optical quantum computing with photonic qubits.** *Rev. Mod. Phys.* 2007, **79**(1):135.
37. Reiserer A, Kalb N, Rempe G, Ritter S: **A quantum gate between a flying optical photon and a single trapped atom.** *Nature* 2014, **508**(7495):237-240.
38. Tiecke T, Thompson J, de Leon N, Liu L, Vuletić V, Lukin M: **Nanophotonic quantum phase switch with a single atom.** *Nature* 2014, **508**(7495):241-244.
39. Doherty AC, Wardrop MP: **Two-qubit gates for resonant exchange qubits.** *Phys. Rev. Lett.* 2013, **111**(5):050503.
40. Taminiau TH, Cramer J, van der Sar T, Dobrovitski VV, Hanson R: **Universal control and error correction in multi-qubit spin registers in diamond.** *Nat. Nanotechnol.* 2014, **9**:171-176.
41. Robledo L, Childress L, Bernien H, Hensen B, Alkemade PFA, Hanson R: **High-fidelity projective read-out of a solid-state spin quantum register.** *Nature* 2011, **477**(7366):574-578.
42. Waldherr G, Wang Y, Zaiser S, Jamali M, Schulte-Herbrüggen T, Abe H, Ohshima T, Isoya J, Du JF, Neumann P, Wrachtrup J: **Quantum error correction in a solid-state hybrid spin register.** *Nature* 2014, **506**(7487):204-207.
43. Borhani M, Loss D: **Cluster states from Heisenberg interactions.** *Phys. Rev. A* 2005, **71**(3):034308.
44. Niskanen AO, Harrabi K, Yoshihara F, Nakamura Y, Lloyd S, Tsai JS: **Quantum coherent tunable coupling of superconducting qubits.** *Science* 2007, **316**(5825):723-726.
45. Stanwix PL, Pham LM, Maze JR, Le Sage D, Yeung TK, Cappellaro P, Hemmer PR, Yacoby A, Lukin MD, Walsworth RL: **Coherence of nitrogen-vacancy electronic spin ensembles in diamond.** *Phys. Rev. B* 2010, **82**(20):201201.
46. Mladenova CD, Mladenov IM: **Vector decomposition of finite rotations.** *Rep. Math. Phys.* 2011, **68**(1):107-117.

12

A Trotter-Kato theorem for quantum Markov limits

Luc Bouten[1], Rolf Gohm[2*], John Gough[2] and Hendra Nurdin[3]

*Correspondence: rog@aber.ac.uk
[2]Dept. for Mathematics and Physics, Aberystwyth University, SY23 3BZ, Wales, United Kingdom
Full list of author information is available at the end of the article

Abstract

Using the Trotter-Kato theorem we prove the convergence of the unitary dynamics generated by an increasingly singular Hamiltonian in the case of a single field coupling. The limit dynamics is a quantum stochastic evolution of Hudson-Parthasarathy type, and we establish in the process a graph limit convergence of the pre-limit Hamiltonian operators to the Chebotarev-Gregoratti-von Waldenfels Hamiltonian generating the quantum Itō evolution.

1 Introduction

In the situation of regular perturbation theory, we typically have a Hamiltonian interaction of the form $H = H_0 + H_{\text{int}}$ with associated strongly continuous one-parameter unitary groups $U_0(t) = e^{-itH_0}$ (the free evolution) and $U(t) = e^{-itH}$ (the perturbed evolution), then we transform to the Dirac interaction picture by means of the unitary family $V(t) = U_0(-t)U(t)$. Although $V(\cdot)$ is strongly continuous, it does not form a one-parameter group but instead yields what is known as a left U_0-cocycle:

$$V(t+s) = U_0(s)^\dagger V(t) U_0(s) V(s). \tag{1}$$

One obtains the interaction picture dynamical equation

$$i\frac{d}{dt}V(t) = \Upsilon(t)V(t), \tag{2}$$

where $\Upsilon(t) = U_0(t)^\dagger H_{\text{int}} U_0(t)$.

More generally, we may have a pair of unitary groups $U(\cdot)$ and $U_0(\cdot)$ with Stone generators H and H_0 respectively, but where the intersection of the domains of the generators are not dense. This is the situation of a singular perturbation. In this case we cannot expect the Dirac picture dynamical equation (2) to be anything but formal since the difference $H_{\text{int}} = H - H_0$ is not densely defined.

Remarkably, the steps above can be reversed even for the situation of singular perturbations. If we assume at the outset a fixed free dynamics $U_0(\cdot)$, with Stone generator H_0, and a strongly continuous unitary left U_0-cocycle $V(\cdot)$, then $U(t) = U_0(t)V(t)$ will then form a strongly continuous one-parameter unitary group with Stone generator H. In practice however the problem of reconstructing H from the prescribed H_0 and $V(\cdot)$ will be difficult.

In the situation of quantum stochastic evolutions introduced by Hudson and Parthasarathy [1], we have a strongly continuous adapted process $V(\cdot)$ satisfying a quantum stochastic differential equation (including Wiener and Poisson noise as special commutative cases) in place of (2), and the solution constitutes a cocycle with respect to the time-shift maps $U_0 \equiv \Theta$ (see below). Nevertheless, $V(\cdot)$ arises as the Dirac picture evolution for a singular perturbation of a unitary $U(\cdot)$ with some generator H with respect to the time-shift: it was a long standing problem to find an explicit form for H which was finally resolved by Gregoratti [2], see also [3].

The purpose of this paper is to approximate the singular perturbation arising in quantum stochastic evolution models by a sequence of regular perturbation models. That is, to construct a sequence of Hamiltonians $H^{(k)} = H_0 + H^{(k)}_{\text{int}}$ yielding a regular perturbation $V^{(k)}(\cdot)$ converging to a singular perturbation $V(\cdot)$ in some controlled way. We exploit the fact that the limit Hamiltonian is now known through the work of Chebotarev [4] and Gregoratti [2]. The strategy is to employ the Trotter-Kato theorem which guarantees strong uniform convergence of the unitaries once graph convergence of the Hamiltonians is established.

1.1 Quantum stochastic evolutions

The seminal work of Hudson and Parthasarathy [1] on quantum stochastic evolutions lead to explicit constructions of unitary adapted quantum stochastic processes V describing the open dynamical evolution of a system with a singular Boson field environment. We fix the system Hilbert space $\mathfrak{h}$ and model the environment as having n channels so that the underlying Fock space is $\mathfrak{F} = \Gamma(\mathbb{C}^n \otimes L^2(\mathbb{R}))$. Here $\Gamma(\mathfrak{H})$ denotes the symmetric (boson) Fock space over a one-particle space $\mathfrak{H}$: we set the inner product as $\langle \Psi | \Phi \rangle = \sum_{m=0}^{\infty} \frac{1}{m!} \langle \Psi_m | \Phi_m \rangle$ and take the exponential vectors to be defined as ($\otimes_s$ denoting a symmetric tensor product)

$$e(f) = (1, f, f \otimes_s f, f \otimes_s f \otimes_s f, \ldots)$$

with test function $f \in \mathfrak{H}$. Here the one particle space is $L^2(\mathbb{R})$, the space of complex-valued square-integrable functions on $\mathbb{R}$. We define the operators

$$\begin{aligned}
\Lambda^{00}(t) &\triangleq t, \\
\Lambda^{10}(t) = A^\dagger(t) &\triangleq a^\dagger(1_{[0,t]}), \\
\Lambda^{01}(t) = A(t) &\triangleq a(1_{[0,t]}), \\
\Lambda^{11}(t) = \Lambda(t) &\triangleq d\Gamma(\chi_{[0,t]}),
\end{aligned}$$

where $1_{[0,t]}$ is the characteristic function of the interval $[0,t]$ and $\chi_{[0,t]}$ is the operator on $L^2(\mathbb{R})$ corresponding to multiplication by $1_{[0,t]}$. Hudson and Parthasarathy [1] have developed a quantum Itō calculus where the basic objects are integrals of adapted processes with respect to the fundamental processes $\Lambda^{\alpha\beta}$. The quantum Itō table is then

$$d\Lambda^{\alpha\beta}(t)\, d\Lambda^{\mu\nu}(t) = \hat{\delta}_{\beta\mu}\, d\Lambda^{\alpha\nu}(t),$$

where $\hat{\delta}_{\alpha\beta}$ is the Evans-Hudson delta defined to equal unity if $\alpha = \beta = 1$ and zero otherwise. This may be written as

$\times$	dA	$d\Lambda$	$dA^\dagger$	dt
dA	0	dA	dt	0
$d\Lambda$	0	$d\Lambda$	dA	0
$dA^\dagger$	0	0	0	0
dt	0	0	0	0

.

In particular, we have the following theorem [1].

Theorem 1 *There exists a unique solution $V(\cdot,\cdot)$ to the quantum stochastic differential equation*

$$V(t,s) = I + \int_s^t dG(\tau)V(\tau,s) \tag{3}$$

$(t \geq s \geq 0)$ *where*

$$dG(t) = G_{\alpha\beta} \otimes d\Lambda^{\alpha\beta}(t)$$

with $G_{\alpha\beta} \in \mathfrak{B}(\mathfrak{h})$. (*We adopt the convention that we sum repeated Greek indices over the range* 0, 1.)

In particular, set $V(t) = V(t,0)$ then we have the quantum stochastic differential equation $dV(t) = dG(t)V(t)$ which replaces the regular Dirac picture dynamical equation (2).

We refer to $\mathbf{G} = [G_{\alpha\beta}] \in \mathfrak{B}(\mathfrak{h} \oplus \mathfrak{h})$, as the *coefficient matrix*, and V as the left process generated by $\mathbf{G}$. The conditions for the process V to be unitary are that $\mathbf{G}$ takes the form, with respect to the decomposition $\mathfrak{h} \oplus \mathfrak{h}$,

$$\mathbf{G} = \begin{bmatrix} -\frac{1}{2}\mathsf{L}^\dagger\mathsf{L} - i\mathsf{H} & -\mathsf{L}^\dagger\mathsf{S} \\ \mathsf{L} & \mathsf{S} - I \end{bmatrix}, \tag{4}$$

where $\mathsf{S} \in \mathfrak{B}(\mathfrak{h})$ is a unitary, $\mathsf{L} \in \mathfrak{B}(\mathfrak{h})$ and $\mathsf{H} \in \mathfrak{B}(\mathfrak{h})$ is self-adjoint. We may write in more familiar notation [1]

$$dG(t) = \left(-\frac{1}{2}\mathsf{L}^\dagger\mathsf{L} - i\mathsf{H}\right) \otimes dt - \mathsf{L}^\dagger\mathsf{S} \otimes dA(t) + \mathsf{L} \otimes dA^\dagger(t) + (\mathsf{S} - I) \otimes d\Lambda(t).$$

We denote the shift map on $L^2(\mathbb{R})$ by θ_t, that is $(\theta_t)f(\cdot) = f(\cdot + t)$ and its second quantization as $\Theta_t = I \otimes \Gamma(\theta_t)$. It then turns out that $\Theta_\tau^\dagger V(t,s)\Theta_\tau = V(t+\tau, s+\tau)$ and so $V(t) = V(0,t)$ is a left unitary Θ-cocycle and that there must exist a self-adjoint operator H such that

$$\Theta_t V(t) \equiv e^{-iHt}$$

for $t \geq 0$. (For $t < 0$ one has $V(-t)^\dagger \Theta_{-t} \equiv e^{-iHt}$.) Here H will be a singular perturbation of generator of the shift, and its characterization was given by Gregoratti [2]. See also [5].

1.2 Physical motivation

As a precursor to and motivation for further approximations, we fix on a simple model of a quantum mechanical system coupled to a boson field reservoir R. In the Markov approximation we assume that the auto-correlation time of the field processes vanishes in the limit: this includes weak coupling (van Hove) and low density limits. The Hilbert space for the field is the Fock space $\mathcal{F}_R = \Gamma(\mathcal{H}_R^1)$ with one-particle space $\mathcal{H}_R^1 = L^2(\mathbb{R})$ taken as the momentum space. (For convenience we consider a one-dimensional situation because this is the setting studied in this paper but of course $\mathbb{R}^3$ is particularly relevant physically.) It is convenient to write annihilation operators formally as $A_R(g) = \int_{\mathbb{R}} g(p)^* a_p \, dp$ where $[a_p, a_{p'}^\dagger] = \delta(p - p')$.

In particular, let us fix a function $g \in L^2(\mathbb{R})$, and set

$$a(t,k) = \sqrt{k} \int e^{-i\omega(p)tk} g(p)^* a_p \, dp,$$

where $\omega = \omega(p)$ is a given function (determining the dispersion relation for the free quanta) and k is a dimensionless parameter rescaling time. We have the commutation relations

$$\left[a(s,k), a(t,k)^\dagger\right] = k\rho\big(k(t-s)\big),$$

where

$$\rho(\tau) \equiv \int |g(p)|^2 e^{i\omega(p)\tau} \, dp.$$

The limit $k \to \infty$ leads to singular commutation relations, and it is convenient to introduce smeared fields

$$A(\varphi, k) = \int \varphi(t)^* a(t,k) \, dt$$

in which case we have the two-point function (and define an operator C_k by)

$$\left[A(\varphi,k), A(\psi,k)^\dagger\right] = \int dt \, dt' \varphi(t)^* k\rho\big(k(t-t')\big)\psi(t') \equiv \langle \varphi | C_k \psi \rangle.$$

For ρ integrable, we expect

$$\lim_{k\to\infty} \left[A(\varphi,k), A(\psi,k)^\dagger\right] = \gamma \int dt \varphi(t)^* \psi(t),$$

where $\gamma = \int_{-\infty}^{\infty} \rho(\tau) \, d\tau = 2\pi \int |g|^2(p)\delta(\omega(p)) \, dp \geq 0$. When $\gamma = 1$, the $A(\varphi,k)$ are smeared versions of the annihilators on $\Gamma(L^2(\mathbb{R}))$.

The limit $k \uparrow \infty$ corresponds to the smeared field becoming singular and this leads to a quantum Markovian approximation. The formulation of such models was first given and treated in a systematic way by Accardi, Frigero and Lu who developed a set of powerful quantum functional central limit theorems including the weak coupling [6] and low density [7] regimes. Theorem 2 is an extension of these which includes both quantum diffusion and jump terms [8, 9].

Theorem 2 *Let* $(\mathsf{E}_{\alpha\beta})$ *be bounded operators on a fixed separable Hilbert space* $\mathfrak{h}$ *labeled by* $\alpha, \beta \in \{0,1\}$ *with* $\mathsf{E}_{\alpha\beta}^{\dagger} = \mathsf{E}_{\beta\alpha}$ *and* $\|E_{11}\| < 2$. *Let*

$$\Upsilon(t,k) = \mathsf{E}_{11} \otimes a(t,k)^{\dagger} a(t,k) + \mathsf{E}_{10} \otimes a(t,k)^{\dagger} + \mathsf{E}_{01} \otimes a(t,k) + \mathsf{E}_{00} \otimes I$$

and

$$e(\varphi,k) = \exp\{A(\varphi,k) - A(\varphi,k)^{\dagger}\}\Omega_R$$

with Ω_R *the Fock vacuum of* $\mathcal{F}_R$. *The solution* $V(t,k)$ *to the equation*

$$\frac{d}{dt}V(t,k) = -i\Upsilon(t,k)V(t,k), \qquad V(0,k) = I,$$

exists and we have the limit

$$\lim_{k\to\infty} \langle u_1 \otimes e(\varphi,k) | V(t,k) | u_2 \otimes e(\psi,k) \rangle = \langle u_1 \otimes e(\varphi) | V(t) | u_2 \otimes e(\psi) \rangle$$

for all $u_1, u_2 \in \mathfrak{h}$ *and* $\varphi, \psi \in L^2(\mathbb{R})$, *where* V *is a unitary adapted process on* $\mathfrak{h} \otimes \Gamma(\mathbb{C}^n \otimes L^2(\mathbb{R}))$ *with coefficient matrix* $\mathbf{G}$ *given by*

$$\mathbf{G} = -i\mathbf{E} - i\frac{1}{2}\mathbf{G}\begin{bmatrix} 0 & 0 \\ 0 & 1 \end{bmatrix}\mathbf{E}, \tag{5}$$

where we assume $\int_{-\infty}^{0} \rho(\tau)\, d\tau = \frac{1}{2}$.

The proof of the theorem is given in [8] and requires a development and a uniform estimation of the Dyson series expansion. Summability of the series requires that $\|E_{11}\| < 2$.

The triple $(\mathsf{S}, \mathsf{L}, \mathsf{H})$ from (4) obtained through (5) is

$$\begin{aligned} \mathsf{S} &= \frac{I - \frac{i}{2}\mathsf{E}_{11}}{I + \frac{i}{2}\mathsf{E}_{11}}, \qquad \mathsf{L} = -\frac{i}{I + \frac{i}{2}\mathsf{E}_{11}}\mathsf{E}_{10}, \\ \mathsf{H} &= \mathsf{E}_{00} + \mathsf{E}_{01}\,\mathrm{Im}\left\{\frac{1}{I + \frac{i}{2}\mathsf{E}_{11}}\right\}\mathsf{E}_{10}. \end{aligned} \tag{6}$$

Our objective is reappraise Theorem 2, where we will prove a related result by an alternative technique. Using the Trotter-Kato theorem, we will establish a stronger mode of convergence (uniformly on compact intervals of time and strongly in the Hilbert space) by means of a graph convergence of the Hamiltonians. The new approach has the advantage of been simpler and is likely to be more readily extended to other cases, for instance a continuum of input channels as originally treated in [1], which cannot be treated by the perturbative techniques used in the proof of Theorem 2.

2 Trotter-Kato theorems for quantum stochastic limits

Our main results will employ the Trotter-Kato theorem, which we recall next in a particularly convenient form. See [10], Theorem 3.17, or [11], Chapter VIII.7.

Theorem 3 (Trotter-Kato) *Let* $\mathcal{H}$ *be a Hilbert space and let* $U^{(k)}(\cdot)$ *and* $U(\cdot)$ *be strongly continuous one-parameter groups of unitaries on* $\mathcal{H}$ *with Stone generators* $H^{(k)}$ *and* H, *respectively. Let* $\mathcal{D}$ *be a core for* H. *The following are equivalent*

1. *For all $f \in \mathcal{D}$ there exist $f^{(k)} \in \mathrm{Dom}(H^{(k)})$ such that*

$$\lim_{k\to\infty} f^{(k)} = f, \qquad \lim_{k\to\infty} H^{(k)} f^{(k)} = Hf.$$

2. *For all $0 \le T < \infty$ and all $f \in \mathcal{H}$ we have*

$$\lim_{k\to\infty} \sup_{0\le t\le T} \left\| \left(\mathcal{U}^{(k)}(t) - \mathcal{U}(t)\right) f \right\| = 0.$$

The theorem yields a strong uniform convergence if we can establish graph convergence of the Hamiltonians. We now present the Trotter-Kato theorems for the class of problems that interest us, treating the first and second quantized problems in sequence.

2.1 First quantization example

Definition 4 Let $g \in C_c^\infty(\mathbb{R})$, i.e., an infinitely differentiable function with compact support, such that $\int_{-\infty}^{\infty} g(s)\,ds = 1$. We define $\rho(t) = \int_{\mathbb{R}} g(s)^* g(s+t)\,ds$. Moreover, for all $k > 0$, we define functions $g^{(k)}$ and $\rho^{(k)}$ by

$$g^{(k)}(t) = kg(kt), \qquad \rho^{(k)}(t) = k\rho(kt), \quad t \in \mathbb{R}.$$

Furthermore, we define two complex numbers by $\kappa_+ := \int_0^\infty \rho(s)\,ds$ and $\kappa_- := \int_{-\infty}^0 \rho(s)\,ds$.

Note that $\kappa_+ + \kappa_- = 1$ and that κ_+ and κ_- are complex conjugate: $\kappa_+ = (\kappa_-)^*$ (substitute $-s$ for s), hence $\kappa_\pm = \frac{1}{2} \pm i\sigma$ with σ real. The choice of ρ is such that $\langle g | g * f\rangle = \langle \rho | f\rangle$, where $(g * f)(t) = \int_{-\infty}^{\infty} g(s) f(t-s)\,ds$ is the usual convolution.

Let $\mathfrak{h}$ be a Hilbert space and let E be a bounded self-adjoint operator on $\mathfrak{h}$. We consider the following family of operators on $L^2(\mathbb{R};\mathfrak{h}) \simeq \mathfrak{h} \otimes L^2(\mathbb{R})$:

$$\begin{aligned} H^{(k)} &= i\partial + \mathsf{E}|g^{(k)}\rangle\langle g^{(k)}| \simeq I \otimes i\partial + E \otimes |g^{(k)}\rangle\langle g^{(k)}|, \\ \mathrm{Dom}\left(H^{(k)}\right) &= W^{1,2}(\mathbb{R};\mathfrak{h}), \end{aligned} \tag{7}$$

where $W^{1,2}(X;\mathfrak{h})$, $X \subseteq \mathbb{R}$, denotes the Sobolev space of $\mathfrak{h}$-valued functions square integrable on X with square integrable weak derivatives on X. It follows easily that $H^{(k)}$ is self-adjoint for every $k > 0$ (for example by the Kato-Rellich theorem, see [12], Theorem X.12). We define a unitary operator on $\mathfrak{h}$ by

$$\mathsf{S} = \frac{I - i\kappa_- \mathsf{E}}{I + i\kappa_+ \mathsf{E}} \tag{8}$$

and an operator H on $L^2(\mathbb{R};\mathfrak{h})$ by

$$\begin{aligned} \mathrm{Dom}(H) &= \left\{ f \in W^{1,2}(\mathbb{R}\backslash\{0\};\mathfrak{h}) : f(0^-) = \mathsf{S} f(0^+) \right\}, \\ Hf &= i\partial f. \end{aligned} \tag{9}$$

It follows easily that H is self-adjoint, compare [11], VIII.2, final example.

Remark Any $f \in W^{1,2}(\mathbb{R}\backslash\{0\};\mathfrak{h})$ is absolutely continuous both on $(-\infty, 0)$ and $(0, \infty)$, see for example [13], 2.6 Ex. 6, but the exclusion of test functions supported at 0 allows jumps at 0. Higher dimensional situations ($\mathbb{R}^n$ with $n > 1$) are more complicated in this respect.

We define strongly continuous one-parameter groups of unitaries on $L^2(\mathbb{R};\mathfrak{h})$ by

$$U^{(k)}(t) = \exp\big(-itH^{(k)}\big), \qquad U(t) = \exp(-itH).$$

We then have the following theorem.

Theorem 5 *Let* $0 \leq T < \infty$. *Then*

$$\lim_{k\to\infty} \sup_{0\leq t\leq T} \big\| \big(U^{(k)}(t) - U(t)\big)f \big\| = 0, \quad \forall f \in L^2(\mathbb{R};\mathfrak{h}).$$

We prove Theorem 5 at the end of this subsection. From the Trotter-Kato Theorem 3, it suffices to find, for every $f \in \mathrm{Dom}(H)$, a sequence $f^{(k)} \in \mathrm{Dom}(H^{(k)})$ that satisfies condition (i) of Theorem 3.

If g is a $\mathbb{C}$-valued function on X and $f \in L^2(X;\mathfrak{h}) \simeq \mathfrak{h} \otimes L^2(X;\mathbb{C})$ then we use the short notation gf for $(I \otimes M_g)f$ where M_g is multiplication by g. With this convention we can also define $g * f \in L^2(X;\mathfrak{h})$ and $\langle g|f\rangle \in \mathfrak{h}$ for suitable functions g, using the same formulas as for $\mathfrak{h} = \mathbb{C}$.

Definition 6 Let f be an element in the domain of H. Define an element $f^{(k)}$ in the domain of $H^{(k)}$ by

$$f^{(k)}(t) = \big(g^{(k)} * f\big)(t) = \int_{-\infty}^{\infty} g^{(k)}(t-s) f(s)\, ds.$$

Lemma 7 *Let* η *be an element of* $C(0,\infty)$ *with compact support and let* h *be an element of* $W^{1,2}((0,\infty);\mathfrak{h}) \cap C^1((0,\infty);\mathfrak{h})$ *such that* $h(0^+) = 0$. *Let* $\eta^{(k)}(x) = k\eta(kx)$ *for all* $x \in (0,\infty)$ *and* $k > 0$. *Then*

$$\big\|\langle \eta^{(k)}|h\rangle\big\|_2 \leq \frac{C}{k}, \quad \forall k > 0,$$

for some positive constant C.

Proof Note that the C^1-function h is Lipschitz on the support of η, that is, there exists a positive constant L such that

$$\big\|h(x) - h(y)\big\|_2 \leq L|x-y|, \quad \forall x, y \in \mathrm{supp}(\eta),$$

where $\mathrm{supp}(\eta)$ denotes the support of η. Taking the limit for y to 0^+ gives

$$\big\|h(x)\big\|_2 \leq L|x|, \quad x \in \mathrm{supp}(\eta).$$

We can define $M := \max_{x\in(0,\infty)} |\eta(x)|$ and let N be a number to the right of the support of η. Now we have

$$\begin{aligned} \big\|\langle \eta^{(k)}|h\rangle\big\|_2 &\leq k\int_0^{\infty} |\eta(kx)|\,\big\|h(x)\big\|_2\, dx \\ &\leq \frac{L}{k}\int_0^{\infty} |\eta(u)|\,u\, du \leq \frac{L}{k}\int_0^{N} Mu\, du = \frac{LMN^2}{2k}. \end{aligned}$$

□

Lemma 8 *If f is in* $\mathrm{Dom}(H) \cap C^\infty(\mathbb{R}\backslash\{0\};\mathfrak{h})$, *and $f^{(k)}$ is given by Definition 6, then we have*

$$1. \quad \lim_{k\to\infty} \|f^{(k)} - f\|_2 = 0, \qquad 2. \quad \lim_{k\to\infty} \|H^{(k)}f^{(k)} - Hf\|_2 = 0.$$

Proof Note that the first limit follows immediately from a standard result on approximations by convolutions, see e.g. [14], Thm. 2.16. For the second limit, note that

$$\partial\left(g^{(k)} * f\right) = g^{(k)} * \partial f + \left(f(0^+) - f(0^-)\right)g^{(k)}, \tag{10}$$

because $\partial f = Hf$ and using [14], Thm. 2.16, once more, we find that

$$\lim_{k\to\infty} g^{(k)} * Hf = Hf.$$

That is, all we need to show is that

$$\lim_{k\to\infty} \left\| \left(if(0^+) - if(0^-) + \mathsf{E}\langle g^{(k)} | g^{(k)} * f\rangle\right) g^{(k)} \right\|_2 = 0. \tag{11}$$

Note that $\langle g^{(k)} | g^{(k)} * f\rangle = \langle \rho^{(k)} | f\rangle$. We can now apply Lemma 7 with $h = f\chi_{(0,\infty)} - f(0^+)$ and $\eta = \rho\chi_{(0,\infty)}$ (resp. $h = f\chi_{(-\infty,0)} - f(0^-)$ and $\eta = \rho\chi_{(-\infty,0)}$) to conclude that

$$\left\langle \rho^{(k)} | f\right\rangle \overset{k\to\infty}{\longrightarrow} (\kappa_-)^* f(0^-) + (\kappa_+)^* f(0^+) = \kappa_+ f(0^-) + \kappa_- f(0^+),$$

with rate $\frac{1}{k}$. Using the boundary condition for f, we therefore find that

$$if(0^+) - if(0^-) + \mathsf{E}\langle g^{(k)} | g^{(k)} * f\rangle \longrightarrow i\left[(I - i\kappa_-\mathsf{E})f(0^+) - (I + i\kappa_+\mathsf{E})f(0^-)\right] = 0,$$

with rate $\frac{1}{k}$. Note that the L^2-norm of $g^{(k)}$ grows with rate $\sqrt{k}$, so that the limit in equation (11) follows. This completes the proof of the lemma. □

Proof of Theorem 5 The theorem follows from a combination of the results in Theorem 3 and Lemma 8 and the fact that $\mathrm{Dom}(H) \cap C^\infty(\mathbb{R}\backslash\{0\};\mathfrak{h})$ is a core for H. The latter follows from [14], Thm. 7.6. □

3 A second quantized model

Let $\mathsf{E}_{\alpha\beta}$ be bounded operators on $\mathfrak{h}$ such that $\mathsf{E}^\dagger_{\alpha\beta} = \mathsf{E}_{\beta\alpha}$ for $\alpha, \beta \in \{0,1\}$. Consider the following family of operators on $\mathfrak{h} \otimes \mathcal{F}$

$$H^{(k)} = i\,d\Gamma(\partial) + \mathsf{E}_{11}A^\dagger\left(g^{(k)}\right)A\left(g^{(k)}\right) + \mathsf{E}_{10}A^\dagger\left(g^{(k)}\right) + \mathsf{E}_{01}A\left(g^{(k)}\right) + \mathsf{E}_{00}, \tag{12}$$

choosing a suitable domain $\mathrm{Dom}(H^{(k)})$ of essential self-adjointness for all $k > 0$. (We conjecture that $\mathfrak{h} \otimes \mathcal{E}(C_c^\infty(\mathbb{R}))$, where $\mathcal{E}(C_c^\infty(\mathbb{R}))$ is the set of exponential vectors $e(f)$ with $f \in C_c^\infty(\mathbb{R})$, is a set of analytic vectors for the $H^{(k)}$ but we haven't been able to prove this rigorously and leave it as an open problem.)

We denote the strongly continuous group of unitaries on $\mathfrak{h} \otimes \mathcal{F}$ generated by the unique self-adjoint extension of $H^{(k)}$ by $U^{(k)}(t)$. Let the triple $(\mathsf{S}, \mathsf{L}, \mathsf{H})$ appearing in (4) be obtained from $\mathbf{E} = (\mathsf{E}_{\alpha\beta})$ through (5): see (6).

The space $\mathfrak{h} \otimes \mathcal{F} = \mathfrak{h} \otimes \Gamma(L^2(\mathbb{R}))$ consists of vectors $\Psi = (\Psi_m)_{m \geq 0}$ which are sequences of symmetric $\mathfrak{h}$-valued functions $\Psi_m(t_1, \ldots, t_m)$ where $t_j \in \mathbb{R}$. Following Gregoratti [2], we define the following spaces (for I a Borel subset of $\mathbb{R}$ and $\mathfrak{H}$ a Hilbert space):

$$\mathcal{H}^\Sigma(I^m, \mathfrak{H}) = \left\{ v \in L^2(I^m, \mathfrak{H}) : \sum_{i=1}^m \partial_i v \in L^2(I^m, \mathfrak{H}) \right\};$$

$$\mathcal{W} = \left\{ \Psi \in \mathfrak{h} \otimes \mathcal{F} : \Psi_m \in \mathcal{H}^\Sigma(\mathbb{R}^m, \mathfrak{h}) : \sum_{m=0}^\infty \frac{1}{m!} \left\| \sum_{i=1}^m \partial_i \Psi_m \right\|^2 < \infty \right\};$$

$$\mathcal{V}_s = \left\{ \Psi \in \mathcal{W} : \sum_{m=0}^\infty \frac{1}{m!} \left\| \Psi_{m+1}(\cdot, t_{m+1} = s) \right\|^2 < \infty \right\};$$

$$\mathcal{V}_{0^\pm} = \mathcal{V}_{0^+} \cap \mathcal{V}_{0^-}.$$

We remark that $\mathcal{W}$ is the natural domain for $d\Gamma(i\partial)$. On $\mathcal{V}_s$ we define the operators

$$\left(a(s)\Psi\right) = \Psi_{n+1}(\cdot, t_{n+1} = s).$$

On the subspace $\mathcal{V}_{0^\pm}$, the operators $d\Gamma(i\partial)$ and $a(0^\pm)$ are all simultaneously defined.

Definition 9 (The Gregoratti Hamiltonian) Define the following operator H on $\mathfrak{h} \otimes \mathcal{F}$

$$H\Phi = d\Gamma(i\partial_{ac})\Phi - i\mathsf{L}^\dagger \mathsf{S} a(0^+)\Phi + \left(\mathsf{H} - \frac{i}{2}\mathsf{L}^\dagger \mathsf{L}\right)\Phi, \tag{13}$$

$$\mathrm{Dom}(H) = \left\{ \Phi \in \mathcal{V}_{0^\pm} : a(0^-)\Phi = \mathsf{S} a(0^+)\Phi + \mathsf{L}\Phi \right\}. \tag{14}$$

It follows from the work of Chebotarev and Gregoratti [2, 4] that the operator H is essentially self-adjoint and its unique self-adjoint extension generates the unitary group $U(t) = \Theta_t V_t$ where V_t is the unitary solution to the following quantum stochastic differential equation (3):

$$\begin{aligned} dV(t) &= \left\{ (\mathsf{S} - 1)\, d\Lambda(t) + \mathsf{L}\, dA^\dagger(t) - \mathsf{L}^\dagger \mathsf{S}\, dA(t) - \frac{1}{2}\mathsf{L}^\dagger \mathsf{L}\, dt - i\mathsf{H}\, dt \right\} V(t), \\ V(0) &= I. \end{aligned} \tag{15}$$

The main result of this section is the following theorem.

Theorem 10 *Let $0 \leq T < \infty$. We have the following*

$$\lim_{k \to \infty} \sup_{0 \leq t \leq T} \left\| (U^{(k)}(t) - U(t))\Phi \right\| = 0, \quad \forall \Phi \in \mathfrak{h} \otimes \mathcal{F}.$$

Before proving the theorem (see the end of this section), we make some preparations. As in the previous section, we would like to use the Trotter-Kato theorem, therefore, for every Φ in a core for $\mathrm{Dom}(H)$, we need to construct an approximating sequence $\Phi^{(k)}$ that satisfies the first condition of Theorem 3. We again employ a smearing through convolution with $g^{(k)}$, this time applied as a second quantization.

Definition 11 Let $g^{(k)}$ be as in Definition 4 and assume further that $g(t) \geq 0$ for all t (hence $\|g\|_1 = 1$). Let $G^{(k)} : L^2(\mathbb{R}) \to L^2(\mathbb{R})$ be the convolution with $g^{(k)}$, i.e.

$$G^{(k)}h = g^{(k)} * h, \quad \forall h \in L^2(\mathbb{R}).$$

Let Φ be an element in $\mathrm{Dom}(H)$. We define an element $\Phi^{(k)}$ in the domain of $H^{(k)}$ by

$$\Phi^{(k)} = \Gamma\left(G^{(k)}\right)\Phi. \tag{16}$$

Here $\Gamma(G^{(k)})$ denotes the second quantization of $G^{(k)}$.

Note that $G^{(k)}$ is a contraction ($\|g^{(k)}\|_1 = 1$, i.e. $\|\hat{g}^{(k)}\|_\infty \leq 1$ with $\hat{g}^{(k)}$ the Fourier transform $\hat{g}^{(k)} = \int_{-\infty}^{\infty} g^{(k)}(t)e^{-i\omega t}\, dt$), so its second quantization is well-defined). The positivity assumption on g implies that $\kappa_+ = \kappa_- = \frac{1}{2}$ (which agrees with Section 1.2).

Lemma 12 *For all $\Phi \in \mathfrak{h} \otimes \mathcal{F}$, we have*

$$\lim_{k\to\infty} \Gamma\left(G^{(k)}\right)\Phi = \Phi.$$

Proof Since the linear span of exponential vectors $v \otimes e(h)$ is dense in $\mathfrak{h} \otimes \mathcal{F}$ and $\Gamma(G^{(k)})$ is bounded, it is enough to prove the lemma for all vectors of the form $\Phi = v \otimes e(h)$. We have

$$\begin{aligned}
&\left\|\Gamma\left(G^{(k)}\right)v \otimes e(h) - v \otimes e(h)\right\|^2 \\
&\quad = \|v\|^2\left[\exp\left(\left\|G^{(k)}h\right\|^2\right) + \exp\left(\|h\|^2\right) - \exp\left(\left\langle G^{(k)}h|h\right\rangle\right) - \exp\left(\left\langle h|G^{(k)}h\right\rangle\right)\right] \to 0,
\end{aligned}$$

where in the last step we used [14], Thm. 2.16. □

We now recall the following result, see for instance [15].

Lemma 13 *Let $C : L^2(\mathbb{R}) \to L^2(\mathbb{R})$ be a contraction. We have for $h \in L^2(\mathbb{R})$*

$$\Gamma(C)\left(\mathrm{Dom}\left(A\left(C^\dagger h\right)\right)\right) \subset \mathrm{Dom}(A(h)).$$

Moreover, on the domain of $A(C^\dagger h)$, we have

$$A(h)\Gamma(C) = \Gamma(C)A\left(C^\dagger h\right).$$

Note that we have the following second quantized version of equation (10):

$$d\Gamma(i\partial)\Phi^{(k)} = \Gamma\left(G^{(k)}\right) d\Gamma(i\partial_{ac})\Phi + iA^\dagger\left(g^{(k)}\right)\Gamma\left(G^{(k)}\right)a_J\Phi,$$

where

$$(a_J\Phi)_m(t_1, \ldots, t_m) = \Phi_{m+1}\left(t_1, \ldots, t_m, 0^+\right) - \Phi_{m+1}\left(t_1, \ldots, t_m, 0^-\right).$$

The action of $H^{(k)}$ on $\Phi^{(k)}$ can now be written as

$$H^{(k)}\Phi^{(k)} = \Gamma(G^{(k)})\, d\Gamma(i\partial_{ac})\Phi + A^{\dagger}(g^{(k)})\Gamma(G^{(k)})(ia_{J}\Phi + \mathsf{E}_{11}A(\rho^{(k)})\Phi + \mathsf{E}_{10}\Phi) + \mathsf{E}_{01}\Gamma(G^{(k)})A(\rho^{(k)})\Phi + \mathsf{E}_{00}\Gamma(G^{(k)})\Phi. \tag{17}$$

Here we have used Lemma 13 and the fact that $A(G^{(k)\dagger}g^{(k)}) = A(\rho^{(k)})$.

Lemma 14 *The singular component of equation* (17) *converges strongly to zero as* $k \to \infty$, *i.e.,*

$$\|A^{\dagger}(g^{(k)})\Gamma(G^{(k)})(ia_{J}\Phi + \mathsf{E}_{11}A(\rho^{(k)})\Phi + \mathsf{E}_{10}\Phi)\|_{2} \overset{k\to\infty}{\longrightarrow} 0,$$

for all Φ *in a core domain* $\mathcal{D}$ *of* H.

We defer the proof of this lemma to the next section.

Using Lemma 12, we find that the first term in equation (17) converges to the first term in the Hamiltonian H given by equation (13), i.e.

$$\lim_{k\to\infty} \|\Gamma(G^{(k)})\, d\Gamma(i\partial_{ac})\Phi - d\Gamma(i\partial_{ac})\Phi\|_{2} = 0.$$

In the proof of the Lemma 14, it is shown that $A(\rho^{(k)})\Phi$ converges in L^2-norm to $\frac{1}{2}a(0^-)\Phi + \frac{1}{2}a(0^+)\Phi$. Therefore, we find for the last line of equation (17)

$$\mathsf{E}_{01}\Gamma(G^{(k)})A(\rho^{(k)})\Phi + \mathsf{E}_{00}\Gamma(G^{(k)})\Phi \longrightarrow \mathsf{E}_{01}\left(\frac{1}{2}a(0^+) + \frac{1}{2}a(0^-)\right)\Phi + \mathsf{E}_{00}\Phi.$$

Employing the boundary condition, we have that

$$\begin{aligned}
&\mathsf{E}_{01}\left(\frac{1}{2}a(0^+) + \frac{1}{2}a(0^-)\right)\Phi + \mathsf{E}_{00}\Phi \\
&= \mathsf{E}_{01}\left(\frac{1}{2}a(0^+)\Phi + \frac{1}{2}[\mathsf{S}a(0^+)\Phi + \mathsf{L}\Phi]\right) + \mathsf{E}_{00}\Phi \\
&\equiv -i\mathsf{L}^{\dagger}\mathsf{S}a(0^+)\Phi + \left(\mathsf{H} - \frac{i}{2}\mathsf{L}^{\dagger}\mathsf{L}\right)\Phi.
\end{aligned}$$

Here we have used the algebraic identities

$$\mathsf{E}_{01}\left(\frac{1}{2} + \frac{1}{2}\mathsf{S}\right) = \mathsf{E}_{01}\left(\frac{1}{2} + \frac{1}{2}\frac{I - i\frac{1}{2}\mathsf{E}_{11}}{I + i\frac{1}{2}\mathsf{E}_{11}}\right) = \mathsf{E}_{01}\frac{1}{I + i\frac{1}{2}\mathsf{E}_{11}} \equiv -iL^{\dagger}\mathsf{S},$$

$$-i\frac{\frac{1}{2}}{I + i\frac{1}{2}\mathsf{E}_{11}} = \frac{1}{2}\operatorname{Im}\left\{\frac{\frac{1}{2}}{I + i\frac{1}{2}\mathsf{E}_{11}}\right\} - \frac{i}{2}\frac{I}{I + i\frac{1}{2}\mathsf{E}_{11}}\frac{I}{I - i\frac{1}{2}\mathsf{E}_{11}}.$$

Applying the Trotter-Kato theorem, this completes the proof of our main result Theorem 10.

4 Proof of Lemma 14

Setting $V^{(k)} = ia_j\Phi + \mathsf{E}_{11}A(\rho^{(k)})\Phi + \mathsf{E}_{10}\Phi$, we see that

$$\begin{aligned}
&\|A^\dagger(g^{(k)})\Gamma(G^{(k)})V^{(k)}\|_2^2\\
&= \langle\Gamma(G^{(k)})V^{(k)}|A(g^{(k)})A^\dagger(g^{(k)})\Gamma(G^{(k)})V^{(k)}\rangle\\
&= \langle\Gamma(G^{(k)})V^{(k)}|(A^\dagger(g^{(k)})A(g^{(k)}) + \|g^{(k)}\|_2^2)\Gamma(G^{(k)})V^{(k)}\rangle\\
&\leq \|A(g^{(k)})\Gamma(G^{(k)})V^{(k)}\|_2^2 + \|g^{(k)}\|_2^2\|V^{(k)}\|_2^2,
\end{aligned}$$

where in the last step we used that $\Gamma(G^{(k)})$ is a contraction. We need to establish two further results: the first is that $V^{(k)}$ goes to 0 sufficiently quickly and we prove this in Lemma 16 below; then we will have to show that this implies that the first term $\|A(g^{(k)})\Gamma(G^{(k)})V^{(k)}\|_2^2$ converges to 0 and we prove this in Lemma 17.

If we accept these results for the moment, then from the boundary conditions we have

$$\begin{aligned}
&ia_j\Phi + \mathsf{E}_{11}\left(\frac{1}{2}a(0^+) + \frac{1}{2}a(0^-)\right)\Phi + \mathsf{E}_{10}\Phi\\
&= i\left(I - i\frac{1}{2}\mathsf{E}_{11}\right)a(0^+)\Phi - i\left(I + i\frac{1}{2}\mathsf{E}_{11}\right)a(0^-)\Phi + \mathsf{E}_{10}\Phi\\
&= i\left(I + i\frac{1}{2}\mathsf{E}_{11}\right)\left[Sa(0^+)\Phi + L\Phi - a(0^-)\Phi\right] = 0
\end{aligned}$$

so that, in fact,

$$V^{(k)} = \mathsf{E}_{11}\left[A(\rho^{(k)})\Phi - \left(\frac{1}{2}a(0^+) + \frac{1}{2}a(0^-)\right)\Phi\right].$$

As $\|g^k\|_2$ grows at rate $\sqrt{k}$, it suffices to show that $A(\rho^{(k)})\Phi - (\frac{1}{2}a(0^+) + \frac{1}{2}a(0^-))\Phi$ goes to 0 in norm with rate faster than $\frac{1}{\sqrt{k}}$. We will now establish this result below, but first we need to recall the definition of a pseudo-exponential vector from [2].

Definition 15 Let $\mathsf{F} : t \mapsto \mathsf{F}_t$ be a function from $\mathbb{R}$ to $\mathfrak{B}(\mathfrak{h})$ and define the corresponding pseudo-exponential vector $\Psi(\mathsf{F}, h)$ as

$$\left[\Psi(\mathsf{F}, h)\right]_m(t_1, \ldots, t_m) = \vec{T}\mathsf{F}_{t_1}\cdots\mathsf{F}_{t_m}h$$

for given $h \in \mathfrak{h}$, where $\vec{T}$ denotes chronological ordering. That is

$$\vec{T}\mathsf{F}_{t_1}\cdots\mathsf{F}_{t_m} = \mathsf{F}_{t_{\sigma(1)}}\cdots\mathsf{F}_{t_{\sigma(m)}},$$

where σ is a permutation for which $t_{\sigma(1)} \geq \cdots \geq t_{\sigma(m)}$.

Lemma 16 *Let $v \in W^{1,2}(\mathbb{R}/\{0\})$ and $u \in W^{1,2}(\mathbb{R}/\{0\})$ with $u|_{\mathbb{R}_+} = 0$ and $u(0^-) = 1$, then define F_t by*

$$\mathsf{F}_t = v(t) + u(t)\left[\mathsf{S}v(0^+) + \mathsf{L} - v(0^-)\right] \tag{18}$$

then the domain $\mathcal{D}$ of such pseudo-exponential vectors $\Phi = \Psi(\mathsf{F}, h)$ is a core for H. Moreover, for each such vector we have

$$\left\| A(\rho^{(k)})\Phi - \left(\frac{1}{2}a(0^+) + \frac{1}{2}a(0^-)\right)\Phi \right\|_2 = O\left(\frac{1}{k}\right).$$

Proof The first part of this lemma is proved by Gregoratti where it is shown that $\mathcal{D}$ is dense, and is contained in $\mathrm{Dom}(H) \cap \mathcal{V}_{0^\pm}$, see [2], Propositions 4 and 5. Note that for $\Phi = \Psi(\mathsf{F}, h)$, by (4) in [2] we have

$$\begin{aligned} a(t)\Phi &= \nu(t)\Phi, \quad t \in \{0^+\} \cup (0, \infty), \\ a(0^-)\Phi &= (\mathsf{S}\nu(0^+) + \mathsf{L})\Phi. \end{aligned}$$

To prove the second part, we begin by setting

$$\begin{aligned} Z_m(t_1, \dots, t_m) &= \left[A(\rho^{(k)})\Phi - \left(\frac{1}{2}a(0^+) + \frac{1}{2}a(0^-)\right)\Phi\right]_m (t_1, \dots, t_m) \\ &= \int_0^\infty \rho^{(k)}(s)\left[\Phi_{m+1}(t_1, \dots, t_m, s) - \Phi_{m+1}(t_1, \dots, t_m, 0^+)\right] ds \\ &\quad + \int_{-\infty}^0 \rho^{(k)}(s)\left[\Phi_{m+1}(t_1, \dots, t_m, s) - \Phi_{m+1}(t_1, \dots, t_m, 0^-)\right] ds \\ &\equiv Z_m^+(t_1, \dots, t_m) + Z_m^-(t_1, \dots, t_m). \end{aligned}$$

We have $\|Z_m\|^2 \le (\|Z_m^+\| + \|Z_m^-\|)^2$ but

$$Z_m^+(t_1, \dots, t_m) = \int_0^\infty \rho^{(k)}(s)\left[\nu(s) - \nu(0^+)\right] ds \Phi_m(t_1, \dots, t_m)$$

and this prefactor is clearly $O(\frac{1}{k})$ from the argument used in Lemma 8.

However, we then have

$$\begin{aligned} &Z_m^-(t_1, \dots, t_m) \\ &= \int_{-\infty}^0 \rho^{(k)}(s)[\mathsf{F}_{t_{\sigma(1)}} \cdots \mathsf{F}_s \cdots \mathsf{F}_{t_{\sigma(m)}} - \mathsf{F}_{0^-}\mathsf{F}_{t_{\sigma(1)}} \cdots \mathsf{F}_{t_{\sigma(m)}}]h\, ds, \end{aligned}$$

where σ is the chronological time ordering permutation.

We note however that $[\mathsf{F}_t, \mathsf{F}_s] = 0$ for all t, s, therefore we have

$$\begin{aligned} Z_m^-(t_1, \dots, t_m) &= \int_{-\infty}^0 \rho^{(k)}(s)[\mathsf{F}_s - \mathsf{F}_{0^-}]\mathsf{F}_{t_{\sigma(1)}} \cdots \mathsf{F}_{t_{\sigma(m)}} h\, ds \\ &= \int_{-\infty}^0 \rho^{(k)}(s)\left[u(s) - u(0^-)\right]\left[\mathsf{S}\nu(0^+) + \mathsf{L} - \nu(0^-)\right] \\ &\quad \times \mathsf{F}_{t_{\sigma(1)}} \cdots \mathsf{F}_{t_{\sigma(m)}} h\, ds, \end{aligned}$$

where we used (18). From the argument in Lemma 8 again, we see that this is $O(\frac{1}{k})$. □

Lemma 17 *For Φ chosen as a pseudo-exponential vector, as in Lemma 16, we have that $\|A(g^{(k)})\Gamma(G^{(k)})V^{(k)}\|_2^2$ converges to 0 as $k \to \infty$.*

Proof We have that

$$A\left(g^{(k)}\right)\Gamma\left(G^{(k)}\right)V^{(k)} = \Gamma\left(G^{(k)}\right)A\left(\rho^{(k)}\right)V^{(k)},$$

with $\Gamma(G^{(k)})$ a contraction. The mth level of the Fock space component of $A(\rho^{(k)})V^{(k)}$ may be written as

$$E_{11}A\left(\rho^{(k)}\right)Z_m^+ + E_{11}A\left(\rho^{(k)}\right)Z_m^-,$$

where we use the same conventions as in Lemma 16. The first term has the explicit components

$$E_{11}\int dt\rho^{(k)}(t)\int_0^\infty \rho^{(k)}(s)\left[v(s)-v\left(0^+\right)\right]ds\Phi_{m+1}(t,t_1,\ldots,t_m)$$
$$= E_{11}\int dt\rho^{(k)}(t)\mathsf{F}_t\int_0^\infty \rho^{(k)}(s)\left[v(s)-v\left(0^+\right)\right]ds\Phi_m(t_1,\ldots,t_m)$$

which is norm bounded by $\|E_{11}\|\int dt\rho^{(k)}(t)\|\mathsf{F}_t\|\|Z_m^+\|$, and we note that in fact $\int dt\rho^{(k)}(t)\times\|\mathsf{F}_t\| = \int d\tau\rho(\tau)\|\mathsf{F}_{\tau/k}\|$. An equivalent bound is easily shown to hold for $E_{11}A(\rho^{(k)})Z_m^-$ and so by an argument similar to lemma 16 we obtain the desired result. □

4.1 Epilogue

After completion of this work, the authors became aware of the book by W. von Waldenfels [16] which gives a complete resolvent analysis of the Chebotarev-Gregoratti-von Waldenfels Hamiltonian, and in the final chapter describes a strong resolvent limit by colored noise approximations. The convergence is comparable to the strong uniform convergence considered here, but the approach is very different.

Competing interests

The authors declare that they have no competing interests.

Authors' contributions

All authors contributed equally to the writing of this paper. All authors read and approved the final manuscript.

Author details

[1]BTA VOF, Slootsekuilen 9, 5986 PE, Beringe, The Netherlands. [2]Dept. for Mathematics and Physics, Aberystwyth University, SY23 3BZ, Wales, United Kingdom. [3]School of Electrical Engineering and Telecommunications, UNSW Australia, Sydney, NSW 2052, Australia.

Acknowledgements

JG and RG are grateful to EPSRC EP/G039275/1 and EP/L006111/1 grant for support. They also wish to thank the Isaac Newton Institute for Mathematical Sciences, Cambridge, for support and hospitality during the programme *Quantum Control Engineering* where work on this paper was completed. HN acknowledges support through a research visit funded through EPSRC EP/H016708/1 and Australian Research Council grants DP0986615 and DP130104191.

References

1. Hudson RL, Parthasarathy KR. Quantum Ito's formula and stochastic evolutions. Commun Math Phys. 1984;93:301-23.
2. Gregoratti M. The Hamiltonian operator associated to some quantum stochastic differential equations. Commun Math Phys. 2001;222:181-200.

3. von Waldenfels W. Symmetric differentiation and Hamiltonian of a quantum stochastic process. Infin Dimens Anal Quantum Probab Relat Top. 2005;8(1):73-116.
4. Chebotarev AM. Quantum stochastic differential equation is unitarily equivalent to a symmetric boundary problem for the Schrödinger equation. Math Notes. 1997;61(4):510-8.
5. Quezada-Batalla R, González-Gaxiola O. On the Hamiltonian of a class of quantum stochastic processes. Math Notes. 2007;81(5-6):734-52.
6. Accardi L, Frigerio A, Lu YG. The weak coupling limit as a quantum functional central limit. Commun Math Phys. 1990;131:537-70.
7. Accardi L, Frigerio A, Lu YG. The low density limit in finite temperature case. Nagoya Math J. 1992;126:25-87.
8. Gough J. Quantum flows as Markovian limit of emission, absorption and scattering interactions. Commun Math Phys. 2005;254:489-512.
9. Gough J. Quantum Stratonovich calculus and the quantum Wong-Zakai theorem. J Math Phys. 2006;47:113509.
10. Davies EB. One-parameter semigroups. London: Academic Press; 1980.
11. Reed M, Simon B. Methods of mathematical physics I: functional analysis. New York: Academic Press; 1980.
12. Reed M, Simon B. Methods of mathematical physics II: Fourier analysis, self-adjointness. New York: Academic Press; 1975.
13. Gustafson KE. Introduction to partial differential equations and Hilbert space methods. 3rd ed. New York: Dover; 1999.
14. Lieb EH, Loss M. Analysis. Providence: Am. Math. Soc.; 1997.
15. Petz D. Invitation to the canonical commutation relations. Leuven: Leuven University Press; 1990.
16. von Waldenfels W. A measure theoretical approach to quantum stochastic processes. Lecture notes in physics. vol. 878. Berlin: Springer; 2014.

13

A coherent perceptron for all-optical learning

Nikolas Tezak* and Hideo Mabuchi

*Correspondence:
ntezak@stanford.edu
Edward L. Ginzton Laboratory,
Stanford University, Stanford, CA
94305, USA

Abstract

We present nonlinear photonic circuit models for constructing programmable linear transformations and use these to realize a coherent perceptron, i.e., an all-optical linear classifier capable of learning the classification boundary iteratively from training data through a coherent feedback rule. Through extensive semi-classical stochastic simulations we demonstrate that the device nearly attains the theoretical error bound for a model classification problem.

Keywords: optical information processing; coherent feedback; machine learning; photonic circuits; nonlinear optics; perceptron

1 Introduction

Recent progress in integrated nanophotonic engineering [1–10] has motivated follow-up proposals [11, 12] of nanophotonic circuits for all-optical information processing. While most of these focus on implementations of digital logic, we present here an approach to all-optical analog, *neuromorphic* computation and propose design schemes for a set of devices to be used as building blocks for large scale circuits.

Optical computation has been a long-time goal [13, 14], with research interest surging regularly after new engineering capabilities are attained [15, 16], but so far the parallel progress and momentum of CMOS based integrated electronics has outperformed all-optical devices.

In recent years we have seen rapid progress in the domain of machine learning, and artificial intelligence in general. Although most current 'big data'-applications are realized on digital computing architectures, there is now an increasing amount of computation done in specialized hardware such as GPUs. Specialized analog computational devices for solving specific subproblems more efficiently than possible with either GPUs or general purpose computers are being considered or already implemented by companies such as IBM, Google and HP and in academia, as well [17–20]. Specifically in the field of neuromorphic computation, there has been impressive progress on CMOS based analog computation platforms [21, 22].

Several neuromorphic approaches to use complex nonlinear optical systems for machine learning applications have recently been proposed [23–26] and some initial schemes have been implemented [9, 27]. So far, however, all of these 'optical reservoir computers' have still required digital computers to prepare the inputs and process the output of these devices with the optical systems only being employed as static nonlinear mappings for dimensional lifting to a high dimensional feature space [28], in which one then applies straightforward linear regression or classification for learning an input-output map [29].

In this work, we address how the final stage of such a system, i.e., the linear classifier could be realized all-optically. We provide a universal scheme, i.e., independent of which particular kind of optical nonlinearity is employed, for constructing *tunable* all-optical, phase-sensitive amplifiers and then outline how these can be combined with self-oscillating systems to realize an optical amplifier with *programmable* gain, i.e., where the gain can be set once and is then fixed subsequently.

Using these as building blocks we construct an all-optical *perceptron* [30, 31], a system that can classify multi-dimensional input data and, using pre-classified training data learn the correct classification boundary 'on-line', i.e., incrementally. The perceptron can be seen as a highly simplified model of a neuron. While the idea of all-optical neural networks has been proposed before [32] and an impressive scheme using electronic, measurement-based feedback for spiking optical signals has been realized [33], to our knowledge, we offer the first complete description for how the synaptic weights can be stored in an optical memory and programmed via feedback.

The physical models underlying the employed circuit components are high intrinsic-Q optical resonators with strong optical nonlinearities. For theoretical simplicity we assume resonators with either a χ_2 or a χ_3 nonlinearity, but the design can be adapted to depend on only one of these two or alternative nonlinearities such as those based on free carrier effects or optomechanical interactions.

The strength of the optical nonlinearity and the achievable Q-factors of the optical resonators determine the overall power scale and rate at which a real physical device could operate. Both a stronger nonlinearity and higher Q allow operating at lower overall power.

We present numerical simulations of the system dynamics based on the semi-classical Wigner-approximation to the full coherent quantum dynamics presented in [10]. For photon numbers as low as (~10-20) this approximation allows us to accurately model the effect of optical quantum shot noise even in large-scale circuits.

In the limit of both very high Q and very strong nonlinearity, we expect quantum effects to become significant as entanglement can arise between the field modes of physically separated resonators. In the Appendix, we provide full quantum models for all basic components of our circuit. The possibility of a quantum speedup is being addressed in ongoing work. Recently, D-Wave Systems has generated a lot of interest in their own superconducting qubit based quantum annealer. Although the exact benefits of quantum dynamics in their machines has not been conclusively established [34], recent results analyzing the role of tunneling in a quantum annealer [35] are intriguing and suggest that quantum effects can be harnessed in computational devices that are not unitary quantum computers.

1.1 The perceptron algorithm

The perceptron is a machine learning algorithm that maps an input $x \in \mathbb{R}^n$ to a single binary class label $\hat{y}_w[x] \in \{0,1\}$. Binary classifiers generally operate by dividing the input space into two disjoint sets and identifying these with the class labels. The perceptron is a linear classifier, meaning that the surface separating the two class label sets is a linear space, a hyperplane, and its output is computed simply by applying a step function $\theta(u) := \mathbb{1}_{u\geq 0}$

to the inner product of a single data point x with a fixed *weight vector* w:

$$\hat{y}_w[x] := \theta\left(w^T x\right) = \begin{cases} 1 & \text{for } w^T x \geq 0, \\ 0 & \text{otherwise.} \end{cases} \tag{1}$$

Geometrically, the weight vector w parametrizes the hyperplane $\{z \in \mathbb{R}^n : w^T z = 0\}$ that forms the decision boundary.

In the above parametrization the decision boundary always contains the origin $z = 0$, but the more general case of an affine decision boundary $\{\tilde{z} \in \mathbb{R}^n : \tilde{w}^T \tilde{z} = b\}$ can be obtained by extending the input vector by a constant $z = (\tilde{z}^T, 1)^T \in \mathbb{R}^{n+1}$ and similarly defining an extended weight vector $w = (\tilde{w}^T, -b)^T$.

The perceptron converges in a finite number of steps for all linearly separable problems [30] by randomly iterating over a set of pre-classified training data $\{(y^{(j)}, x^{(j)}) \in \{0,1\} \otimes \mathbb{R}^n, j = 1, 2, \ldots, M\}$ and imparting a small weight correction $w \to w + \Delta w$ for each falsely classified training example $x^{(j)}$

$$\Delta w = \tilde{\alpha}\left(y^{(j)} - \hat{y}_w\left[x^{(j)}\right]\right)x^{(j)}. \tag{2}$$

The *learning rate* $\tilde{\alpha} > 0$ determines the magnitude of the correction applied for each training example. The expression in parentheses can only take on the values $\{0, -1, 1\}$ with the zero corresponding to a correctly classified example and the non-zero values corresponding to the two different possible classification errors.

Usually there exist many separating hyperplanes for a given linear binary classification problem. The standard perceptron is only guaranteed to find one that works for the training set. It is possible to introduce a notion of optimality to this problem by considering the minimal distance ('margin') of the training data to the found separating hyperplane. Maximization of this margin naturally leads to the 'support vector machine' (SVM) algorithm [36]. Although the SVM outperforms the perceptron in many classification tasks it does not lend itself to a hardware implementation as readily because it cannot be trained incrementally. It is this that makes the perceptron algorithm especially suited for a hardware implementation: We can convert the discrete update rule (2) to a differential equation

$$\dot{w}(t) = \alpha\left\{y(t) - \hat{y}_{w(t)}(t)\right\}x(t), \tag{3}$$

and then construct a physical system that realizes these dynamics. In this continuous-time version the inputs are piece-wise constant $x(t) = x^{(j_t)}$, $y(t) = y^{(j_t)}$ and take on the same discrete values as above indexed by $j_t := \lceil \frac{t}{\Delta t} \rceil \in \{1, 2, \ldots, M = \frac{T}{\Delta t}\}$.

1.2 The circuit modeling framework

Circuits are fully described via Quantum Hardware Description Language (QHDL) [37] based on Gough and James' SLH-framework [38, 39]. To carry out numerical simulations for large scale networks, we derive a system of semi-classical Langevin equations based on the Wigner-transformation as described in [10]. Note that there is a perfect one-to-one correspondence between nonlinear cavity models expressed via SLH and the Wigner method as long as the nonlinearities involve only oscillator degrees of freedom. There is ongoing research in our group to establish similar results for more general nonlinearities [40].

Both the Wigner method and the more general SLH framework can be used to model networks of quantum systems where the interconnections are realized through bosonic quantum fields. The SLH framework describes a system interacting with n independent input fields in terms of a unitary scattering matrix S parametrizing direct field scattering, a coupling vector $L = (L_1, L_2, \ldots, L_n)^T$ parametrizing how external fields couple into the system and how the system variables couple to the output and a Hamilton operator inducing the internal dynamics. We summarize these objects in a triplet (S, L, H). L and H are sufficient to parametrize any Schrödinger picture simulation of the quantum dynamics, e.g., the master equation for a mixed system state ρ is given by

$$\dot{\rho} = -i[H, \rho] + \sum_{j=1}^{n} \left(L_j \rho L_j^\dagger - \frac{1}{2} \{ L_j^\dagger L_j, \rho \} \right). \tag{4}$$

The scattering matrix S is important when composing components into a network. In particular, the input-output relation in the SLH framework is given by

$$dA_{\text{out}} = S\, dA_{\text{in}} + L\, dt, \tag{5}$$

where the $dA_{\text{in/out},j}$, $j = 1, 2, \ldots, n$ are to be understood as quantum stochastic processes whose differentials can be manipulated via a quantum Ito calculus [38]. The Wigner method provides a simplified, approximate description which is valid when all nonlinear resonator modes are in strongly displaced states [10]. The simulations presented here were carried out exclusively at energy scales for which the Wigner method is valid, allowing us to scale to much larger system sizes than we could in a full SLH-based quantum simulation. This is because the computational complexity of the Wigner method scales at most quadratically (and in sparsely interconnected systems nearly linearly) with the number of components as opposed to the exponential state space scaling of a quantum mechanical Hilbert space. We nonetheless provide our models in both Wigner-method form and SLH form in anticipation that our component models will also be extremely useful in the full quantum regime.

In the Wigner-based formalism, a system is described in terms of time-dependent complex coherent amplitudes $\alpha(t) = (\alpha_1(t), \alpha_2(t), \ldots, \alpha_m(t))^T$ for the internal cavity modes and external inputs $\beta_{\text{in}}(t) = (\beta_{\text{in},1}(t), \beta_{\text{in},2}(t), \ldots, \beta_{\text{in},n}(t))^T$. These amplitudes relate to quantum mechanical expectations as $\langle \alpha_j \rangle \approx \langle a_j \rangle_{\text{QM}}$, where $\langle \cdot \rangle$ denotes the expectation with respect to the Wigner quasi distribution and $\langle \cdot \rangle_{\text{QM}}$ a quantum mechanical expectation value. See [10] for the corresponding relations of higher order moments.

To simplify the analysis, we exclusively work in a rotating frame with respect to all driving fields. As in the SLH case we define output modes $\beta_{\text{out}}(t)$ that are algebraically related to the inputs and the internal modes. The full dynamics of the internal and external modes are then governed by a multi-dimensional Langevin equation

$$\dot{\alpha}(t) = \left[\mathbf{A}\alpha(t) + \mathbf{a} + A_{\text{NL}}(\alpha, t) \right] + \mathbf{B}\beta_{\text{in}}(t), \tag{6}$$

as well as a purely algebraic, linear input-output relationship

$$\beta_{\rm out}(t) = \left[\mathbf{C}\alpha(t) + \mathbf{c}\right] + \mathbf{D}\beta_{\rm in}(t). \tag{7}$$

The complex matrices **A**, **B**, **C**, **D** as well as the constant bias input vectors **a** and **c** parametrize the linear dynamics, whereas the function $A_{\rm NL}(\alpha, t)$ gives the nonlinear contribution to the dynamics of the internal cavity modes.

Each input consists of a coherent, deterministic part and a stochastic contribution $\beta_{{\rm in},j}(t) = \bar{\beta}_{{\rm in},j}(t) + \eta_j(t)$. The stochastic terms $\eta_j(t) = \eta_{j,1}(t) + i\eta_{j,2}(t)$ are assumed to be independent complex Gaussian white noise processes with correlation function $\langle \eta_{j,s}(t)\eta_{k,r}(t')\rangle = \frac{1}{4}\delta_{jk}\delta_{sr}\delta(t-t')$.

The linearity of the input-output relationship in either framework (5) and (7) in the external degrees of freedom leads to algebraic rules for deriving reduced models for whole circuits of nonlinear optical resonators by concatenating component models and algebraically solving for their interconnections [10, 39]. To see the basic component models used in this work see the Appendix. Netlists for composite components and the whole circuit will be made available at [41].

2 The coherent perceptron circuit

The full perceptron's circuit is visualized in Figure 1. The input data x to the perceptron circuit is encoded in the real quadrature of N coherent optical inputs. Equation (3) informs us what circuit elements are required for a hardware implementation by decomposing the necessary operations:

1 Each input x_j is multiplied by a weight w_j.
2 The weighted inputs are coherently added.
3 The sum drives a thresholding element to generate the estimated class label $\hat{y}$.
4 In the training phase (input $T = 1$) the estimated class label $\hat{y}$ is compared with the true class label (input Y) and based on the outcome, feedback is applied to modify the weights $\{w_j\}$.

The most crucial element for this circuit is the system that multiplies an input x_j with a programmable weight w_j. This not only requires having a linear amplifier with tunable

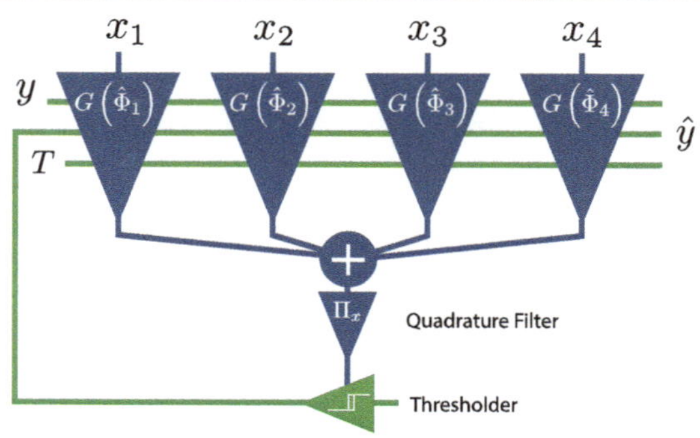

Figure 1 An example perceptron circuit consisting of $N = 4$ programmable amplifiers for the coherent input vector $x = (x_1, x_2, x_3, x_4)^T$, a static mixing element that sums their output, a quadrature filter to remove the imaginary quadrature and a final thresholding element to generate the estimated binary class label $\hat{y}$. The additional binary input T controls whether the system is in *training mode*, in which case the estimated class label $\hat{y}$ is compared to the true class label Y which is provided as an additional input. When they differ, the programmable amplifiers receive a feedback signal to adjust their internal weights.

gain, but also a way to encode and store the continuous weights w_j. In the following we outline one way how such systems can be constructed from basic nonlinear optical cavity models: Section 2.1 presents an elegant way to construct a phase sensitive linear optical amplifier where the gain can be tuned by changing the amplitude of a bias input. In Section 2.2 we propose using an above threshold non-degenerate optical parametric amplifier to store a continuous variable in the output phase of the signal (or idler) mode. In Section 2.3 these systems are combined to realize an optical amplifier with *programmable* gain, i.e., a control input can program its gain, which then stays constant even after the control has been turned off. Finally, we present a simple model for all-optical switches based on a cavity with two modes that interact via a cross-Kerr-effect in Section 2.4. This element is used both for the feedback logic as well as the thresholding function to generate the class label $\hat{y}$.

2.1 Tunable gain Kerr-amplifier

A single mode Kerr-nonlinear resonator driven by an appropriately detuned coherent drive ϵ can have a strongly nonlinear dependence of the intra-cavity energy on the drive power. When the drive of a single resonator is given by the sum of a constant large bias amplitude and a small signal $\epsilon = \frac{1}{\sqrt{2}}(\epsilon_0 + \delta\epsilon)$, the steady state reflected amplitude is $\epsilon' = \frac{1}{\sqrt{2}}(\eta\epsilon_0 + g_-(\epsilon_0)\delta\epsilon + g_+(\epsilon_0)\delta\epsilon^*) + O(\delta\epsilon^2)$, where $|\eta| \leq 1$ with equality for the ideal case of negligible intrinsic cavity losses. The small signal thus experiences phase sensitive gain dependent on the bias amplitude and phase. We provide analytic expressions for the gain in Appendix A.2.1.

Placing two identical resonators in the arms of an interferometer allows for isolating the signal and bias outputs even if their amplitudes vary by canceling the scattered bias in one output and the scattered signal in the other (cf. Figure 2). This highly symmetric construction, which generalizes to any other optical nonlinearity, ensures that the signal output is linear in $\delta\epsilon$ up to third order.[a] If the system parameters are well-chosen, the amplifier gain depends very strongly on small variations of the bias amplitude. This allows to tune the gain from close to unity to its maximum value, which, for a given waveguide coupling κ and Kerr coefficient χ depends on the drive detuning from cavity. For Kerr-nonlinear resonators there exists a critical detuning beyond which the system becomes

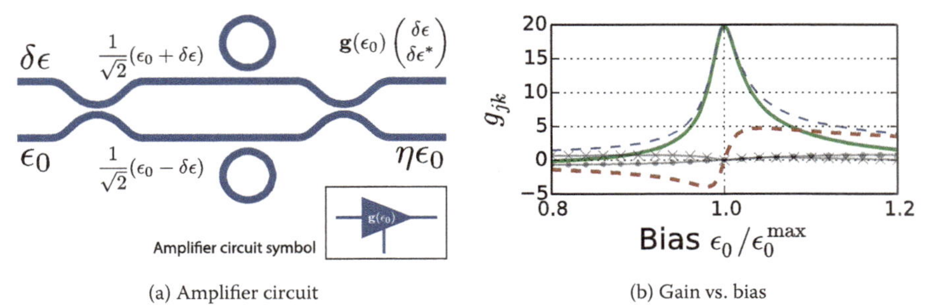

Figure 2 Amplifier circuit and bias-gain characteristic. Part (**a**) shows two identical single mode Kerr-nonlinear optical resonators symmetrically placed in the two arms of an interferometer. Part (**b**) gives the phase sensitive amplifier gain $g_{rr}(\epsilon_0)$ (green, solid) and the $g_{ir}(\epsilon_0)$ (red, dashed) as a function of the bias photon input rate normalized by the drive power at which dynamic resonance occurs. For completeness we also provide g_{ri} (black ×'s) and g_{ii} (black dots). The detuning has been chosen such that $g_{rr}^{\max} = g_{rr}(\epsilon_0^{\max}) = 20$. The dashed blue envelope gives the maximal input output gain achievable between any two signal quadratures at that bias. Note that g_{rr} vanishes at $\epsilon_0/\epsilon_0^{\max} \approx 0.8$.

bi-stable and exhibits hysteresis. This can be used for thresholding type behavior though as shown in [42] in this case it may be advantageous to reduce the symmetry of the circuit. It is convenient to engineer the relative propagation phases such that at maximum gain, a real quadrature input signal $x \in \mathbb{R}$ leads to an amplified output signal $x' = g_{rr}^{\max} x$ with no imaginary quadrature component (other than noise and higher order contributions). However, for different bias input amplitudes and consequently lower gain values the output will generally feature a linear imaginary quadrature component $x' = [g_{rr}(\epsilon_0) + ig_{ir}(\epsilon_0)]x$ as well. Figure 2(b) demonstrates this for a particular choice of maximal gain. We note that there exist previous proposals of using nonlinear resonator pairs inside interferometers to achieve desirable input-output behavior [42], but to our knowledge, no one has proposed using these for signal/bias isolation and tunable gain. To first order the linearized Kerr model is actually identical to a sub-threshold degenerate OPO model. This implies that it can be used to generate squeezed light and also that one could replace the Kerr-model by an OPO model.

An almost identical circuit, but featuring resonators with additional internal loss equal to the wave-guide coupling[b] and constantly biased to *dynamic resonance* $\langle|\alpha|^2\rangle_{ss} = -\Delta/\chi$ can be used to realize a *quadrature filter*, i.e., an element that has unity gain for the real quadrature and zero for the imaginary one. Now the quadrature filtered signal still has an imaginary component, but to linear order this only consists of transmitted noise from the additional internal loss. While it would be possible to add one of these downstream of every tunable Kerr amplifier, in our specific application it is more efficient to add just a single one downstream of where the individual amplifier outputs are summed (cf. Section 2.5). This also reduces the total amount of additional noise introduced to the system.

2.2 Encoding and storing the gain

In the preceding section we have seen how to realize a tunable gain amplifier, but for programming and *storing* this gain (or equivalently its bias amplitude) an additional component is needed. Although it is straightforward to design a multi-stable system capable of outputting a discrete set of different output powers to be used as the amplifier bias, such schemes would likely require multiple nonlinear resonators and it would be more cumbersome to drive transitions between the output states.

An alternative to such schemes is given by systems that have a continuous set of stable states. Recent analysis of continuous time recurrent neural network models trained for complex temporal information processing tasks has revealed multi-dimensional stable attractors in the internal network dynamics that are used to store information over time [43].

A simple semi-classical nonlinear resonator model to exhibit this is given by a non-degenerate optical parametric oscillator (NOPO) pumped above threshold; for low pump input powers this system allows for parametric amplification of a weak coherent signal (or idler) input. In this case vacuum inputs for the signal and idler lead to outputs with zero expected photon number. Above a critical threshold pump power, however, the system down-converts pump photons into pairs of signal and idler photons.

Due to an internal $U(1)$ symmetry of the underlying Hamiltonian (cf. Appendix A.2.3), the signal and idler modes spontaneously select phases that are dependent on each other but independent of the pump phase. This implies that there exists a whole manifold of fix-points related to each other via the symmetry transformation $(\alpha_s, \alpha_i) \to (\alpha_s e^{i\phi}, \alpha_i e^{-i\phi})$,

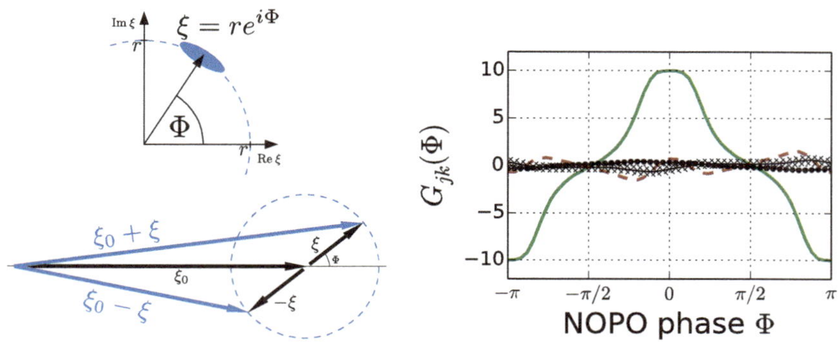

(a) Combined bias (b) Gain vs. OPO phase

Figure 3 The NOPO's signal output $\xi = \sqrt{\kappa}\alpha_s$ lives on a circular manifold parametrized by Φ (a, upper figure). Vacuum input shot noise leads to small fluctuations perpendicular to the manifold and diffusion along it. Mixing this signal output with a constant bias offset on a beamsplitter produces two outputs with anti-correlated total amplitude (**a**, lower figure). When both outputs are used to drive a complementary pair of tunable amplifiers whose outputs are subtracted, the overall real-to-real quadrature gain (green) of the system varies from positive to negative values (**b**). We can also see that the real-to-imaginary gain (dashed red) stays small for all NOPO phases, which allows us to efficiently subtract it downstream by the quadrature filter. The imaginary to real and imaginary gains are also plotted.

where α_s and α_i are the rotating frame signal and idler mode amplitudes, respectively. Consequently the signal output of an above threshold NOPO lives on a circular manifold (cf. Figure 3).

Vacuum shot noise on the inputs leads to phase diffusion with a rate of $\gamma_\Phi = \frac{\kappa}{8n_0}$, where κ is the signal and idler line width and n_0 is the steady state intra cavity photon number in either mode. We point out that this diffusion rate does not directly depend on the strength of the nonlinearity which only determines how strongly the system must be pumped to achieve a given intra cavity photon number n_0.

A weak external signal input breaks the symmetry and biases the signal output phase towards the external signal's phase. This allows for changing the programmed phase value.

Finally, we note that parametric oscillators can also be realized in materials with vanishing χ_2 nonlinearity. They have been successfully realized via four-wave mixing (i.e., exploiting a χ_3 nonlinearity) in [1, 2, 44] and even in opto-mechanical systems [8] in which case the idler mode is given by a mechanical degree of freedom.

In principle any nonlinear optical system that has a stable limit cycle could be used to store and encode a continuous value in its oscillation phase. Non-degenerate parametric oscillators stand out because of their theoretical simplicity allowing for a 'static' analysis inside a rotating frame.

2.3 Programmable gain amplifier

Combining the circuits described in the preceding sections allows us to construct a fully programmable phase sensitive amplifier. In Figure 2(b) we see that there exists a particular bias amplitude at which the real to real quadrature gain vanishes $g_{rr}(\epsilon_0^{\min}) = 0$. We combine the NOPO signal output $\xi = re^{i\Phi}$ with a constant phase bias input ξ_0 (cf. Figure 3(a)) on a beamsplitter such that the outputs vary between zero gain and the maximal gain bias values $|\frac{\xi_0 \pm re^{i\Phi}}{\sqrt{2}}| \in [\epsilon_0^{\min}, \epsilon_0^{\max}]$. To realize both positive and negative gain, we use the second output of that beamsplitter to bias another tunable amplifier. The two amplifiers are always

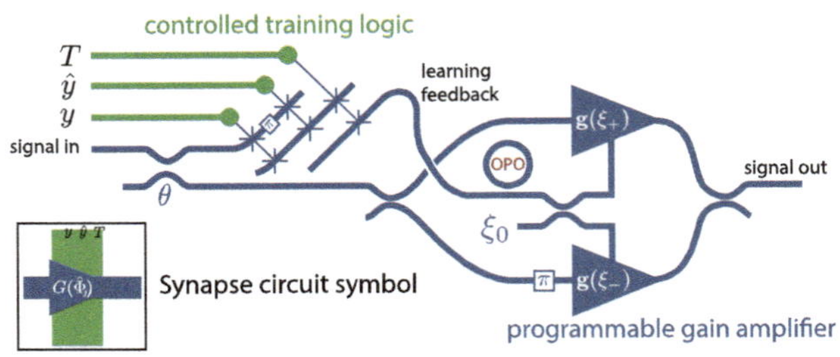

Figure 4 Synapse circuit composed of a programmable amplifier and feedback logic (cf. Section 2.4) that implements the perceptron learning feedback (3) for a single weight. The upper amplifier when biased optimally leads to positive gain whereas the lower amplifier leads to negative gain due to the additional π phase shift.

biased oppositely meaning that one will have maximal gain when the other's gain vanishes and vice versa. The overall input signal is split and sent through both amplifiers and then re-combined with a relative π phase shift. This complementary setup leads to an overall effective gain tunable within $G_{rr}(\Phi) \in [-\frac{g_{rr}^{\max}}{2}, \frac{g_{rr}^{\max}}{2}]$ (cf. Figure 3(b)).

In Figure 4 we present both the complementary pair of amplifiers and the NOPO used for storing the bias as well as some logic elements (described in Section 2.4) used for implementing conditional training feedback. We call the full circuit a synapse because it features programmable gain and implements the perceptron's conditional weight update rule.

The resulting synapse model is quite complex and certainly not optimized for a minimal component number but rather the ease of theoretical analysis. A more resource efficient programmable amplifier could easily be implemented using just two or three nonlinear resonators. E.g., inspecting the real to imaginary quadrature gain $g_{ir}(\epsilon_0)$ in Figure 2(b) we see that close to $\epsilon_0^{\max}$ it passes through zero fairly linearly and with an almost symmetric range. This indicates that we could use a single tunable amplifier to realize both positive and negative gain. Using only a single resonator for the tunable amplifier could work as well, but it would require careful interferometric bias cancellation and more tedious upfront analysis. We do not think it is feasible to use just a single resonator for both the parametric oscillator and the amplifier because any amplified input signal would have an undesirable back-action on the oscillator phase.

2.4 Optical switches

The feedback to the perceptron weights (cf. Equation (3)) is conditional on the binary values of the given and estimated class labels y and $\hat{y}$, respectively. The logic necessary for implementing this can be realized by means of all-optical switches. There have been various proposals and demonstrations [7, 45] of all-optical gates/switches and quantum optical switches [46].

The model that we assume here (cf. Figure 5) is to use two different modes of a resonator that interact via a cross-Kerr-effect, i.e., power in the control mode leads to a refractive index shift (or detuning) for the signal mode. The index shift translates to a control mode dependent phase shift of a scattered signal field yielding a controlled optical phase modulator. Wrapping this phase modulator in a Mach-Zehnder interferometer then realizes a controlled switch: If the control mode input is in one of two different states $|\xi| \in 0, \xi_0$, the signal inputs are either passed through or switched. This operation is often referred to as a *controlled swap* or Fredkin gate [47] which was originally proposed for realizing

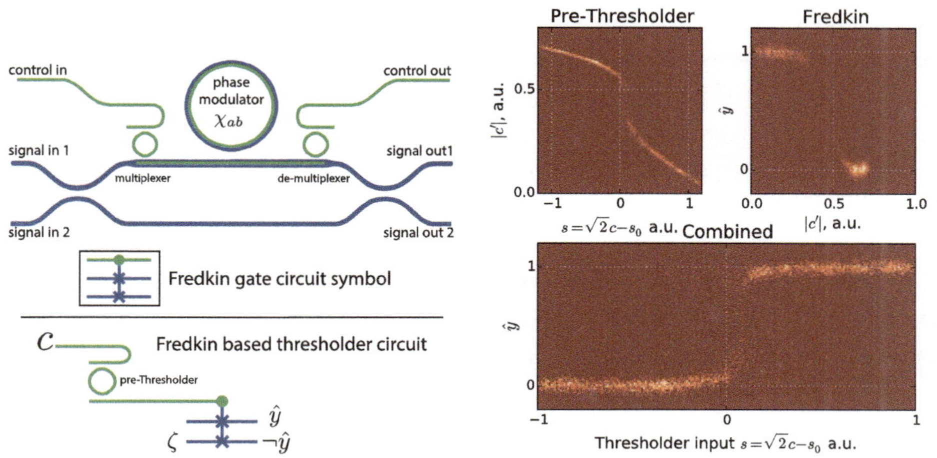

(a) Fredkin gate and thresholder (b) Thresholder input/output

Figure 5 Fredkin gate and thresholder circuit. In the upper graphic of (**a**) we present a schematic for Fredkin gate based on a two mode cross-Kerr-nonlinear resonator. The lower graphic shows how this circuit can be pre-pended with a single mode nonlinear resonator to better approximate a thresholding response. In (**b**) we present the input-output characteristic of the prepended resonator (upper left), the Fredkin gate (upper right) and the combined input output relationship between the inner product amplitude s and the estimated state label $\hat{y}$.

reversible computation. This dispersive model has the advantage that the control input signal can be reused.

Note that at control input amplitudes significantly different from the two control levels the outputs are coherent mixtures of the inputs, i.e., the switch then realizes a tunable beamsplitter.

Finally, we point out that using two different (frequency non-degenerate) resonator modes has the advantage that the interaction between control and signal inputs is phase insensitive which greatly simplifies the design and analysis of cascaded networks of such switches.

2.5 Generation of the estimated label

The estimated classifier label $\hat{y}$ should be a step function applied to the inner product of the weight vector and the input. In the preceding sections we have shown how individual inputs x_j can be amplified with programmable gain to give $\tilde{s}_j = \tilde{G}(\Phi_j)x_j$, thus realizing the individual contributions to the inner product. These are then summed on an n-port beamsplitter that has an output which gives the uniformly weighted sum $\tilde{s} := \frac{1}{\sqrt{N}}\sum_{k=1}^{N}\tilde{G}(\Phi_k)x_k$.

The gain factors $\tilde{G}(\Phi_k) = G_{rr}(\Phi_k) + iG_{ir}(\Phi_k)$ generally have an unwanted imaginary part which we subtract by passing the summed output through a *quadrature filter* circuit (cf. the last paragraph of Section 2.1), which has unit gain for the real quadrature and zero gain for the imaginary quadrature leading to an overall output $s = \mathrm{Re}\,\tilde{s} = \frac{1}{\sqrt{N}}\sum_{k=1}^{N}G_{rr}(\Phi_k)x_k$. The thresholding circuit should now produce a high output if $s > 0$ and a zero output if $s \leq 0$.

It turns out that the optical Fredkin gate described in the previous section already works almost as a two mode thresholder, where the control input leads to a step-like response in the signal outputs: A constant signal input amplitude which encodes the logical '1' state is applied to one of the signal inputs. When the control input amplitude is varied from zero to ξ_0, the signal output turns on fairly abruptly at some threshold $\xi_{th} < \xi_0$. To make the thresholding phase sensitive, the control input is given by the sum of s and a constant offset s_0 that provides a phase reference: $c = \frac{1}{\sqrt{2}}(s + s_0)$.

For a Fredkin gate operated with continuous control inputs the signal output is almost zero for a considerable range of small control inputs. However, for very high control inputs, i.e., significantly above ξ_0, the signal output decreases instead of staying constant as would be desirable for a step-function like profile. We found that this issue can be addressed by transmitting the control input through a single mode Kerr-nonlinear cavity, with resonance frequency chosen such that the transmission gain $|c'/c|$ is peaked close to $c' = \xi_0$. For input amplitudes larger than c, the transmission gain is lower (although $|c'|$ still grows monotonically with $|c|$) which extends the input range over which the subsequent Fredkin gate stays in the on-state.

3 Results

The perceptron's SDEs where simulated using a newly developed custom software package named QHDLJ [48] implemented in Julia [49] which allows for dynamic compilation of circuit models to LLVM [50] bytecode that runs at speed comparable to C/C++. All individual simulations can be carried out on a laptop, but the results in Figure 8 were obtained by averaging over the results of 100 stochastic simulation run on an HP ProLiant server with 80 cores. The current version of QHDLJ uses one process per trajectory, but the code could easily be vectorized.

In Figure 6 we present an example of a single application of an $N = 8$ perceptron including both a learning stage with pre-labeled training data and a classification testing stage in which the perceptron's estimated class labels are compared with their correct values. The data to be classified here are sampled from a different 8-dimensional Gaussian distribution for each class label with their mean vectors separated by a distance $\|\mu_1 - \mu_0\|_2/\sigma = 2$ relative to the standard deviation of both individual clusters. For each sample the input was held constant for a duration $\Delta t = 2\kappa^{-1}$ where κ is the NOPO signal and idler line width. The perceptron was first trained with $M_{\text{train}} = 100$ training examples and subsequently tested on $M_{\text{test}} = 100$ test examples with the learning feedback turned off.

In Figure 7 we visualize linear projections of the testing data as well as the estimated classification boundaries. We can see that the classifier performs very well far away from the decision boundary. Close to the decision boundary there are some misclassified examples. We proceed to compare the performance of the classifier to the theoretically optimal performance achievable by any classifier and with the optimal classifier for this scenario, Gaussian Discriminant Analysis (GDA) [51, 52], implemented in software. Using the identical perceptron model as above and an identical training/testing procedure, we estimate the error rate $p_{\text{err}} = \mathbb{P}[y \neq \hat{y}]$ of the trained perceptron as a function of the cluster separation $\|\mu_1 - \mu_0\|_2/\sigma$. The results are presented in Figure 8(a). Identically distributed training and testing data was used to evaluate the performance of the GDA algorithm and both results are compared to the theoretically optimal error rate for this discrimination task, which can be computed analytically to be $p_{\text{err, optim.}} = \frac{1}{2}\,\text{erfc}(\frac{\|\mu_1-\mu_0\|_2}{\sqrt{8}\sigma})$, where $\text{erfc}(x) = \frac{2}{\sqrt{\pi}}\int_x^\infty e^{-u^2}du$ is the complementary error function. We see that the all-optical perceptron's performance is comparable to GDA's performance for this problem and both algorithms attain performance close to the theoretical optimum.

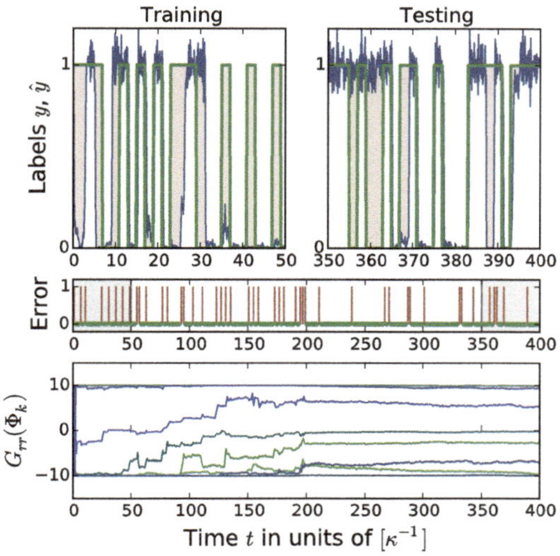

Figure 6 Single trajectory divided into a training interval $0 \le t \le M_{\mathrm{train}}\Delta t$ during which the learning feedback is active and a test interval $M_{\mathrm{train}}\Delta t < t \le M_{\mathrm{test}}\Delta t$. During training and testing, respectively, the system is driven by $M_{\mathrm{train}} = M_{\mathrm{test}} = 100$ separate input states which are held constant for an interval $\Delta t = 2\kappa^{-1}$. The estimated class label is discretized by averaging the output intensity over each input interval, dividing the result by the intensity $|\zeta|^2$ corresponding to the logical '1' output state and rounding. The upper panel compares the correct class label y (green) with the estimated class label $\hat{y}$ (black) during training and testing, respectively. The area between them indicates errors or at least lag of the estimator and is shaded in light red. The second panel shows occurrences of classification errors (red vertical bars). The slight shading near the beginning and the end of the trajectory in the second panel visualizes the segments corresponding to the upper left and right panel, respectively. The third panel shows the learned linear amplitude gains for each synapse. After the learning feedback is turned off at $t = M_{\mathrm{train}}\Delta t$, they diffuse slightly due to optical shot noise.

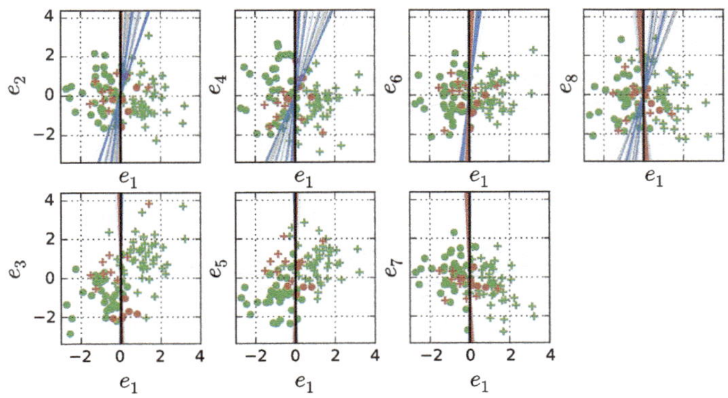

Figure 7 Projection of training data and classification boundaries. The data has been rotated such that the s_1 coordinate lines up with the learned normal vector of the separating hyperplane. Incorrectly classified data are plotted in red. The faint blue (red) lines visualize the evolution of the classifier boundary during training (testing).

The learning rate of the perceptron is determined by two things, the overall strength of the learning feedback as well as the time for which each example is presented to the circuit. In Figure 8(b) we plot the estimated error rate for varying feedback strength and duration. As can be expected intuitively, we find that there are trade-offs between speed (smaller Δt preferable) and energy consumption (smaller α preferable).

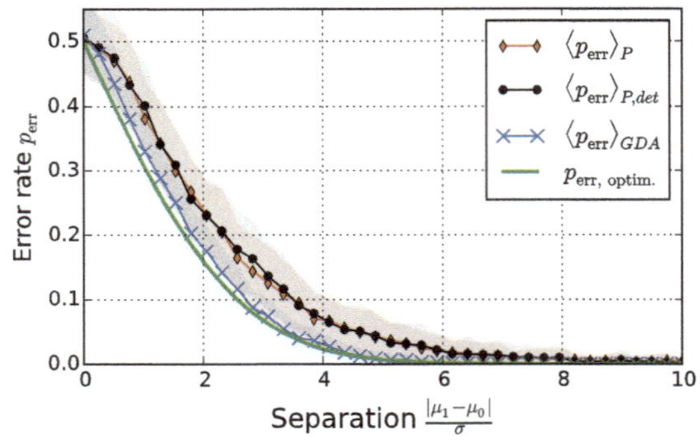

(a) Error rate vs. hardness

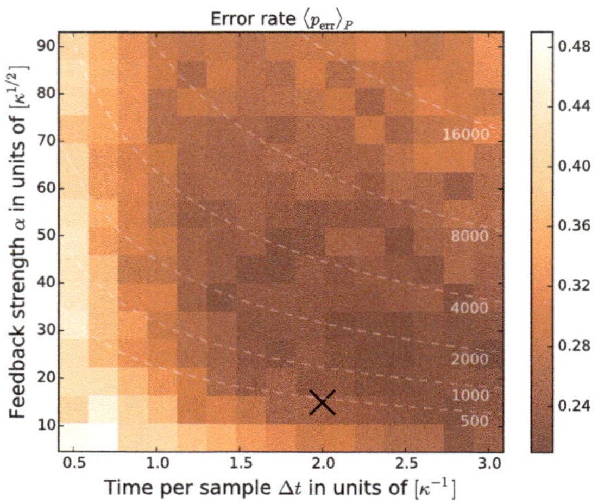

(b) Error rate vs. learning parameters

Figure 8 The perceptron's error rate vs the difficulty of the classification task and as a function of the parameters determining the learning rate. In part (**a**) we compare the unoptimized performance of the perceptron circuit (red diamonds) to the optimal performance bound (solid, green) as well as a GDA (blue ×'s) trained on the same number of training examples. We show averages over 100 trials at each cluster separation. The GDA data was similarly averaged over 100 trials. The transparent envelopes indicate the sample standard deviation. The black dots show the perceptron performance when simulated without shot noise. We see that the shot noise has very little effect. In part (**b**) we plot the average error rate (averaged over 50 trials) at fixed cluster separation $\|\mu_1 - \mu_0\|_2/\sigma = 2$ for various values of the time interval Δt for which each data sample is presented to the circuit as well as the strength of the training feedback α. The total number of feedback photons $N_{fb} = |\alpha|^2 \Delta t$ per sample is constant along the faint dashed lines and the actual value is indicated on the right. A good choice of parameters is characterized both by low feedback power (small $|\alpha|^2$) and high input rate (low sample time Δt) while still resulting in a low classification error rate. The × marks the parameters used for the results in (a) and the previous figures.

3.1 Time scales and power budget

Here we roughly estimate the power consumption of the whole device and discuss how to scale it up to a higher input dimension.

Any real-world implementation will depend strongly on the engineering paradigm, i.e., the choice of material/nonlinearity as well as the engineering precision, but based on recently achieved progress in nonlinear optics we will estimate an order of magnitude range for the input power.

The signal and feedback input power to the circuit will scale linearly in the number of synapses N.

The bias inputs for the amplifiers has to be larger than the signal to ensure linearly operation, but it should be expected that some of the scattered bias amplitudes can be reused to power multiple synapses.

In our models we have defined all rates relative to the line width of the signal and idler mode of the NOPO, because this is the component that should necessarily have the smallest decay rate to ensure a long lifetime for the memory.

All other resonators are employed as nonlinear input-output transformation devices and therefore a high bandwidth (corresponding to much lower loaded quality factor) is necessary for achieving a high bit rate. For our simulations we typically assumed quality factors that were lower than the NOPO's by 1-2 orders of magnitude. Based on self-oscillation threshold powers reported in [1, 2, 4, 53] and the switching powers of [7] we estimate the necessary power per synapse to be in the range of ~10-100 μWatt. By re-using the scattered pump and bias fields it should be possible to reduce the power consumption per amplifier even further. Even for the continuous wave signal paradigm we have assumed (as opposed to pulsed/spiking signals such as considered in [25]) the devices proposed here could be competitive with the current state of the art CMOS-based neuromorphic electrical circuits [22].

In the simulations for the 8-dimensional perceptron our input rate for training data was set to $\Delta t^{-1} = \frac{\kappa}{2}$. This value corresponds to roughly ten times the average feedback delay time between arrival of an input pattern and the conditional switching of the feedback logic upon arrival of the generated estimated state label $\hat{y}$. This time can be estimated as $\tau_{fb}(n) \approx G_{\max}\kappa_A^{-1} + \kappa_{QF}^{-1} + \kappa_{\text{thresh}}^{-1} + n\kappa_F^{-1}$, where n is the index of the synaptic weight, $G_{\max}$ is the amplifier gain range and κ_A, κ_{QF}, κ_{thresh} and κ_F are the line widths of the amplifier, quadrature filter, the combined thresholding circuit (cf. Figure 5) and the feedback Fredkin gates. There is a contribution scaling with n because the feedback traverses the individual weights sequentially to save power.

When scaling up the perceptron to a higher dimension while retaining approximately the same input signal powers, it is intuitively clear that the combined 'inner product' signal amplitude s scales as $s \propto \sqrt{N}s_1$, where s_1 is the signal amplitude for a single input. This allows to similarly scale up the amplitude ζ_0 of the signal encoding the generated estimated state label $\hat{y}$ and consequently the bandwidth of the feedback Fredkin gates that it drives. A detailed analysis reveals that the Fredkin gate threshold scales as $\sqrt{N}$, in particular we find that $\sqrt{|\chi|}\zeta_0 \propto \kappa_F \propto \sqrt{|\chi|}\xi_0 \propto \kappa_{\text{thresh}} \propto \sqrt{|\chi|s} \propto \sqrt{N|\chi|s_1}$. The first two scaling relationships are due to the constraints on the Fredkin gate construction (cf. Appendix A.2.2), the next two scaling relationships follow from demanding that the additional thresholding resonator be approximately dynamically resonant at the highest input level (cf. Appendices A.2.1 and A.2.2). The last proportionality is simply due to the amplitude summation at the N-port beamsplitter.

This reveals that when increasing N the perceptron as constructed here would have to be driven at a lower input bit rate scaling as $\Delta t^{-1} \propto N^{-\frac{1}{2}}$ or alternatively be driven with higher signal input powers. A possible solution that could greatly reduce the difference in arrival time $\sim \kappa_F^{-1}$ at each synapse could be to increase the waveguide-coupling to the control signal and thus decrease the delay per synapse. The resulting increase in the required control amplitude ζ_0 can be counter-acted with feedback, i.e., by effectively creating a large cavity around the control loop. When even this strategy fails one could add fan-out stages for $\hat{y}$ which introduce a delay that grows only logarithmically with N.

Finally, we note that the bias power of all the Kerr-effect based models considered here scales inversely with the respective nonlinear coefficient $\{|\zeta_0|^2, |s|^2\} \times |\chi| \sim$ const when keeping the bandwidth fixed. This implies that improvements in the nonlinear coefficient translate to lower power requirements or alternatively a faster speed of operation.

4 Conclusion and outlook

In conclusion we have shown how to design an all-optical device that is capable of supervised learning from input data, by describing how tunable gain amplifiers with signal/bias isolation can be constructed from nonlinear resonators and subsequently combined with self-oscillating resonators to encode the programmed amplifier gain in their oscillation phase. By considering a few additional nonlinear devices for thresholding and all-optical switching we then show how to construct a perceptron, including the perceptron feedback rule. To our knowledge this is the first end-to-end description of an all-optical circuit capable of learning from data. We have furthermore demonstrated that despite optical shot-noise it nearly attains the performance of the optimal software algorithm for the classification task that we considered. Finally, we have discussed the relevant time-scales and pointed out how to scale the circuit up to large input dimensions while retaining the signal processing bandwidth and a low power consumption per input.

Possible applications of an all-optical perceptron are as the trainable output filter of an optical reservoir computer or as a building block in a multi-layer all-optical neural network.

The programmable amplifier could be used as a building block to construct other learning models that rely on continuously tunable gain such as Boltzmann machines and hardware implementations of message passing algorithms.

An interesting next step would be to design a perceptron that can handle inputs at different carrier frequencies. In this case wavelength division multiplexing (WDM) might allow to significantly reduce the physical footprint of the device.

A simple modification of the perceptron circuit could autonomously learn to invert linear transformations that were applied to its input signals. This could be used for implementing a circuit capable of solving linear regression problems. In combination with a multi-mode optical fibers such a device could also have applications for all-optical sensing.

Finally, an extremely interesting question is whether harnessing quantum dynamics could lead to a performance increase. We hope to address these ideas in future work.

Appendix: Basic component models

Here we present the component models used to build the perceptron circuit. We will first describe the static components such as beamsplitters, phase shifts and coherent displacements, then proceed to describe the different Kerr-nonlinear models and finally the NOPO model.

A.1 Static, linear circuit components

All of these components have in common that they have no internal dynamics, implying that the A, B and C matrices and the a-vector have zero elements, and A_{NL} is not defined.

A.1.1 Constant laser source

The simplest possible static component is given by single input/output coherent displacement with coherent amplitude η. This model is employed to realize static coherent input amplitudes. The D matrix is trivially given by $D = (1)$ and the coherent amplitude is encoded in $c = (\eta)$. This leads to the desired input-output relationship $\beta_{\text{out}} = \eta + \beta_{\text{in}}$. For completeness we also provide the SLH [38] model $((1), (\eta), 0)$.

A.1.2 Static phase shifter

The static single input/outputs phase shifter has $D = (e^{i\phi})$ and $c = (0)$, leading to an input output relationship of $\beta_{\text{out}} = e^{i\phi}\beta_{\text{in}}$. Its SLH model is $((e^{i\phi}), (0), 0)$.

A.1.3 Beamsplitter

The static beamsplitter mixes (at least) two input fields and can be parametrized by a mixing angle θ. It has $D = \begin{pmatrix} \cos\theta & -\sin\theta \\ \sin\theta & \cos\theta \end{pmatrix}$ and $c = (0, 0)^T$. This leads to an input output relationship

$$\begin{pmatrix} \beta_{\text{out},1} \\ \beta_{\text{out},2} \end{pmatrix} = \begin{pmatrix} \cos\theta & -\sin\theta \\ \sin\theta & \cos\theta \end{pmatrix} \begin{pmatrix} \beta_{\text{in},1} \\ \beta_{\text{in},2} \end{pmatrix}. \tag{8}$$

Its SLH model is $\left(\begin{pmatrix} \cos\theta & -\sin\theta \\ \sin\theta & \cos\theta \end{pmatrix}, \begin{pmatrix} 0 \\ 0 \end{pmatrix}, 0\right)$.

A.2 Resonator models

We consider resonator models with m internal modes and n external inputs and outputs. We assume for simplicity that $a = \mathbf{0}$ and $c = \mathbf{0}$ meaning that we will model all coherent displacements explicitly in the fashion described above. We also assume that their scattering matrices are trivially given by $D = \mathbf{1}_n$ which means that far off-resonant input fields are simply reflected without a phase shift. Furthermore, none of our assumed models feature *linear* coupling between the internal cavity modes. This implies that the A-matrix is always diagonal. We are always working in a rotating frame.

A.2.1 Single mode Kerr-nonlinear resonator

A Kerr-nonlinearity is modeled by the nonlinear term $A_{\text{NL}}^{\text{Kerr}}(\alpha) = -i\chi|\alpha|^2\alpha$ which can be understood as an intensity dependent detuning. The A-matrix is given by $(-\frac{\kappa_T}{2} - i\Delta)$, its B-matrix is $-(\sqrt{\kappa_1}, \sqrt{\kappa_2}, \ldots, \sqrt{\kappa_n})$, where the total line width is given by $\sum_{j=1}^{n} \kappa_j = \kappa_T$ and the cavity detuning from any external drive is given by Δ. The C-matrix is given by $C = -B^T$. The corresponding SLH model is

$$\left(\mathbf{1}_n, \begin{pmatrix} \sqrt{\kappa_1}a \\ \vdots \\ \sqrt{\kappa_n}a \end{pmatrix}, \tilde{\Delta}a^\dagger a + \frac{\chi}{2}a^{2\dagger}a^2\right), \tag{9}$$

where the detuning differs slightly $\tilde{\Delta} = \Delta + \chi$ as can be shown in the derivation of the Wigner-formalism [10].

The special case of a single mirror with coupling rate κ and negligible internal losses is of interest for constructing the phase sensitive amplifier described in Section 2.1. Considering again an input given by a large static bias and a small signal $\epsilon = \frac{1}{\sqrt{2}}(\epsilon_0 + \delta\epsilon)$, the steady state reflected amplitude is to first order

$$\epsilon' \approx \frac{1}{\sqrt{2}}\left[\eta\epsilon_0 + g_-(\epsilon_0)\delta\epsilon + g_+(\epsilon_0)\delta\epsilon^*\right]. \tag{10}$$

For negligible internal losses we can give provide exact expressions for η, g_+ and g_-. Rather than parametrizing these by the bias ϵ_0 we parametrize them by the mean coherent intra-cavity amplitude α_0. When the system is not bi-stable (see below) relationship (14) defines a one-to-one map between ϵ_0 and α_0.

$$\eta = -\frac{\kappa/2 - i(\Delta + \chi|\alpha_0|^2)}{\kappa/2 + i(\Delta + \chi|\alpha_0|^2)} \quad \Rightarrow \quad |\eta| = 1, \tag{11}$$

$$g_- = 1 + \frac{\kappa[-\frac{\kappa}{2} + i\Delta + 2i\chi|\alpha_0|^2]}{(\frac{\kappa}{2})^2 + (\Delta + 2\chi|\alpha_0|^2)^2 - |\chi|^2|\alpha_0|^4}, \tag{12}$$

$$g_+ = \frac{i\kappa\chi\alpha_0^2}{(\frac{\kappa}{2})^2 + (\Delta + 2\chi|\alpha_0|^2)^2 - |\chi|^2|\alpha_0|^4}, \tag{13}$$

$$\epsilon_0 = -\frac{1}{\sqrt{\kappa}}\left[\frac{\kappa}{2} + i\left(\Delta + i\chi|\alpha_0|^2\right)\right]\alpha_0. \tag{14}$$

The Kerr cavity exhibits bistability for a particular interval of bias amplitudes if and only if $\Delta/\chi < 0$ and $|\Delta| \geq \frac{\sqrt{3}\kappa}{2} = \Delta_{\text{th}}$.

At any fixed bias amplitude and corresponding internal steady state mode amplitude the maximal gain experienced by a small signal is given by $g^{\max} = |g_-| + |g_+|$. Here maximal means that we maximize over all possible signal input phases relative to the bias input. To experience this gain, the signal has to be in an appropriate quadrature defined by $\arg\delta\epsilon = \frac{\arg g_- - \arg g_+}{2}$. The orthogonal quadrature is then maximally de-amplified by a gain of $||g_-| - |g_+||$ and it is possible to show that for negligible losses the perfect squeezing relationship $(|g_-| + |g_+|)||g_-| - |g_+|| = ||g_-|^2 - |g_+|^2| = 1$ holds for any bias amplitude. Furthermore, for fixed cavity parameters $g^{\max}$ is maximized at a particular non-zero intra-cavity photon amplitude

$$|\alpha_0^{\max}|^2 = \sqrt{\frac{\Delta^2 + \frac{\kappa^2}{4}}{3\chi^2}} \tag{15}$$

$$\Rightarrow \quad g^{\max} = \sqrt{\frac{\sqrt{f} + \kappa}{\sqrt{f} - \kappa}}, \quad \text{with } f = 28\Delta^2 + 4\kappa^2 - 8\Delta\sqrt{12\Delta^2 + 3\kappa^2}. \tag{16}$$

Note that the maximal gain does not depend on the strength of the nonlinearity. The relationship between $g^{\max}$ and Δ can be inverted:

$$\Delta = \frac{\sqrt{3}\kappa}{2}\frac{(g^{\max} - \sqrt{3})(g^{\max} - \frac{1}{\sqrt{3}})}{g^{\max 2} - 1}. \tag{17}$$

Using all this it is straightforward to construct a tunable Kerr-amplifier. The symmetric construction proposed in Section 2.1 provides the additional advantage that one does not have to cancel the scattered bias. It is also convenient to prepend and append phase shifters to the signal input and output that ensure $g_-, g_+ \in \mathbb{R}_{>0}$ at maximum gain.

The quadrature filter construction relies on the presence of additional cavity losses that are equal to the input coupler $\kappa_2 = \kappa_1 = \kappa$. In this case the gain coefficients for reflection of the first port are given by

$$g_- = 1 + \frac{\kappa[-\kappa + i\Delta + 2i\chi|\alpha_0|^2]}{\kappa^2 + (\Delta + 2\chi|\alpha_0|^2)^2 - |\chi|^2|\alpha_0|^4}, \tag{18}$$

$$g_+ = \frac{i\kappa\chi\alpha_0^2}{\kappa^2 + (\Delta + 2\chi|\alpha_0|^2)^2 - |\chi|^2|\alpha_0|^4}, \tag{19}$$

$$\epsilon_0 = -\frac{1}{\sqrt{\kappa}}\left[\kappa + i\left(\Delta + i\chi|\alpha_0|^2\right)\right]\alpha_0 \tag{20}$$

and one may easily verify that for dynamic resonance, i.e., $\chi|\alpha_0|^2 = -\Delta$, the gain coefficients are equal in magnitude $|g_-| = |g_+|$ which implies that there exists an input phase for which the reflected signal vanishes.

A.2.2 Two mode Kerr-nonlinear resonator

We label the mode amplitudes as α_1 and α_2. In this case the nonlinearity includes a cross-mode induced detuning

$$A_{\rm NL}^{\rm Kerr2}(\alpha) = \begin{pmatrix} -i\chi_a|\alpha_1|^2\alpha_1 - i\chi_{ab}|\alpha_2|^2\alpha_1 \\ -i\chi_{ab}|\alpha_1|^2\alpha_2 - i\chi_b|\alpha_2|^2\alpha_2 \end{pmatrix}. \tag{21}$$

The model matrices are

$$A = \begin{pmatrix} -\frac{\kappa_{a,T}}{2} - i\Delta_a & 0 \\ 0 & -\frac{\kappa_{b,T}}{2} - i\Delta_b \end{pmatrix}, \tag{22}$$

$$B = -\begin{pmatrix} \sqrt{\kappa_{a,1}} & \sqrt{\kappa_{a,2}} & \dots & \sqrt{\kappa_{a,n_a}} & 0 & \dots & 0 \\ 0 & 0 & \dots & 0 & \sqrt{\kappa_{b,1}} & \sqrt{\kappa_{b,2}} & \dots & \sqrt{\kappa_{b,n_b}} \end{pmatrix}, \tag{23}$$

$$C = -B^T, \tag{24}$$

and the corresponding SLH model is

$$\left(\mathbf{1}_{n_a+n_b}, C\begin{pmatrix} a \\ b \end{pmatrix}, \tilde{\Delta}_a a^\dagger a + \tilde{\Delta}_b b^\dagger b + \frac{\chi_a}{2}a^{2\dagger}a^2 + \frac{\chi_b}{2}b^{2\dagger}b^2 + \chi_{ab}a^\dagger a b^\dagger b\right), \tag{25}$$

with $\tilde{\Delta}_{a/b} = \Delta_{a/b} + \chi_{a/b} + \frac{\chi_{ab}}{2}$ and where the Wigner-correspondence[c] is $\langle\alpha_1\rangle_{\rm W} = \langle a\rangle$, $\langle\alpha_2\rangle_{\rm W} = \langle b\rangle$.

We briefly summarize how to construct a controlled phase shifter using an ideal two-mode Kerr cavity with a single input coupling to each mode and negligible additional internal losses. We exploit that in this case the reflected steady state signal amplitude ζ' is identical to the input amplitude ζ up to a power dependent phase shift

$$\zeta' = -\frac{\frac{\kappa_a}{2} - i(\Delta_a + i\chi_a|\alpha_0|^2 + i\chi_{ab}|\beta_0|^2)}{\frac{\kappa_a}{2} + i(\Delta_a + i\chi_a|\alpha_0|^2 + i\chi_{ab}|\beta_0|^2)}\zeta \quad \Rightarrow \quad |\zeta'| = |\zeta|. \tag{26}$$

We assume that the control input amplitude takes on two discrete values $\xi = 0$ or $\xi = \xi_0$ and that variations of the signal input amplitude are small $|\zeta| \approx |\zeta_0|$. In this case a good choice of detunings and coupling rates is given by

$$\Delta_a = \frac{\kappa_a}{2} - \frac{2\chi_a|\zeta_0|^2}{\kappa_a}, \tag{27}$$

$$\Delta_b = \frac{\kappa_a\chi_b}{\chi_{ab}} - \frac{2\chi_{ab}|\zeta_0|^2}{\kappa_a}, \tag{28}$$

$$\xi_0 = \frac{\sqrt{\kappa_a \kappa_b}}{2\sqrt{|\chi_{ab}|}} \tag{29}$$

in addition to two inequality constraints

$$\Delta_a \leq \sqrt{3}\frac{\kappa_a}{2}, \tag{30}$$

$$\Delta_b \leq \sqrt{3}\frac{\kappa_b}{2} \tag{31}$$

that ensure that the system is stable. This construction ensures that $\frac{\zeta'|_{\xi=\xi_0}}{\zeta'|_{\xi=0}} = -1$ and in fact it can easily be generalized to the more realistic case of non-negligible internal losses.

Finally note that the inequality constraints imply that the lower bounds for the input couplings scale as $\kappa_a^{\min}, \kappa_b^{\min} \propto |\zeta_0|$ which is important for our power analysis in Section 3.1. This, in turn implies that $\xi_0 \propto |\zeta_0|$ which is a fairly intuitive result.

The controlled phase shifter can now be included in one arm of a Mach-Zehnder interferometer to create a Fredkin gate (cf. Section 2.4).

To realize a thresholder, the control mode input is prepended with a two port Kerr-cavity with parameters chosen such that it becomes dynamically resonant with maximal differential transmission gain close to where its output gives the correct high control input ξ_0.

Overall, we remark that even when we account for the prepended cavity, the relationship $c \propto |\zeta_0|$ still holds, where c is the input to the thresholder. To see how the total decay rate of the thresholding cavity κ_{thresh} scales consider first that to get maximum differential gain or contrast, we ought pick a detuning right at or below the Kerr stability threshold $\Delta \approx \Delta_{\text{th}} = \sqrt{3}\kappa_{\text{thresh}}/2$.

We choose the maximum input amplitude such that it approximately achieves dynamic resonance within the prepended thresholding cavity. This occurs when $\Delta = -\chi|\alpha_0|^2$ (cf. Appendix A.2.1) and at an input amplitude of $c \propto \sqrt{\kappa_{\text{thresh}}|\frac{\Delta}{\chi}|} \propto \kappa_{\text{thresh}}$.

A.2.3 NOPO model

The NOPO model has consists of three modes, the signal and idler modes α_s, α_i and the pump mode α_p. We assume a triply resonant model[d] and that $\omega_s + \omega_i = \omega_p$, allowing for resonant conversion of pump photons into pairs of signal and idler photons and vice versa. The nonlinearity is given by

$$A_{\text{NL}}^{\text{NOPO}}(\alpha) = \begin{pmatrix} \chi\alpha_i^*\alpha_p \\ \chi\alpha_s^*\alpha_p \\ -\chi\alpha_s\alpha_i \end{pmatrix} \tag{32}$$

and the model matrices are

$$A = \begin{pmatrix} -\frac{\kappa}{2} & 0 & 0 \\ 0 & -\frac{\kappa}{2} & 0 \\ 0 & 0 & -\frac{\kappa_p}{2} \end{pmatrix}, \quad B = -\begin{pmatrix} \sqrt{\kappa} & 0 & 0 \\ 0 & \sqrt{\kappa} & 0 \\ 0 & 0 & \sqrt{\kappa_p} \end{pmatrix}, \tag{33}$$

$$C = -B^T. \tag{34}$$

Here, the SLH model is given by

$$\left(\mathbf{1}_3, C\begin{pmatrix} a \\ b \\ c \end{pmatrix}, i\chi\left(abc^\dagger - a^\dagger b^\dagger c\right)\right), \tag{35}$$

where now a, b and c correspond to α_s, α_i and α_p.

A steady state analysis of the system driven only by a pump input amplitude ϵ reveals that below a critical threshold $|\epsilon| < \epsilon_{\rm th} = \frac{\kappa\sqrt{\kappa_p}}{4\chi}$ the system as a unique fixpoint with $\alpha_s = \alpha_i = 0$ and $\alpha_p = -\frac{2\epsilon}{\sqrt{\kappa_p}}$. Above threshold $|\epsilon| \geq \epsilon_{\rm th}$, the intra-cavity pump amplitude stays constant at the threshold value $\alpha_p = -\frac{2\epsilon_{\rm th}\epsilon/|\epsilon|}{\sqrt{\kappa_p}} = -\frac{\kappa\epsilon/|\epsilon|}{2\chi}$ and the signal and idler mode obtain non-zero magnitude

$$|\alpha_s| = |\alpha_i| = \sqrt{\frac{4\epsilon_{\rm th}}{\kappa}\left(|\epsilon| - \epsilon_{\rm th}\right)}. \tag{36}$$

As an interesting consequence of the model's symmetry there exists not a single above threshold state but a whole manifold of fixpoints parametrized by a correlated signal and idler phase

$$\alpha_s = \sqrt{\frac{4\epsilon_{\rm th}}{\kappa}\left(|\epsilon| - \epsilon_{\rm th}\right)}e^{i\phi + i\phi_0}, \tag{37}$$

$$\alpha_i = \sqrt{\frac{4\epsilon_{\rm th}}{\kappa}\left(|\epsilon| - \epsilon_{\rm th}\right)}e^{-i\phi + i\phi_0}, \tag{38}$$

where the common phase ϕ_0 is fixed by the pump input phase via

$$\alpha_s\alpha_i = -\frac{4\epsilon_{\rm th}}{\kappa}\left(|\epsilon| - \epsilon_{\rm th}\right)\frac{\epsilon}{|\epsilon|}. \tag{39}$$

In particular, for $\epsilon < 0$ we have $\alpha_i = \alpha_s^*$. Above threshold the system will rapidly converge to a fixpoint of well-defined phase ϕ. Without quantum shot noise ϕ would remain constant. With noise, however, the system can freely diffuse along the manifold. When the pump bias input is sufficiently large compared to threshold and consequently there are many signal and idler photons present in the cavity at any given time ($|\alpha_{s/i}|^2 \gg 1$) one can analyze the dynamics along the manifold and of small orthogonal deviations from the manifold. In the symmetric case considered here where signal and idler have equal decay rates, the differential phase degree of freedom $\phi = \frac{\arg\alpha_i - \arg\alpha_s}{2}$ decouples from all other variables and approximately obeys the SDE

$$d\phi = \sqrt{\gamma_\phi}\, dW_t, \qquad dW_t^2 = dt \tag{40}$$

$$\text{with } \gamma_\phi = \frac{\kappa}{8|\alpha_s|^2} = \frac{\kappa^2}{32\epsilon_{\rm th}\left(|\epsilon| - \epsilon_{\rm th}\right)}. \tag{41}$$

It is relatively straightforward to generalize these results to a less symmetric model with different signal and idler couplings and even non-zero detunings, but for a given nonlinearity the model considered here provides the smallest phase diffusion and thus the best analog memory. For a very thorough analysis of this model we refer to [54].

A.3 Composite component models

Due to the scope of this article, we will refrain from including the full net lists for the composite component models in this article and instead publish them online at [41]. We remark that composing a photonic circuit from the above described nonlinear photonic models is often complicated by the fact that the steady state input-output relationships are hard or even impossible to invert analytically. A systematic approach to optimizing component parameters would be highly desirable.

Competing interests

The authors declare that they have no competing interests.

Authors' contributions

All authors contributed equally to the writing of this paper. All authors read and approved the final manuscript.

Acknowledgements

This work is supported by DARPA-MTO under award no. N66001-11-1-4106. NT acknowledges support from a Stanford Graduate Fellowship. We would also like to thank Ryan Hamerly, Jeff Hill, Peter McMahon and Amir Safavi-Naeini for helpful discussion.

Endnotes

[a] One can easily convince oneself that all even order contributions are scattered into the bias output.

[b] In the photonics community this is referred to as *critically coupled*, whereas the amplifier circuit would ideally be strongly *overcoupled* such that additional internal losses are negligible.

[c] In this appendix we denote expectations with respect to the Wigner function as $\langle\cdot\rangle_W$ and quantum mechanical expectations as $\langle\cdot\rangle$.

[d] It is possible to drop this resonance assumption for the pump.

References

1. Kippenberg T, Spillane S, Vahala K. Kerr-nonlinearity optical parametric oscillation in an ultrahigh-Q toroid microcavity. Phys Rev Lett. 2004;93(8):83904.
2. Del'Haye P, Schliesser A, Arcizet O, Wilken T, Holzwarth R, Kippenberg TJ. Optical frequency comb generation from a monolithic microresonator. Nature. 2007;450(7173):1214-7.
3. Levy M. Nanomagnetic route to bias-magnet-free, on-chip Faraday rotators. J Opt Soc Am B. 2005;22(1):254-60.
4. Razzari L, Duchesne D, Ferrera M, Morandotti R, Chu S, Little BE, Moss DJ. CMOS-compatible integrated optical hyper-parametric oscillator. Nat Photonics. 2009;4(1):41-5.
5. Englund D, Faraon A, Fushman I, Stoltz N, Petroff P, Vučković J. Controlling cavity reflectivity with a single quantum dot. Nature. 2007;450(7171):857-61.
6. Fushman I, Englund D, Faraon A, Stoltz N, Petroff P, Vuckovic J. Controlled phase shifts with a single quantum dot. Science. 2008;320(5877):769-72.
7. Nozaki K, Tanabe T, Shinya A, Matsuo S. Sub-femtojoule all-optical switching using a photonic-crystal nanocavity. Nat Photonics. 2010;4:477-83.
8. Cohen JD, Meenehan SM, MacCabe GS, Gröblacher S, Safavi-Naeini AH, Marsili F, Shaw MD, Painter O. Phonon counting and intensity interferometry of a nanomechanical resonator. p. 1-10. arXiv preprint, arXiv:1410.1047 (2014).
9. Vandoorne K, Mechet P, Van Vaerenbergh T, Fiers M, Morthier G, Verstraeten D, Schrauwen B, Dambre J, Bienstman P. Experimental demonstration of reservoir computing on a silicon photonics chip. Nat Commun. 2014;5:1-6.
10. Santori C, Pelc JS, Beausoleil RG, Tezak N, Hamerly R, Mabuchi H. Quantum noise in large-scale coherent nonlinear photonic circuits. Phys Rev Applied. 2014;1:054005.
11. Mabuchi H. Nonlinear interferometry approach to photonic sequential logic. Appl Phys Lett. 2011;99(15):153103.
12. Pavlichin DS, Mabuchi H. Photonic circuits for iterative decoding of a class of low-density parity-check codes. New J Phys. 2013;16:105017. doi:10.1088/1367-2630/16/10/105017.
13. Abraham E, Smith SD. Optical bistability and related devices. Rep Prog Phys. 1982;45:815.
14. Smith SD. Optical bistability: towards the optical computer. Nature. 1984;307(5949):315-6.
15. Miller DAB. Physical reasons for optical interconnection. Int J Optoelectron. 1997;11(3):155-68.
16. Miller DAB. Are optical transistors the logical next step? Nat Photonics. 2010;4(1):3-5.
17. Ananthanarayanan R, Esser SK, Simon HD, Modha DS. The cat is out of the bag. In: Proceedings of the conference on high performance computing networking, storage and analysis - SC '09 (c); 2009.
18. Neven H. Hardware initiative at quantum artificial intelligence lab. 2014. http://googleresearch.blogspot.com/2014/09/ucsb-partners-with-google-on-hardware.html.
19. Strukov DB, Snider GS, Stewart DR, Williams RS. The missing memristor found. Nature. 2008;453(7191):80-3.
20. Wang Z, Marandi A, Wen K, Byer RL, Yamamoto Y. Coherent Ising machine based on degenerate optical parametric oscillators. Phys Rev A. 2013;88(6):063853.
21. Choudhary S, Sloan S, Fok S, Neckar A, Trautmann E, Gao P, Stewart T, Eliasmith C, Boahen K. Silicon neurons that compute. In: Artificial neural networks and machine learning ICANN 2012. Lecture notes in computer science vol. 7552. Berlin: Springer; 2012. p. 121-8.

22. Cassidy AS, Alvarez-Icaza R, Akopyan F, Sawada J, Arthur JV, Merolla PA, Datta P, Tallada MG, Taba B, Andreopoulos A, Amir A, Esser SK, Kusnitz J, Appuswamy R, Haymes C, Brezzo B, Moussalli R, Bellofatto R, Baks C, Mastro M, Schleupen K, Cox CE, Inoue K, Millman S, Imam N, Mcquinn E, Nakamura YY, Vo I, Guok C, Nguyen D, Lekuch S, Asaad S, Friedman D, Jackson BL, Flickner MD, Risk WP, Manohar R, Modha DS. Real-time scalable cortical computing at 46 giga-synaptic OPS/Watt with ~100× speedup in time-to-solution and ~100,000× reduction in energy-to-solution. In: Proceedings of the international conference for high performance computing, networking, storage and analysis; 2014. p. 27-38.
23. Duport F, Schneider B, Smerieri A, Haelterman M, Massar S. All-optical reservoir computing. Opt Express. 2012;20(20):22783.
24. Vandoorne K, Dambre J, Verstraeten D, Schrauwen B, Bienstman P. Parallel reservoir computing using optical amplifiers. IEEE Trans Neural Netw. 2011;22(9):1469-81.
25. Van Vaerenbergh T, Fiers M, Mechet P, Spuesens T, Kumar R, Morthier G, Schrauwen B, Dambre J, Bienstman P. Cascadable excitability in microrings. Opt Express. 2012;20(18):20292.
26. Dejonckheere A, Duport F, Smerieri A, Fang L, Oudar J-L, Haelterman M, Massar S. All-optical reservoir computer based on saturation of absorption. Opt Express. 2014;22(9):10868.
27. Larger L, Soriano MC, Brunner D, Appeltant L, Gutierrez JM, Pesquera L, Mirasso CR, Fischer I. Photonic information processing beyond turing: an optoelectronic implementation of reservoir computing. Opt Express. 2012;20(3):3241-9.
28. Cover TM. Geometrical and statistical properties of systems of linear inequalities with applications in pattern recognition. IEEE Trans Electron Comput. 1965;EC-14(3):326-34.
29. Verstraeten D. Reservoir computing: computation with dynamical systems. PhD thesis (2010).
30. Rosenblatt F. The perceptron - a perceiving and recognizing automaton. Report 85, Cornell Aeronautical Laboratory; 1957.
31. Rosenblatt F. The perceptron: a probabilistic model for information storage and organization in the brain. Psychol Rev. 1958;65:386-408.
32. Miller DAB. Novel analog self-electrooptic-effect devices. IEEE J Quantum Electron. 1993;29:678-98.
33. Fok MP, Tian Y, Rosenbluth D, Prucnal PR. Pulse lead/lag timing detection for adaptive feedback and control based on optical spike-timing-dependent plasticity. Opt Lett. 2013;38(4):419-21.
34. Boixo S, Rønnow TF, Isakov SV, Wang Z, Wecker D, Lidar DA, Martinis JM, Troyer M. Evidence for quantum annealing with more than one hundred qubits. Nat Phys. 2014;10(3):218-24.
35. Boixo S, Smelyanskiy VN, Shabani A, Isakov SV, Dykman M, Denchev VS, Amin M, Smirnov A, Mohseni M, Neven H. Computational role of collective tunneling in a quantum annealer. arXiv:1411.4036v1 (2014).
36. Cortes C, Vapnik V. Support-vector networks. Mach Learn. 1995;20(3):273-97.
37. Tezak N, Niederberger A, Pavlichin DS, Sarma G, Mabuchi H. Specification of photonic circuits using Quantum Hardware Description Language. Philos Trans R Soc A, Math Phys Eng Sci. 2012;370(1979):5270-90.
38. Gough J, James MRR. The series product and its application to quantum feedforward and feedback networks. IEEE Trans Autom Control. 2009;54(11):2530-44.
39. Gough J, James MR. Quantum feedback networks: Hamiltonian formulation. Commun Math Phys. 2008;287(3):1109-32.
40. Hamerly R, Mabuchi H. Quantum noise of free-carrier dispersion in semiconductor optical cavities. arXiv:1504.04409 (2015).
41. Tezak N. Perceptron-files. https://github.com/ntezak/perceptron-files (2014).
42. Tait AN, Shastri BJ, Fok MP, Nahmias MA, Prucnal PR. The DREAM: an integrated photonic thresholder. J Lightwave Technol. 2013;31(8):1263-72.
43. Sussillo D, Barak O. Opening the black box: low-dimensional dynamics in high-dimensional recurrent neural networks. Neural Comput. 2013;25(3):626-49.
44. Savchenkov A, Matsko A, Strekalov D, Mohageg M, Ilchenko V, Maleki L. Low threshold optical oscillations in a whispering gallery mode CaF_2 resonator. Phys Rev Lett. 2004;93(24):243905.
45. Poustie AJJ, Blow KJJ. Demonstration of an all-optical Fredkin gate. Opt Commun. 2000;174(1-4):317-20.
46. Milburn G. Quantum optical Fredkin gate. Phys Rev Lett. 1989;62(18):2124-7.
47. Fredkin E, Toffoli T. Conservative logic. Int J Theor Phys. 1982;21(3/4):219-53.
48. Tezak N. QHDLJ. https://bitbucket.org/ntezak/qhdlj.jl (2014).
49. Bezanson J, Edelman A, Karpinski S, Shah VBJ. Julia: A fresh approach to numerical computing. arXiv:1411.1607 (2014).
50. Lattner C, Adve V. LLVM: a compilation framework for lifelong program analysis and transformation. In: International symposium on code generation and optimization, 2004. CGO 2004; 2014. p. 75-86.
51. Fisher RA. The use of multiple measurements in taxonomic problems. Ann Eugen. 1936;7(2):179-88.
52. McLachlan GJ. Discriminant analysis and statistical pattern recognition. Wiley series in probability and statistics. Hoboken: Wiley; 1992.
53. Levy JS, Gondarenko A, Foster MA, Turner-Foster AC, Gaeta AL, Lipson M. CMOS-compatible multiple-wavelength oscillator for on-chip optical interconnects. Nat Photonics. 2009;4(1):37-40.
54. Graham R, Haken H. The quantum-fluctuations of the optical parametric oscillator. I. Z Phys. 1968;210(3):276-302.

14

Fermionic models with superconducting circuits

Urtzi Las Heras[1*], Laura García-Álvarez[1], Antonio Mezzacapo[1], Enrique Solano[1,2] and Lucas Lamata[1]

*Correspondence: ulasheras@gmail.com
[1] Department of Physical Chemistry, University of the Basque Country UPV/EHU, Apartado 644, 48080 Bilbao, Spain
Full list of author information is available at the end of the article

Abstract

We propose a method for the efficient quantum simulation of fermionic systems with superconducting circuits. It consists in the suitable use of Jordan-Wigner mapping, Trotter decomposition, and multiqubit gates, be with the use of a quantum bus or direct capacitive couplings. We apply our method to the paradigmatic cases of 1D and 2D Fermi-Hubbard models, involving couplings with nearest and next-nearest neighbours. Furthermore, we propose an optimal architecture for this model and discuss the benchmarking of the simulations in realistic circuit quantum electrodynamics setups.

PACS Codes: 03.67.Ac; 42.50.Pq; 85.25.Cp; 05.30.Fk

Keywords: quantum information; quantum simulation; superconducting circuits

1 Background

Quantum simulations are one of the most promising research fields in quantum information, allowing the possibility of solving problems exponentially faster than classical computers [1]. In those cases in which analog quantum simulation is hard or impossible, one may decompose the simulated quantum dynamics in terms of discrete quantum gates through a technique known as digital quantum simulation [2–4]. Problems involving interacting fermions are frequently intractable for classical computers due to, among other features, the exponential growth of the Hilbert space dimension with the size of the system. Moreover, standard numerical methods such as quantum Monte Carlo algorithms, do not converge for fermionic systems. Indeed, neither fermionic models in more than one dimension nor systems with the well-known sign problem [5] can be efficiently simulated employing a classical computer. Quantum simulations allow us the reproduction and study of complex systems by means of the use of minimal experimental resources and going beyond mean field approximations in numerical calculations.

Circuit quantum electrodynamics (cQED) [6] is one of the most advanced quantum technologies in terms of coherent control and scalability aspects. Several analog quantum simulations have been proposed in this quantum platform, e.g., spin models [7], quantum phase transitions [8], spin glasses [9], disordered systems [10], metamaterials [11], time symmetry breaking [12], topological order [13], atomic physics [14], open systems [15], dynamical gauge theories [16], and fermionic models in one dimension [17], among others. Furthermore, digital quantum simulations have been recently proposed for superconducting circuits [18, 19] and two pioneering experiments have been performed [20, 21].

In this article, we present a method for encoding the simulation of fermionic systems for arbitrary spatial dimensions, long range or short range couplings, and highly nonlinear interactions, in superconducting circuits. For this purpose, we differentiate two kinds of cQED setups, those employing pairwise capacitive qubit interactions [22], and the ones employing microwave resonators as quantum buses [23]. Our method can be summarized in three steps. The first step consists in mapping a set of N fermionic modes to N spin operators via the Jordan-Wigner transformation [24]. Then, we make use of the Trotter expansion [2–4] to decompose the unitary evolution of the simulated system in a sequence of quantum gates. Finally, many-body interactions [25, 26] are implemented either with a sequence of capacitive two-qubit gates or by fast multiqubit gates mediated by resonators [27]. Our method allows to implement highly nonlinear and long-range interactions employing only polynomial resources, which makes it suitable for simulating complex physical problems intractable for classical computers. To this extent, we analyze the simulation of the Fermi-Hubbard model with different cQED architectures, considering couplings with nearest neighbours and next-nearest neighbours in two-dimensional fermionic lattices. The structure of the article is the following. In Section 2, we explain the method for decomposing a fermionic dynamics via digital techniques. In Section 3, we describe the proposal for implementing the Fermi-Hubbard model in two distinct situations, either with pairwise capacitive couplings or via resonators. Finally, in Section 4 we give our conclusions.

2 Jordan-Wigner mapping and Trotter expansion

The Jordan-Wigner (JW) transformation allows one to map fermionic creation and annihilation operators onto spin operators. When the fermionic lattice is two or three-dimensional, it is possible that local fermionic interactions are mapped onto nonlocal spin ones. The JW mapping reads $b_k^\dagger = I_N \otimes I_{N-1} \otimes \cdots \otimes \sigma_k^+ \otimes \sigma_{k-1}^z \otimes \cdots \otimes \sigma_1^z$, and $b_k = (b_k^\dagger)^\dagger$, where $b_k(b_k^\dagger)$ are the fermionic annihilation and creation operators and σ_i^α are the spin operators of the ith site, being σ^α for $\alpha = x, y, z$ the Pauli matrices and $\sigma^+ = (\sigma^x + i\sigma^y)/2$.

Often, the simulating system cannot provide in a simple manner the dynamical structure of the simulated systems. Therefore, one may feel compelled to use digital methods for implementing unnatural interactions in the controllable system, based on the decomposition of the exact unitary evolution into a sequence of discrete gates [2]. In this sense, one can use the Trotter formula [3] in order to obtain a polynomial sequence of efficiently implementable gates. The Trotter formula is an approximation of the unitary evolution e^{-iHt}, where H is the simulated Hamiltonian, consisting of M quantum gates e^{-iH_it} that fulfill the condition $H = \sum_i^M H_i$, being H_i the natural interaction terms of the controllable system. The Trotter expansion can be written as ($\hbar = 1$)

$$e^{-iHt} \simeq \left(e^{-iH_1t/l} \cdots e^{-iH_Mt/l}\right)^l + \sum_{i<j} \frac{[H_i, H_j]t^2}{2l}. \tag{1}$$

Here, e^{-iH_it} are the gates that can be implemented in the controllable system and l is the total number of Trotter steps. By shortening the execution times of the gates and applying the protocol repeatedly, the digitized unitary evolution becomes more accurate. As can be seen in Eq. (1), the error estimate in this approximation scales with t^2/l, such that the longer the simulated time is, the more digital steps we need to apply in order to get good fidelities.

3 Circuit QED implementation

3.1 Fermi-Hubbard model: small lattices with pairwise interactions

In this section, we present a cQED encoding of the Fermi-Hubbard model, as an example of a fermionic model with nearest-neighbour pairwise interactions, which hence employs only pairwise capacitive spin-spin interactions. Although we focus on a model with three fermionic modes, for the sake of clarity, these techniques are straightforwardly extendable to arbitrary number of fermionic modes in two and three spatial dimensions. These cases are in general mapped into multi-qubit gates that can be always polynomially decomposed into sets of two-qubit gates, as shown below in Eq. (7). In the last part of Section 3, we focus on another cQED platform that uses resonators instead of direct qubit couplings to mediate the interactions.

The Fermi-Hubbard dynamics is a condensed matter model describing traveling electrons in a lattice. The model captures the competition between the kinetic energy of the electrons, discretized in a lattice and encoded in a hopping term, with their Coulomb interaction that is expressed by a nonlinear term. We begin by considering a small lattice realizable with current technology. We consider the Fermi-Hubbard-like model for three spinless fermions with open boundary conditions,

$$H = -h\left(b_1^\dagger b_2 + b_2^\dagger b_1 + b_2^\dagger b_3 + b_3^\dagger b_2\right) + U\left(b_1^\dagger b_1 b_2^\dagger b_2 + b_2^\dagger b_2 b_3^\dagger b_3\right). \tag{2}$$

Here, $b_m^\dagger$ and b_m are fermionic creation and annihilation operators for the site m.

We will use the Jordan-Wigner transformation in our derivation to encode the fermionic operators into tensor products of Pauli matrices. We will show below that the latter may be efficiently implemented in superconducting circuits. The Jordan-Wigner mapping reads,

$$\begin{aligned} b_1^\dagger &= \mathbb{I} \otimes \mathbb{I} \otimes \sigma^+, \\ b_2^\dagger &= \mathbb{I} \otimes \sigma^+ \otimes \sigma^z, \\ b_3^\dagger &= \sigma^+ \otimes \sigma^z \otimes \sigma^z. \end{aligned} \tag{3}$$

Afterwards, we rewrite the Hamiltonian in Eq. (2) in terms of spin-1/2 operators,

$$\begin{aligned} H = &\frac{h}{2}\left(\mathbb{I} \otimes \sigma^x \otimes \sigma^x + \mathbb{I} \otimes \sigma^y \otimes \sigma^y + \sigma^x \otimes \sigma^x \otimes \mathbb{I} + \sigma^y \otimes \sigma^y \otimes \mathbb{I}\right) \\ &+ \frac{U}{4}\left(\mathbb{I} \otimes \sigma^z \otimes \sigma^z + \mathbb{I} \otimes \sigma^z \otimes \mathbb{I} + \mathbb{I} \otimes \mathbb{I} \otimes \sigma^z + \sigma^z \otimes \sigma^z \otimes \mathbb{I}\right. \\ &\left. + \sigma^z \otimes \mathbb{I} \otimes \mathbb{I} + \mathbb{I} \otimes \sigma^z \otimes \mathbb{I}\right). \end{aligned} \tag{4}$$

Here, the different interactions can be simulated via digital techniques using a specific sequence of gates. We will first consider the associated Hamiltonian evolution in terms of $\exp(-i\phi\sigma^z \otimes \sigma^z)$ interactions. These can be implemented in small steps of CZ_ϕ gates, where an average single-qubit gate and two-qubit gate fidelities of 99.92% and up to 99.4%, respectively, have been recently achieved [22]. One can then use the following relations,

$$\begin{aligned} \sigma^x \otimes \sigma^x &= R_y(\pi/2)\sigma^z \otimes \sigma^z R_y(-\pi/2), \\ \sigma^y \otimes \sigma^y &= R_x(-\pi/2)\sigma^z \otimes \sigma^z R_x(\pi/2), \end{aligned} \tag{5}$$

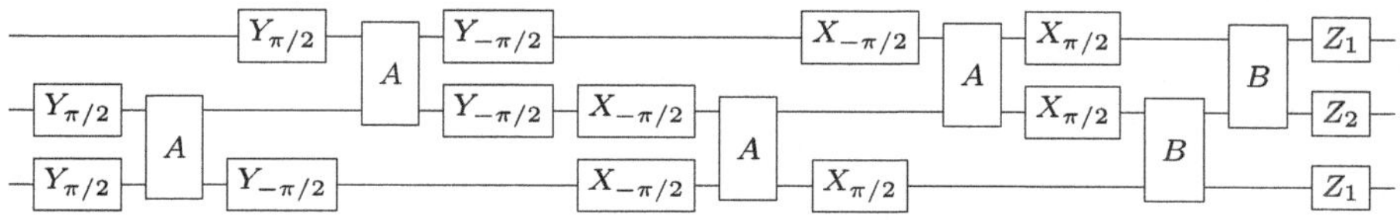

Figure 1 Sequence of gates for one Trotter step of Hamiltonian 2.

where $R_j(\theta) = \exp(-i\frac{\theta}{2}\sigma^j)$ denote local rotations along the jth axis of the Bloch sphere acting on both qubits.

The evolution operator associated with the Hamiltonian in Eq. (4) can be expressed in terms of $\exp(-i\phi\sigma^z \otimes \sigma^z)$ interactions. Moreover, the operators may be rearranged in a more suitable way in order to optimise the number of gates and eliminate global phases,

$$\begin{aligned}\exp(-iHt) \approx \Bigg[& R'_y(\pi/2)\exp\left(-i\frac{h}{2}\mathbb{I}\otimes\sigma^z\otimes\sigma^z\frac{t}{n}\right)R'_y(-\pi/2)R_y(\pi/2) \\ & \times \exp\left(-i\frac{h}{2}\sigma^z\otimes\sigma^z\otimes\mathbb{I}\frac{t}{n}\right)R_y(-\pi/2)R'_x(-\pi/2) \\ & \times \exp\left(-i\frac{h}{2}\mathbb{I}\otimes\sigma^z\otimes\sigma^z\frac{t}{n}\right)R'_x(\pi/2)R_x(-\pi/2) \\ & \times \exp\left(-i\frac{h}{2}\sigma^z\otimes\sigma^z\otimes\mathbb{I}\frac{t}{n}\right)R_x(\pi/2)\exp\left(-i\frac{U}{4}\mathbb{I}\otimes\sigma^z\otimes\sigma^z\frac{t}{n}\right) \\ & \times \exp\left(-i\frac{U}{2}\mathbb{I}\otimes\sigma^z\otimes\mathbb{I}\frac{t}{n}\right)\exp\left(-i\frac{U}{4}\mathbb{I}\otimes\mathbb{I}\otimes\sigma^z\frac{t}{n}\right) \\ & \times \exp\left(-i\frac{U}{4}\sigma^z\otimes\sigma^z\otimes\mathbb{I}\frac{t}{n}\right)\exp\left(-i\frac{U}{4}\sigma^z\otimes\mathbb{I}\otimes\mathbb{I}\frac{t}{n}\right)\Bigg]^n, \end{aligned} \tag{6}$$

where we use the prime notation in the rotation to distinguish between gates applied on different qubits, since R_i acts on the first and second qubits, while R'_i acts on the second and the third. If we consider that $R_j(\alpha)R_j(\beta) = R_j(\alpha + \beta)$, the sequence of gates for one Trotter step in the digital simulation of the Hubbard model with three qubits is shown in Figure 1. There, gates A and B are two-qubit gates written in terms of $\exp(-i\phi\sigma^z \otimes \sigma^z)$ interactions, $A = \exp(-i\frac{h}{2}\sigma^z \otimes \sigma^z \frac{t}{n})$ and $B = \exp(-i\frac{U}{4}\sigma^z \otimes \sigma^z \frac{t}{n})$. Z_1 and Z_2 are single-qubit phases, $Z_1 = \exp(-i\frac{U}{4}\sigma^z\frac{t}{n})$ and $Z_2 = \exp(-i\frac{U}{2}\sigma^z\frac{t}{n})$, while X_α and Y_α are rotations along the x and y axis, respectively.

The $\exp(-i\phi\sigma^z \otimes \sigma^z)$ interaction can be implemented in small steps with optimized CZ_ϕ gates,

$$\exp\left(-i\frac{\phi}{2}\sigma^z\otimes\sigma^z\right) = \begin{pmatrix} 1 & 0 & 0 & 0 \\ 0 & e^{i\phi} & 0 & 0 \\ 0 & 0 & e^{i\phi} & 0 \\ 0 & 0 & 0 & 1 \end{pmatrix}.$$

Quantum circuits for simulating these gates are shown in Figures 2 and 3. In Figures 1-3, X and Y are π pulses.

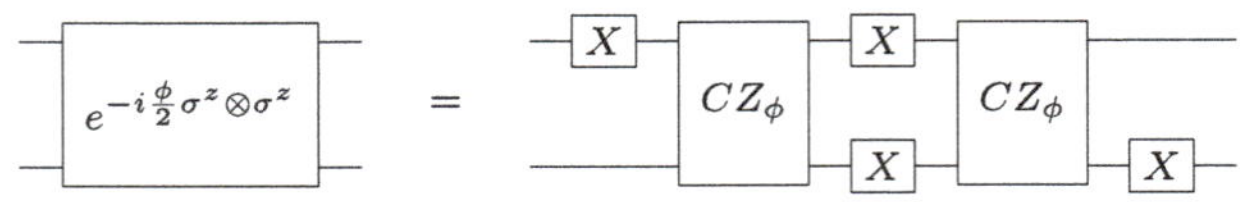

Figure 2 **Two-qubit gates in terms of the optimized CZ_ϕ gate and X π pulses.**

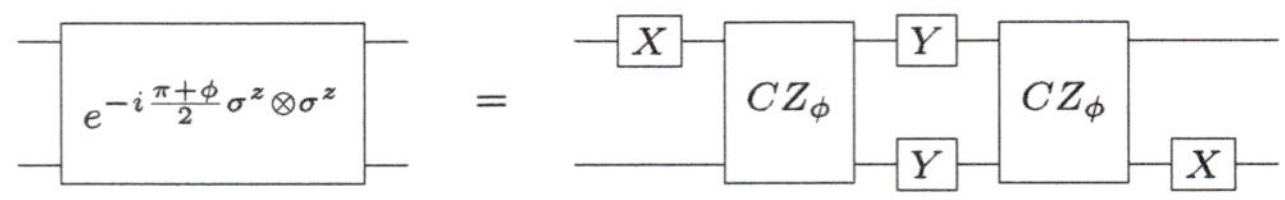

Figure 3 **Two-qubit gates in terms of the optimized CZ_ϕ gate and Y π pulses.**

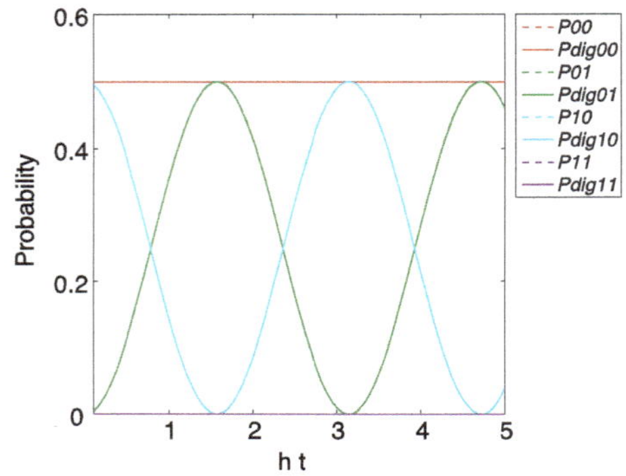

Figure 4 **Probability *Pij* for each state $|ij\rangle$ in the Hubbard model with two fermionic modes.** We obtain the same dynamics for hopping $h = 1$, and values of the potential $U = 1$ and $U = 0.5$. We also consider in both cases different number of Trotter steps, $n = 4$ and $n = 10$, and observe the same result with no Trotter error. The initial state is in all cases $(|00\rangle + |10\rangle)/\sqrt{2}$. Dashed lines refer to numerical solutions without Trotter approximation, and solid lines to numerical solutions with Trotter approximation. The absence of Trotter error comes from the fact that the second term in the Trotter expansion in Eq. (1), i.e., the term proportional to the sum of commutators, is zero for this specific model, allowing us to perform the simulation in a single Trotter step.

3.1.1 Numerical analysis of the errors

In this section, we present numerical simulations for specific values of model parameters, that is, the time t, the hopping coefficient h, and nonlinear coupling U. We compute numerically the results for the proposed model with three fermionic modes, as well as the equivalent one with two fermionic modes, for the sake of completeness. In Figures 4 and 5, we show the results of the Fermi-Hubbard model with two and three fermionic modes, respectively, for $n = 4$ and $n = 10$ Trotter steps. As shown below, the achieved fidelities can be large at the end of each digital protocol.

In Figure 6, we plot the fidelities of the digitally-evolved state with respect to the ideal dynamics associated with Eq. (2), where $\theta \equiv Ut$, for $n = 4$ Trotter steps. The fidelities are defined as $F = |\langle\Psi_T|\Psi\rangle|^2$, being $|\Psi\rangle$ and $|\Psi_T\rangle$ the states evolved with the exact unitary evolution and the digital one, respectively. Fidelities well above 90% can be achieved for a large fraction of the considered period.

Summarizing, we have analized the digital quantum simulation of the Fermi-Hubbard model with three fermionic modes in terms of simulatable spin operators with nearest neighbour interactions. Furthermore, we have considered the digital steps involving optimized gates (CZ_ϕ).

3.2 Large lattices and collective gates mediated via quantum bus

Quantum simulations of fermionic and bosonic models, as well as quantum chemistry problems, have been recently proposed in trapped ions [28–31]. In these proposals, the use

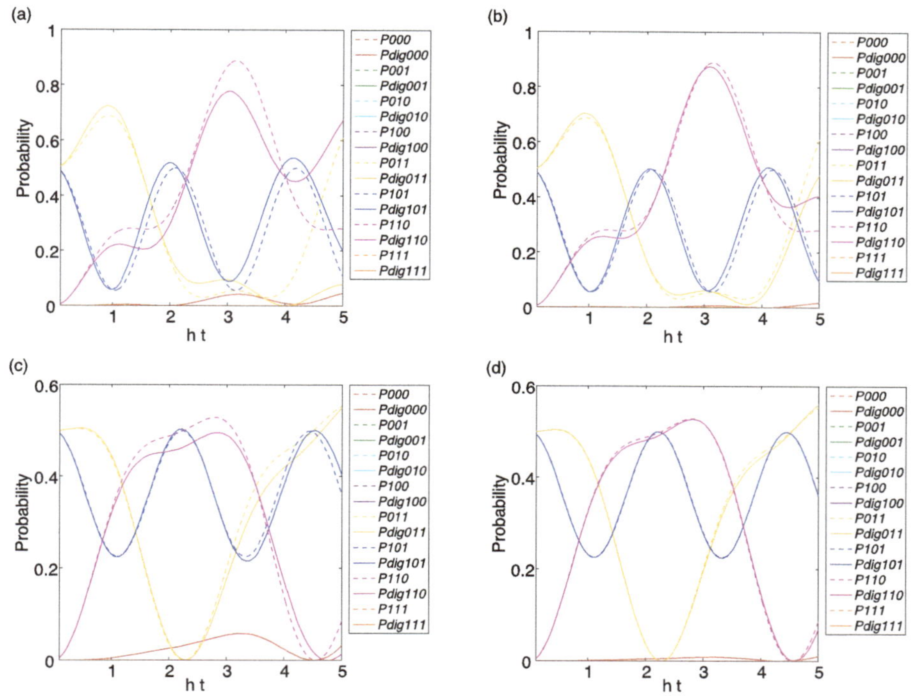

Figure 5 Probability *Pijk* for each state $|ijk\rangle$ in the Hubbard model with three fermionic modes. The physical parameters used are hopping $h = 1$, together with (**a**) $U = 1$ and number of Trotter steps $n = 4$, (**b**) $U = 1$ and $n = 10$, (**c**) $U = 0.5$ and $n = 4$, and (**d**) $U = 0.5$ and $n = 10$. The initial state is in all cases $(|011\rangle + |101\rangle)/\sqrt{2}$. Dashed lines refer to numerical solutions without Trotter approximation, and solid lines to numerical solutions with Trotter approximation.

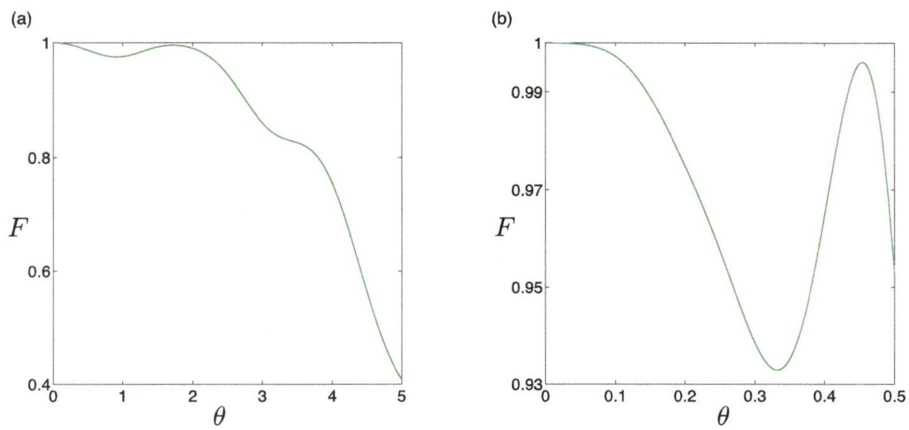

Figure 6 Fidelities obtained for the dynamics of Eq. (2), where $\theta \equiv Ut$, for $n = 4$ Trotter steps. The physical parameters used are hopping $h = 1$, together with (**a**) $U = 1$, and (**b**) $U = 0.5$. The initial state is in both cases $(|011\rangle + |101\rangle)/\sqrt{2}$.

of nonlocal interactions via a quantum bus, together with digital expansion techniques, which have been implemented in recent ion-trap experimental setups [32, 33], allows for the retrieval of arbitrary fermionic dynamics. Most current superconducting circuit setups are composed of superconducting qubits and transmission line resonators [6]. A resonator is a useful tool with several applications such as single-qubit rotations, two-qubit

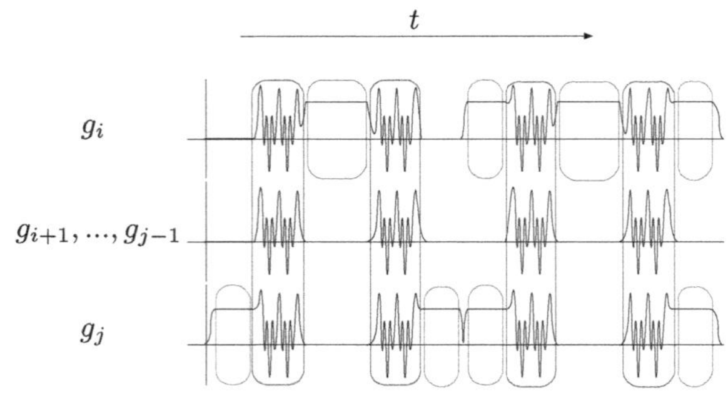

Figure 7 Scheme of the magnitude of the couplings, g_l, of the superconducting qubits $i, \ldots, j$ with the transmission line resonator as a function of time for the simulation of fermionic hopping terms. This sketch shows how sequences of rotations and nonlocal multiqubit gates gives place to interactions of the form $b_i^\dagger b_j + b_j^\dagger b_i$, which can be written in terms of spin operators as $-(\sigma_i^x \otimes \sigma_{i+1}^z \otimes \cdots \otimes \sigma_{j-1}^z \otimes \sigma_j^x + \sigma_i^y \otimes \sigma_{i+1}^z \otimes \cdots \otimes \sigma_{j-1}^z \otimes \sigma_j^y)/2$. Multiqubit gates are marked with red color where all the couplings suffer a frequency modulation [27]. Single-qubit rotations are implemented by coupling a single qubit to the resonator. They are marked with green color for a phase of $\pi/4$ and with blue for a phase-dependent single-qubit rotation, $U_{\sigma_y}(\phi)$, where the phase ϕ is proportional to the simulated time evolution of the hopping term. Note that all the qubits between sites i and j play a role in this interaction in order to fulfill the Jordan-Wigner mapping.

gates between distant spins, and dispersive qubit readout [34, 35]. In this section, we analyze how a resonator permits the efficient reproduction of the dynamics of 2D and 3D fermionic systems.

Recently, engineering of fast multiqubit interactions with tunable transmon-resonator couplings has been proposed [27]. These many-body interactions allow for the realization of multipartite entanglement [36], topological codes [37], and as we show below, simulation of fermionic systems. Employing two multiqubit gates and a single-qubit rotation, the unitary evolution associated with a tensor product of spin operators can be constructed,

$$U = U_{S_z^2} U_{\sigma_y}(\phi) U_{S_z^2}^\dagger = \exp\left[i\phi\sigma_1^y \otimes \sigma_2^z \otimes \sigma_3^z \otimes \cdots \otimes \sigma_k^z\right], \tag{7}$$

where $U_{S_z^2} = \exp[-i\pi/4 \sum_{i<j} \sigma_i^z \sigma_j^z]$ and $U_{\sigma_y}(\phi) = \exp[-i\phi' \sigma_1^{y(x)}]$ for odd (even) k. The phase ϕ' also depends on the number of qubits, i.e., $\phi' = \phi$ for $k = 4n + 1$, $\phi' = -\phi$ for $k = 4n - 1$, $\phi' = -\phi$ for $k = 4n - 2$, and $\phi' = \phi$ for $k = 4n$, where n is a positive integer. Making use of this unitary evolution and introducing single qubit rotations, it is possible to generate any tensor product of Pauli matrices during a controlled phase that is given in terms of ϕ.

In Figure 7, we show how to implement the ith-site hopping terms of a system made of N fermionic sites onto N superconducting qubits coupled to a quantum bus. Notice that local interactions between nearest and next-nearest neighbours in the square lattice involve several qubits in the experimental setup.

In order to benchmark our protocol with a specific example, we consider the Hamiltonian of the Fermi-Hubbard model with both nearest and next-nearest neighbour couplings,

$$H = \sum_{\langle i,j \rangle} \left[-h(b_i^\dagger b_j + \text{H.c.}) + U\left(n_i - \frac{1}{2}\right)\left(n_j - \frac{1}{2}\right)\right]$$
$$+ \sum_{\langle\langle i,j \rangle\rangle} \left[-h'(b_i^\dagger b_j + \text{H.c.}) + U'\left(n_i - \frac{1}{2}\right)\left(n_j - \frac{1}{2}\right)\right], \tag{8}$$

where $\langle i,j \rangle$ (resp., $\langle\langle i,j \rangle\rangle$) denote sum extended to nearest (next-nearest) neighbours, h (h') is the hopping parameter and U (U') is the interaction for nearest (next-nearest) neighbours. Here, b_i ($b_i^\dagger$) is the fermionic annihilation (creation) operator for site i, that satisfies the anticommutation relation $\{b_i, b_j^\dagger\} = \delta_{i,j}$, and $n_i = b_i^\dagger b_i$ is the fermionic number operator.

Employing the method introduced before, it is possible to simulate any fermionic dynamics. Let us analyze the interactions we need to simulate in a superconducting qubit platform considering a two-dimensional lattice of 4×4 sites. Taking as an example the 6th site in Figure 8, the simulation of hopping terms with sites 5 and 7 requires only two-qubit gates, since they are nearest neighbours in the order chosen for the mapping. However, the simulation of hopping terms between sites 2 and 6 involves 5 superconducting qubits, $b_6^\dagger b_2 + b_2^\dagger b_6 = -(\sigma_2^x \otimes \sigma_3^z \otimes \sigma_4^z \otimes \sigma_5^z \otimes \sigma_6^x + \sigma_2^y \otimes \sigma_3^z \otimes \sigma_4^z \otimes \sigma_5^z \otimes \sigma_6^y)/2$. The same thing happens for next-nearest neighbour interactions, which are simulated employing multiqubit gates made of either 4 or 6 spin operators. On the other hand, interaction terms between qubits i and j can be implemented by evolving the system with a global interaction involving all the qubits with labels between i and j, decoupling the rest of the qubits from the resonator.

The number of gates needed for realizing this simulation depends linearly on the number of qubits. Assuming that N is the number of fermionic sites in a 2D square lattice, the number of hopping and interaction terms that we need to simulate is $2\sqrt{N}(\sqrt{N}-1)$ for nearest neighbours and $2(\sqrt{N}-1)^2$ for next-nearest neighbours. As can be seen in Figure 7, every hopping term involving qubits with distant labels is made of 8 single-qubit rotations and 4 multiqubit gates. On the other hand, interaction terms can be simulated applying just one multiqubit gate.

The superconducting setup that we are considering for this quantum simulation is composed of N transmon qubits coupled to a single resonator. In order to perform highly nonlocal interactions between two distant qubits, every qubit with label inside the interval spanned by them should interact with the same resonator. Coupling several qubits to just one resonator can be a difficult task wherever the number of simulated sites is large enough. Therefore, we propose an optimized architecture for the simulation of Fermi-Hubbard model with up to next-nearest neighbours in 2D. As it is shown in Figure 8, we propose a setup with N superconducting circuits distributed in a square lattice [38]. Sequentially coupling two rows by a single transmission line resonator, one can reduce the number of qubits coupled to a single resonator. Nevertheless, all the interactions needed for satisfying the Jordan-Wigner mapping can be simulated with this architecture. Furthermore, one can achieve a speedup of the protocol by performing interactions that involve different qubits in a parallel way, e.g. the interaction between qubits 2 and 3 and the one between qubits 5 and 9 can be performed simultaneously using resonators 1 and 2, respectively.

In order to benchmark our protocol, we study its efficiency by computing the error associated with a digital decomposition. To this extent, we analyze the occupation of the fermionic sites in a 3×3 lattice. In Figure 9, we show a plot of these populations considering a perfect unitary evolution of the Fermi-Hubbard model versus the evolution associated with the digital decomposition, where l is the number of Trotter steps. As l increases, the fidelity $F = |\langle \Psi_T | \Psi \rangle|^2$ improves, being $|\Psi\rangle$ and $|\Psi_T\rangle$ the states evolved with the exact unitary evolution and the digital one, respectively.

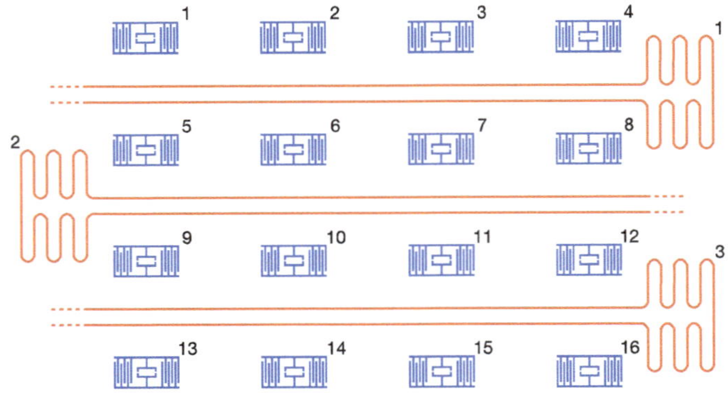

Figure 8 Architecture of superconducting qubits coupled to microwave resonators optimized for the simulation of a square lattice of 4 × 4 fermionic sites with Fermi-Hubbard interactions between nearest and next-nearest neighbours. By the use of resonators as quantum buses in the dispersive regime, several qubits are coupled allowing the implementation of single and many-body gates, which are necessary for the simulation of fermionic operators. Coupling two subsequent rows of superconducting qubits via a resonator allows to implement all the interactions required for the simulation. In order to scale the system, one needs to couple two more qubits to every resonator for simulating one more column of sites, or make use of one more resonator for connecting another row. This scheme shows an optimized architecture for the simulation of fermionic models, and further resonators would be required for the read-out and single-control of the qubits.

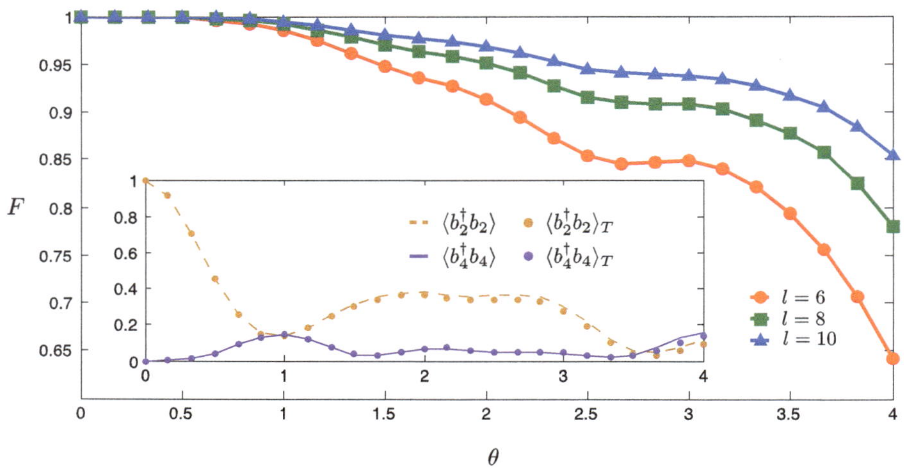

Figure 9 Simulation of a 2D lattice of 3 × 3 sites where the coupling ratios are: $U/t = 2$, $U/t' = 10$ and $U/U' = 5$. The principal plot shows the fidelity of the evolved state with digital methods for a phase of $\theta \equiv Ut = 4$ applying different numbers of Trotter steps, l. The inset shows the population of sites 2 and 4 being the initial state $|\Psi_I\rangle = b_2^\dagger|0\rangle$. The markers denote the digital evolution with 10 Trotter steps while the lines show the exact evolution.

4 Conclusions

We have presented a method for the digital quantum simulation of many-body fermionic systems in superconducting circuits with polynomial resources. Moreover, we have analyzed the efficiency of this method for the simulation of the Fermi-Hubbard model in 1D and 2D with different superconducting platforms. Finally, we have proposed an optimized circuit QED architecture where our ideas may be implemented. This work paves the way for the quantum simulation of complex fermionic dynamics in superconducting circuits.

Competing interests
The authors declare that they have no competing interests.

Authors' contributions
All the authors contributed to the realization and writing of the manuscript.

Author details
[1]Department of Physical Chemistry, University of the Basque Country UPV/EHU, Apartado 644, 48080 Bilbao, Spain. [2]IKERBASQUE, Basque Foundation for Science, Maria Diaz de Haro 3, 48013 Bilbao, Spain.

Acknowledgements
The authors acknowledge useful discussions with John Martinis group at Google/UCSB. We also acknowledge funding from Basque Government IT472-10 Grant, Spanish MINECO FIS2012-36673-C03-02, Ramón y Cajal Grant RYC-2012-11391, UPV/EHU PhD grant, UPV/EHU Project No. EHUA14/04, UPV/EHU UFI 11/55, CCQED, PROMISCE and SCALEQIT European projects.

References

1. Feynman RP. Simulating physics with computers. Int J Theor Phys. 1982;21:467.
2. Lloyd S. Universal quantum simulators. Science. 1996;273:1073.
3. Suzuki M. Fractal decomposition of exponential operators with applications to many-body theories and Monte Carlo simulations. Phys Lett A. 1990;146:319.
4. Berry DW, Ahokas G, Cleve R, Sanders BC. Efficient quantum algorithms for simulating sparse Hamiltonians. Commun Math Phys. 2007;270:359.
5. Troyer M, Wiese U-J. Computational complexity and fundamental limitations to fermionic quantum Monte Carlo simulations. Phys Rev Lett. 2005;94:170201.
6. Devoret MH, Schoelkopf RJ. Superconducting circuits for quantum information: an outlook. Science. 2013;339:1169.
7. Tsomokos DI, Ashhab S, Nori F. Using superconducting qubit circuits to engineer exotic lattice systems. Phys Rev A. 2010;82:052311.
8. van Oudenaarden A, Mooij JE. One-dimensional Mott insulator formed by quantum vortices in Josephson junction arrays. Phys Rev Lett. 1996;76:4947.
9. Tsomokos DI, Ashhab S, Nori F. Fully connected network of superconducting qubits in a cavity. New J Phys. 2008;10:113020.
10. García-Ripoll JJ, Solano E, Martin-Delgado MA. Quantum simulation of Anderson and Kondo lattices with superconducting qubits. Phys Rev B. 2008;77:024522.
11. Rakhmanov AL, Zagoskin AM, Savel'ev S, Nori F. Quantum metamaterials: electromagnetic waves in a Josephson qubit line. Phys Rev B. 2008;77:144507.
12. Koch J, Houck AA, Hur KL, Girvin SM. Time-reversal-symmetry breaking in circuit-QED-based photon lattices. Phys Rev A. 2010;82:043811.
13. You JQ, Shi X-F, Hu X, Nori F. Quantum emulation of a spin system with topologically protected ground states using superconducting quantum circuits. Phys Rev B. 2010;81:014505.
14. Wallraff A, Schuster DI, Blais A, Frunzio L, Huang R-S, Majer J, Kumar S, Girvin SM, Schoelkopf RJ. Strong coupling of a single photon to a superconducting qubit using circuit quantum electrodynamics. Nature. 2004;431:162.
15. Li J, Silveri MP, Kumar KS, Pirkkalainen J-M, Vepsäläinen A, Chien WC, Tuorila J, Sillanpää MA, Hakonen PJ, Thuneberg EV, Paraoanu GS. Motional averaging in a superconducting qubit. Nat Commun. 2013;4:1420.
16. Marcos D, Rabl P, Rico E, Zoller P. Superconducting circuits for quantum simulation of dynamical gauge fields. Phys Rev Lett. 2013;111:110504.
17. Stojanovic VM, Vanevic M, Demler E, Tian L. Transmon-based simulator of nonlocal electron-phonon coupling: a platform for observing sharp small-polaron transitions. Phys Rev B. 2014;89:144508.
18. Las Heras U, Mezzacapo A, Lamata L, Filipp S, Wallraff A, Solano E. Digital quantum simulation of spin systems in superconducting circuits. Phys Rev Lett. 2014;112:200501.
19. Mezzacapo A, Las Heras U, Pedernales JS, DiCarlo L, Solano E, Lamata L. Digital quantum Rabi and Dicke models in superconducting circuits. Sci Rep. 2014;4:7482.
20. Barends R, Lamata L, Kelly J, García-Álvarez L, Fowler AG, Megrant A, Jeffrey E, White TC, Sank D, Mutus JY, Campbell B, Chen Y, Chen Z, Chiaro B, Dunsworth A, Hoi I-C, Neill C, O'Malley PJJ, Quintana C, Roushan P, Vainsencher A, Wenner J, Solano E, Martinis JM. Digital quantum simulation of fermionic models with a superconducting circuit. arXiv:1501.07703.
21. Salathé Y, Mondal M, Oppliger M, Heinsoo J, Kurpiers P, Potočnik A, Mezzacapo A, Las Heras U, Lamata L, Solano E, Filipp S, Wallraff A. Digital quantum simulation of spin models with circuit quantum electrodynamics. arXiv:1502.06778.
22. Barends R, Kelly J, Megrant A, Veitia A, Sank D, Jeffrey E, White TC, Mutus J, Fowler AG, Campbell B, Chen Y, Chen Z, Chiaro B, Dunsworth A, Neill C, O'Malley P, Roushan P, Vainsencher A, Wenner J, Korotkov AN, Cleland AN, Martinis JM. Superconducting quantum circuits at the surface code threshold for fault tolerance. Nature. 2014;508:500.
23. Steffen L, Salathe Y, Oppliger M, Kurpiers P, Baur M, Lang C, Eichler C, Puebla-Hellmann G, Fedorov A, Wallraff A. Deterministic quantum teleportation with feed-forward in a solid state system. Nature. 2013;500:319.
24. Jordan P, Wigner E. Über das Paulische Äquivalenzverbot. Z Phys. 1928;47:631.
25. Mølmer K, Sørensen A. Multiparticle entanglement of hot trapped ions. Phys Rev Lett. 1999;82:1835.
26. Sørensen A, Mølmer K. Entanglement and quantum computation with ions in thermal motion. Phys Rev A. 2000;62:022311.
27. Mezzacapo A, Lamata L, Filipp S, Solano E. Many-body interactions with tunable-coupling transmon qubits. Phys Rev Lett. 2014;113:050501.

28. Casanova J, Mezzacapo A, Lamata L, Solano E. Quantum simulation of interacting fermion lattice models in trapped ions. Phys Rev Lett. 2012;108:190502.
29. Mezzacapo A, Casanova J, Lamata L, Solano E. Digital quantum simulation of the Holstein model in trapped ions. Phys Rev Lett. 2012;109:200501.
30. Lamata L, Mezzacapo A, Casanova J, Solano E. Efficient quantum simulation of fermionic and bosonic models in trapped ions. EPJ Quantum Technol. 2014;1:9.
31. Yung M-H, Casanova J, Mezzacapo A, McClean J, Lamata L, Aspuru-Guzik A, Solano E. From transistor to trapped-ion computers for quantum chemistry. Sci Rep. 2014;4:3589.
32. Müller M, Hammerer K, Zhou YL, Roos CF, Zoller P. Simulating open quantum systems: from many-body interactions to stabilizer pumping. New J Phys. 2011;13:085007.
33. Barreiro JT, Müller M, Schindler P, Nigg D, Monz T, Chwalla M, Hennrich M, Roos CF, Zoller P, Blatt R. An open-system quantum simulator with trapped ions. Nature. 2011;470:486.
34. Blais A, Huang RS, Wallraff A, Girvin SM, Schoelkopf RJ. Cavity quantum electrodynamics for superconducting electrical circuits: an architecture for quantum computation. Phys Rev A. 2004;69:062320.
35. Blais A, Gambetta J, Wallraff A, Schuster DI, Girvin SM, Devoret MH, Schoelkopf RJ. Quantum-information processing with circuit quantum electrodynamics. Phys Rev A. 2007;75:032329.
36. Mlynek JA, Abdumalikov AA Jr, Fink JM, Steffen L, Baur M, Lang C, van Loo AF, Wallraff A. Demonstrating W-type entanglement of Dicke states in resonant cavity quantum electrodynamics. Phys Rev A. 2012;86:053838.
37. Kitaev AY. Fault-tolerant quantum computation by anyons. Ann Phys. 2003;303:2.
38. Helmer F, Mariantoni M, Fowler AG, von Delft J, Solano E, Marquardt F. Cavity grid for scalable quantum computation with superconducting circuits. Europhys Lett. 2009;85:50007.

15

Walsh-synthesized noise filters for quantum logic

Harrison Ball[1,2] and Michael J Biercuk[1,2*]

*Correspondence: michael.biercuk@sydney.edu.au
[1] ARC Centre for Engineered Quantum Systems, School of Physics, The University of Sydney, Sydney, NSW 2006, Australia
[2] National Measurement Institute, West Lindfield, Sydney, NSW 2070, Australia

Abstract

We study a novel class of open-loop control protocols constructed to perform arbitrary nontrivial single-qubit logic operations robust against time-dependent non-Markovian noise. Amplitude and phase modulation protocols are crafted leveraging insights from functional synthesis and the basis set of Walsh functions. We employ the experimentally validated generalized filter-transfer function formalism in order to find optimized control protocols for target operations in SU(2) by defining a cost function for the filter-transfer function to be minimized through the applied modulation. Our work details the various techniques by which we define and then optimize the filter-synthesis process in the Walsh basis, including the definition of specific analytic design rules which serve to efficiently constrain the available synthesis space. This approach yields modulated-gate constructions consisting of chains of discrete pulse-segments of arbitrary form, whose modulation envelopes possess intrinsic compatibility with digital logic and clocking. We derive novel families of Walsh-modulated noise filters designed to suppress dephasing and coherent amplitude-damping noise, and describe how well-known sequences derived in NMR also fall within the Walsh-synthesis framework. Finally, our work considers the effects of realistic experimental constraints such as limited modulation bandwidth on achievable filter performance.

Keywords: decoherence suppression; error correction; open-loop control; dynamic error suppression; quantum control; quantum logic; qubit; Walsh function; functional analysis

1 Introduction

In realistic laboratory settings, decoherence in quantum systems is dominated by *time-dependent* non-Markovian noise processes with long correlations, frequently characterized by low-frequency dominated noise power spectra [1–5]. These may arise either from environmental fluctuations or - in the important case of *driven* quantum systems - from noise in the control device itself [6]. In either case, the result is a reduction in the fidelity of a target control operation, including both memory and nontrivial operations. These phenomena present a major challenge as quantum devices move from proof of principle demonstrations to realistic applications, where performance demands on the quantum devices are frequently extreme. Accordingly, finding ways to control quantum systems efficiently and effectively in the presence of noise is a central task in quantum control theory [7–10].

A range of techniques relying on both open- and closed-loop control have been devised to address this challenge [11–13] at various levels in a layered architecture for quantum computing [14]. In particular, open-loop dynamical error suppression strategies (without the need for measurement or feedback) such as dynamic decoupling (DD) [15–18], dynamically corrected gates (DCGs) [19–25], and composite pulsing [26–28], have emerged as *resource-efficient* approaches for physical-layer decoherence control. They are joined by a new class of continuously modulated ('always-on') dynamical decoupling and dynamically protected gate schemes [29–36] inspired by well established techniques in NMR.

These schemes all address the question of decoherence mitigation, but looking across their breadth, have both benefitted and suffered from reliance on a wide range of theoretical techniques. Unfortunately the analytic tools for crafting control protocols employed in any particular setting do not necessarily translate equivalently between approaches, nor do the methods generally employed for evaluating efficacy easily translate to experimentally measured characteristics of the environment. This is a major challenge for experimentalists or systems designers attempting to determine which of the many open-loop control schemes to employ in a particular experiment. As an example, the powerful group theoretic insights and consideration of time-varying environments that permit the construction of error-robust, bounded-strength SU(2) operations for quantum information in Viola's DCG framework are quite different from the geometric considerations and quasi-static noise assumptions widely employed in NMR composite pulsing. This issue has been highlighted recently as new work has revealed striking differences between the time-domain noise sensitivity of control protocols as compared to longstanding notions of error cancellation in the Magnus expansion [37, 38].

A unified and experimentally relevant framework for *devising* and *evaluating* error-suppressing gates in realistic noise environments is therefore needed to secure the role of dynamical error suppression in systematic designs of quantum technologies including fault-tolerant quantum computers. Kurizki provided a promising path towards this end with his seminal work framing the problem of finding decoherence-suppressing control protocols by considering appropriate frequency-domain modification of the system-environment coupling [39, 40]. Residual errors could be calculated through overlap integrals of the power spectrum describing the environmental noise, and functions capturing the frequency-domain response of any applied control. This framework - effectively a quantum generalization of transfer functions widely used in control engineering [41] - provides a simple heuristic approach to understanding the performance of an arbitrary control protocol in an arbitrary noise environment. Stated simply, effective error-suppressing control protocols 'filter' the noise over a user-defined band, therefore mitigating decoherence in the quantum system [42].

Early demonstrations of this framework applied to the simple case of implementing the protected identity operator to qubits by dynamical decoupling [15, 43–47], where the filter functions could be calculated for pure dephasing in a straightforward manner using concepts of *linear control* [42]. Expanding significantly beyond this work, the challenge of crafting generalized analytic forms for the transfer functions describing arbitrary single-qubit control compatible with universal non-commuting noise (a problem in *nonlinear* control) has recently been addressed theoretically [28, 38, 48, 49], and validated in experiment [37]. Further theoretical extensions of filter-transfer functions to two-qubit gates highlight the breadth of applicability of this approach to quantum control [50–52].

Beyond its simple intuitive nature, the power of the filter transfer function approach comes from the fact that it can in principle be applied to studying dynamic-error-suppression control protocols derived through any manner of analytic approach. It permits the application of well tested engineering concepts for control systems design; the complex physics associated with quantum dynamics in time-dependent environments with non-commuting noise and control Hamiltonians is relegated to the calculation of the generalized filter transfer functions themselves, and once derived these may be deployed in block-diagram systems analyses [41].

With these significant advances and the promise of applying the suite of insights from control theory to the quantum regime, the noise-filtering approach to quantum control has leapt to the fore, providing a unifying framework applicable over a wide parameter range of interest to real experimental settings. Nonetheless, outstanding challenges remain in how to leverage the generalized filter-transfer-function framework [48, 49] to systematically craft effective error-suppressing gate constructions while also heeding realistic system constraints imposed by hardware systems. For instance, the presence of finite timing precision and limited classical communication bandwidth between the physical (quantum) layer and a classical controller [14] impose new constraints not generally captured when solely considering quantum dynamical evolution of an individual state.

We address this challenge, introducing a quantum control toolkit permitting the realization of physical-layer error-suppressing control protocols that are simultaneously effective in suppressing error and compatible with a variety of major hardware restrictions. We leverage the generalized filter-transfer function formalism as a unifying theoretical construct, and employ techniques from functional analysis in order to realize appropriate modulation protocols applied to a near-resonant carrier frequency for enacting high-fidelity quantum control operations on single qubits [30, 53]. Our work identifies the Walsh functions - square-wave analogues of the sines and cosines - as natural building blocks for constructing the modulation protocols designed to filter time-varying noise over a user defined band while enacting a nontrivial qubit rotation. The Walsh functions are defined in a uniform piecewise-constant fashion, building intrinsic compatibility with discrete clocking [54] and classical digital logic, and have previously been identified as providing a powerful mathematical framework in the context of quantum control sequencing [51]. Moreover, they may be arbitrarily combined using Fourier-like synthesis using techniques for arbitrary waveform generation well established in digital signal processing.

We treat a *Walsh-modulated* driven qubit system weakly interacting with both dephasing and coherent amplitude-damping noise processes. The task of finding Walsh-synthesized modulation patterns that produce effective filters is reduced to minimizing a cost function measuring the extent to which noise over a user-defined spectral band is filtered by the applied control. The performance of resulting control protocols is completely characterized by their *Walsh spectra*, facilitating intuitive analytic design rules based on symmetry and spectral properties of the Walsh basis. Our work details the various techniques and mathematical constructs through which we define and then optimize the filter-synthesis process in the Walsh basis, and considers the effects of realistic experimental constraints such as limited modulation bandwidth.

With these insights, we derive novel families of Walsh-modulated noise filters designed to suppress dephasing and coherent amplitude-damping noise, and describe their properties. Modulation protocols are tailored to a particular operation on SU(2), but are otherwise largely *model-robust* (being tailored to suppress noise over a frequency band rather than to a specific time-domain noise signal), and *portable* between different qubit technologies. Combined with the discovery, presented here, that several prominent composite pulse protocols derived in NMR actually fall within the Walsh-synthesis basis - mirroring similar insights in the context of dynamical decoupling [51] - this work positions the Walsh functions as a natural basis for crafting physical-layer error suppression strategies for scalable quantum technologies.

The remainder of this paper is organized as follows. In Section 2 we describe our model quantum system by defining relevant control and noise Hamiltonians. In Section 3 we review the generalized filter-transfer function formalism used to derive a spectral representation of the operational infidelity. Notation for defining and parameterizing the control space is introduced and explicit expressions for computing corresponding filter functions are presented. Section 4 provides a formal definition of a filter cost function used for optimizing operational fidelity over the control space and deriving useful filters. Performance characteristics of these filters are discussed and interpreted, with care taken to differentiate filter order from Magnus order. In Section 5 physically motivated constraints on the control space are established by synthesizing control waveforms as superpositions of functions in the Walsh basis, bounding the dimensionality of the filter-optimization task. Two useful representations of the Walsh basis - Paley ordering and the Hadamard representation - are introduced. We then develop a range of analytic filter-design rules for efficient filter construction based on the symmetry and spectral properties of the Walsh functions. In Sections 6-9 we apply the above framework to derive several novel families of noise filters implementing nontrivial logic gates. These include filters for dephasing and coherent amplitude-damping noise in addition to concatenated filters for universal noise. In Section 10 we study how relaxing the assumption of perfect square pulses reduces the performance of filters optimized in the Walsh basis, and demonstrate that these filter properties may be recovered in general by simply re-optimizing under the assumption of non-square pulses. We then close with a brief summary and outlook, followed by a number of appendices containing detailed derivations of relevant quantities used in the main text.

2 Physical setting

We begin by establishing the Hamiltonian framework for the control and noise interactions treated in this paper. This is necessary background in order to study noise filtering via Walsh-synthesized control fields implementing logic gates. We consider a model quantum system consisting of an ensemble of identically prepared noninteracting qubits immersed in a weakly interacting noise bath and driven by an external control device. Working in the interaction picture with respect to the qubit splitting, state transformations are represented as unitary rotations of the Bloch vector. In this interaction picture the generalized time-dependent Hamiltonian is written

$$H(t) = H_c(t) + H_0(t), \tag{1}$$

where $H_c(t)$ describes perfect control of the qubit state, e.g. via an ideal external driving field, and the noise Hamiltonian $H_0(t)$ captures undesirable interactions with a time-varying non-Markovian noise environment. The full qubit dynamics are governed by the

Schrodinger equation $i\dot{U}(t,0) = H(t)U(t,0)$ where the time-evolution operator $U(t,0)$ transforms an initial state $|\psi(0)\rangle$ to the final $U(\tau,0)|\psi(0)\rangle$ after an interaction of duration τ.

In the absence of noise the total Hamiltonian reduces to $H(t) = H_c(t)$, in which case time-evolution is determined purely by control operations according to $i\dot{U}_c(t,0) = H_c(t)U_c(t,0)$. An *intended* evolution path under ideal control is therefore described by the *control propagator* $U_c(t,0) = \mathcal{T}\exp(-i\int_0^t H_c(t')\,dt')$, with $\mathcal{T}$ denoting the time-ordering operator. For a single qubit the time-dependent control Hamiltonian may in general be written $H_c(t) = \Omega(t)\hat{\mathbf{n}}(t)\cdot\boldsymbol{\sigma}/2$. Here $\hat{\mathbf{n}}(t)\cdot\boldsymbol{\sigma} \equiv n_x\hat{\sigma}_x + n_y\hat{\sigma}_y + n_z\hat{\sigma}_z$ is the rotation generator, $\hat{\mathbf{n}}(t) \in \mathbb{R}$ is a unit vector defining the instantaneous axis of rotation, and $\Omega(t)$ is the instantaneous rate of rotation (Rabi rate) for the Bloch vector.

Environmental interactions are modeled semi-classically, with stochastic noise processes expressed in terms of time-dependent fluctuating classical noise fields. We consider time-dependent dephasing (detuning) and coherent amplitude-damping processes, captured respectively through (stochastic) rotations about $\hat{\sigma}_z$ and about the instantaneous direction of control $\hat{\mathbf{n}}(t)\cdot\boldsymbol{\sigma}$. The universal noise Hamiltonian thus takes the form $H_0(t) = H_0^{(z)}(t) + H_0^{(\Omega)}(t)$ where $H_0^{(z)}(t)$ and $H_0^{(\Omega)}(t)$ denote noise interactions in the dephasing and amplitude noise quadratures respectively. Dephasing noise thus contributes the additive term

$$H_0^{(z)}(t) = \beta_z(t)\hat{\sigma}_z, \tag{2}$$

where $\beta_z(t)$ describes a time-varying noise field. Coherent amplitude-damping noise contributes the multuplicative term

$$H_0^{(\Omega)}(t) = \frac{\beta_\Omega(t)\Omega(t)}{2}\hat{\mathbf{n}}(t)\cdot\boldsymbol{\sigma} = \beta_\Omega(t)H_c(t). \tag{3}$$

Including this term is equivalent to making the substitution $\Omega(t) \longrightarrow \Omega(t)(1+\beta_\Omega(t))$ in the control Hamiltonian, where $\beta_\Omega(t)$ describes a (multiplicative) noise source in the amplitude of the driving field. Inclusion of this term in the noise Hamiltonian enables us to go beyond previous studies where attention has been restricted to dephasing processes. This novel approach is important for most realistic experimental situations where correctable non-Markovian amplitude-damping errors arise from noise in the control system itself (for example, fluctuations in the strength of the driving field).

In our model both $\beta_z(t)$ and $\beta_\Omega(t)$ are assumed to be classical random variables with zero mean and non-Markvovian power spectra. We also assume they are *wide sense stationary* and *independent*.[a] The former implies the autocorrelation functions $\langle\beta_i(t_1)\beta_i(t_2)\rangle$, $i \in \{z,\Omega\}$, depend only on the time *difference* $t_1 - t_2$. The latter implies the cross-correlation functions vanish. That is, $\langle\beta_i(t_1)\beta_j(t_2)\rangle = 0$ where $i,j \in \{z,\Omega | i \neq j\}$. The angle brackets denote a time average of the random variables. Finally, our model permits access to a wide range of parameter regimes, from quasistatic (noise slow compared to $H_c(t)$) to the limit in which the noise fluctuates on timescales comparable to or faster than $H_c(t)$.

These noise Hamiltonians generate uncontrolled rotations in the qubit dynamics, leading to errors in the evolution path (and hence the final state) relative to the target transformation intended under $H_c(t)$. An estimate for this error is derived in the next section using the generalized filter-transfer function formalism.

3 Building noise filters

Overall, our objective is to craft control protocols such that the deleterious effects of time-dependent noise on the intended evolution of an arbitrary qubit transformation are suppressed - *filtered* by the control. Accordingly, we require a measure for the operational fidelity in the presence of both noise and the relevant control. For this we employ the method developed by *Green et al.* [49]. In this framework the error contributed by the noise fields over the duration of the control is approximated, to first order, via a truncated Magnus expansion. Each noise field then contributes a term to the gate infidelity in the spectral domain expressed as an overlap integral between the noise power spectrum and an appropriate generalized filter-transfer function. We describe this in detail below.

3.1 Calculating operational fidelity

In the absence of noise interactions, state evolution is determined by $i\dot{U}_c(t) = H_c(t)U_c(t)$ with $U_c(t)$ the ideal evolution operator describing the target operation. Including the effects of noise, however, time evolution is determined by the operator $U(t)$ satisfying $i\dot{U}(t) = (H_c(t) + H_0^{(z)}(t) + H_0^{(\Omega)}(t))U(t)$. Our measure for operational fidelity is given by $\mathcal{F}_{av}(\tau) = \frac{1}{4}\langle |\operatorname{Tr}(U_c^\dagger(\tau)U(\tau))|^2\rangle$, effectively measuring the extent to which the intended and realized operators 'overlap', as captured by the Hilbert-Schmidt inner product [55]. Computing the evolution dynamics, however, is very challenging since the control and noise Hamiltonians do not commute at different times; sequential application of the resulting time-dependent, non-commuting operations gives rise to both dephasing and depolarization errors, mandating approximation methods.

Our error model assumes non-dissipative qubit evolution with both control and noise interactions resulting in unitary rotations. Hence we approximate the evolution operator as a unitary using a Magnus expansion [56, 57]. This involves moving to a frame co-rotating with the control known as the *toggling frame*, originally appearing in the development of average Hamiltonian theory [58]. This approach allows us to separate the part of the system evolution due solely to the control from the part affected by environmental coupling, and is standard in the study of coherent control in NMR [58, 59] and quantum information.

Defining the *error propagator* $\tilde{U}(t) \equiv U_c^\dagger(t)U(t)$, the total evolution operator is written $U(t) = U_c(t)\tilde{U}(t)$. In this case the realized evolution operator approaches the target operation as $\tilde{U}(\tau) \to \mathbf{I}$, establishing the condition for suppression of noisy evolution dynamics. However, moving to the toggling frame defined by *toggling frame Hamiltonian* $\tilde{H}_0(t) \equiv U_c^\dagger(t)H_0(t)U_c(t)$, the error propagator satisfies the Schrodinger equation $i\dot{\tilde{U}}(t) = \tilde{H}_0(t)\tilde{U}(t)$. Performing a Magnus expansion in this frame - assuming convergence of the series [57] - we may write $\tilde{U}(\tau) = \exp[-i\sum_{\mu=1}^{\infty}\mathbf{a}_\mu(\tau)\cdot\boldsymbol{\sigma}]$ where the *error vectors* $\mathbf{a}_\mu(\tau)$ determine expansion coefficients of the Magnus series operators $\Phi_\mu(\tau)$ expressed in the basis of Pauli matrices (see Appendix A). We may then in principle approximate $\tilde{U}(t)$ to arbitrary accuracy by truncating the infinite series at an appropriate order.

The operational fidelity $\mathcal{F}_{av}(\tau) = \frac{1}{4}\langle |\operatorname{Tr}(\tilde{U}(\tau))|^2\rangle$ may now be fully expressed as an infinite power series over the ensemble-averaged magnitudes of the expansion vectors $\mathbf{a}_\mu(\tau)$. In the limit of sufficiently weak noise,[b] however, it is appropriate to truncate the expansion to first-order yielding $\mathcal{F}_{av}(\tau) \approx 1 - \langle a_1^2\rangle$ with $\langle a_1^2\rangle \equiv \langle \mathbf{a}_1(\tau)\mathbf{a}_1^T(\tau)\rangle$ defining the *first order infidelity*. Now, as set out in Appendix A the first-order error vector is related to the first-order Magnus term according to Eq. (70), yielding $\mathbf{a}_1(\tau)\cdot\boldsymbol{\sigma} = \Phi_1(\tau) = \int_0^\tau dt\tilde{H}_0(t)$.

That is, the first-order infidelity $\langle a_1^2 \rangle$ is associated with the time-average of the toggling frame Hamiltonian over the total sequence duration.

Expressing $\tilde{H}_0(t) \equiv \vec{\mathcal{R}}(t) \cdot \boldsymbol{\sigma}$ in the Pauli basis, where the expansion vector $\vec{\mathcal{R}}(t)$ is some convolution of both control and noise fields, we obtain the computational expression $\mathbf{a}_1(\tau) = \int_0^\tau dt \vec{\mathcal{R}}(t)$. Using the noise model assumptions outlined in Section 2, and performing a number of Fourier-like transforms (see Appendix A for full details), we obtain a spectral representation of the form

$$\langle a_1^2 \rangle = \frac{1}{2\pi} \int_{-\infty}^{\infty} \frac{d\omega}{\omega^2} S_z(\omega) F_z(\omega) + \frac{1}{2\pi} \int_{-\infty}^{\infty} \frac{d\omega'}{\omega'^2} S_\Omega(\omega') F_\Omega(\omega'). \tag{4}$$

Here $S_z(\omega)$ and $S_\Omega(\omega)$ denote the dephasing and amplitude noise power-spectral densities (PSDs). The dephasing $F_z(\omega)$ and amplitude $F_\Omega(\omega)$ filter-transfer functions, on the other hand, capture the spectral response of the control sequence. Moving forward, we will present the mathematical framework that permits calculation of these quantities for arbitrary control protocols.

3.2 Defining the control space

In order to realize specific noise filters, characterized by the filter-transfer functions introduced above, we require a simple framework to define the time-domain control operations that can be applied to the qubit. Representing the qubit state on the Bloch sphere, state manipulation maps to a rotation in $\mathbb{R}^3$ of the Bloch vector associated with a unitary transformation $\mathcal{U}(\theta, \hat{\sigma}_{\hat{\mathbf{n}}}) \equiv \exp[-i(\boldsymbol{\sigma} \cdot \hat{\mathbf{n}})\theta/2]$, reflecting the homeomorphism between $SU(2)$ and $SO(3)$. The rotation generator $\hat{\sigma}_{\hat{\mathbf{n}}} \equiv \hat{\mathbf{n}} \cdot \boldsymbol{\sigma} \equiv n_x\hat{\sigma}_x + n_y\hat{\sigma}_y + n_z\hat{\sigma}_z$ produces a rotation though an angle θ about the axis defined by the unit vector $\hat{\mathbf{n}} \in \mathbb{R}^3$.

We treat control protocols taking the form of an *n-segment sequence* of such unitaries, executed over the time period $[0, \tau]$. This implies a natural partition of the total sequence duration τ into n subintervals $I_l = [t_{l-1}, t_l]$, $l \in \{1, \ldots, n\}$, such that the lth control unitary has duration $\tau_l = t_l - t_{l-1}$. Here t_{l-1} and t_l are the start and end times of the lth rotation respectively, and we define $t_0 \equiv 0$ and $t_n \equiv \tau$. In particular we consider control unitaries of the form

$$P_l \equiv \mathcal{U}(\Omega_l \tau_l, \hat{\sigma}_{\phi_l}) = \exp\left[-i\frac{\Omega_l}{2}\tau_l\hat{\sigma}_{\phi_l}\right], \tag{5}$$

$$\hat{\sigma}_{\phi_l} \equiv \cos(\phi_l)\hat{\sigma}_x + \sin(\phi_l)\hat{\sigma}_y, \tag{6}$$

corresponding to the experimentally relevant case of a resonantly driven qubit. Here Ω_l is the Rabi rate during the lth time interval $[t_{l-1}, t_l]$, and is assumed constant over the duration τ_l of the associated control interaction. During this interaction the rotation generator $\hat{\sigma}_{\phi_l}$, parameterized by $\phi_l \in [0, 2\pi]$, thus generates a rotation of the Bloch vector through an angle $\theta_l \equiv \Omega_l \tau_l$ about the axis $\hat{\mathbf{n}}_l \equiv (\cos(\phi_l), \sin(\phi_l), 0)$ in the xy plane.[c] The control Hamiltonian associated with this n-segment sequence takes the form

$$H_c(t) = \sum_{l=1}^{n} G^{(l)}(t) \frac{\Omega_l}{2} \hat{\sigma}_{\phi_l}, \tag{7}$$

where the function $G^{(l)}(t)$ is 1 if $t \in I_l$ and zero otherwise. Controlled evolution *during* the lth time interval is, under this Hamiltonian, consequently described by the unitary

$$U_c(t, t_{l-1}) = \exp\left[-i\frac{\Omega_l}{2}\hat{\sigma}_{\phi_l}(t - t_{l-1})\right]. \tag{8}$$

That is, implementation of the lth completed rotation is equivalently denoted by the operator $P_l = U_c(t_l, t_{l-1})$. For compactness we define the *cumulative operator*

$$Q_l := P_l P_{l-1} \cdots P_0, \quad P_0 := \mathbf{I} \tag{9}$$

to capture the cumulative action of the first l sequentially competed rotations. Hence the control propagator at any time t may be written

$$U_c(t, 0) = \sum_{l=1}^{n} G^{(l)}(t) U_c(t, t_{l-1}) Q_{l-1}. \tag{10}$$

$H_c(t)$ is thus completely described by the sequence of n triples $\{(\Omega_l, \tau_l, \phi_l)\}_{l=1}^{n}$, and each control operation is completely parameterized by the control variables according to $P_l = P_l(\theta_l, \Omega_l, \tau_l, \phi_l)$. Although not strictly an independent parameter it is useful to include $\theta_l = \Omega_l \tau_l$ in the argument to distinguish different *realizations* of the same net rotation for different choices of $\Omega_l \tau_l$. We define the $(n \times 4)$ n-segment matrix

$$\boldsymbol{\Gamma}_n = \begin{array}{c} \\ P_1 \\ P_2 \\ \vdots \\ P_n \end{array} \begin{array}{c} \begin{array}{cccc} \Omega_l & \tau_l & \theta_l & \phi_l \end{array} \\ \left[\begin{array}{cccc} \Omega_1 & \tau_1 & \theta_1 & \phi_1 \\ \Omega_2 & \tau_2 & \theta_2 & \phi_2 \\ \vdots & \vdots & \vdots & \vdots \\ \Omega_n & \tau_n & \theta_n & \phi_n \end{array}\right] \end{array} \tag{11}$$

to compactly describe any arbitrary n-segment unitary control sequence (see Figure 1). The entire space $\mathfrak{C}_n$ of such control forms, referred to the *n-segment control space*, and written formally

$$\mathfrak{C}_n := \left\{ \boldsymbol{\Gamma}_n \middle| \Omega_l, \theta_l, \tau_l > 0, \phi_l \in [0, 2\pi], l \in 1, \ldots, n, \sum_{l}^{n} \tau_l = \tau \right\} \tag{12}$$

thus corresponds to an infinite set of $\boldsymbol{\Gamma}_n$ matrices ranging continously over all possible values taken by the control variables. This general class of control, consisting of bounded-strength unitary sequences, includes familiar *composite-pulse sequences* in NMR and DCGs in quantum information. We use the more general control space, however, to construct novel qubit gates specifically designed to filter non-Markovian noise.

3.3 Generalized filter-transfer functions

We now present the computational forms of the filter-transfer functions $F_z(\omega)$ and $F_\Omega(\omega)$ introduced in Eq. (4) for arbitrary n-segment control protocols implemented by Eq. (7). As outlined above, the filter-transfer functions are completely parametrized by the control

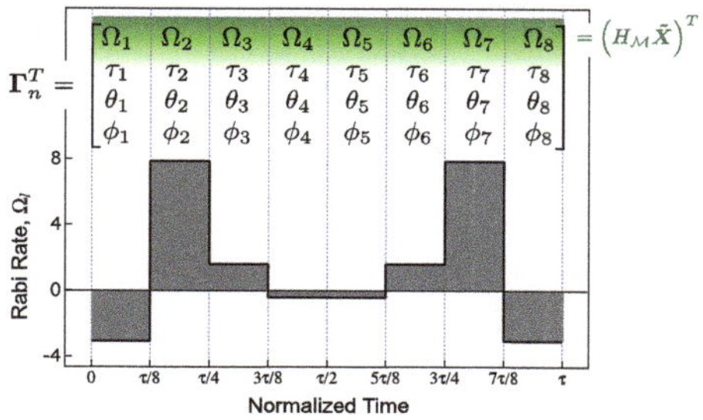

Figure 1 Filter construction. Visualization of the available control space for an $n = 8$ segment control sequence. The filter is synthesized over the parameters presented in $\mathbf{\Gamma}_n$, whose transpose corresponds to the discrete time segments in the time-domain filter. As an illustration, a time-varying Rabi rate (arbitrary units) is presented for each of the l segments. Synthesis of this waveform may be constructed in the Walsh basis using the Hadamard transform (notation upper right), as will be discussed in Section 5.

variables $\{(\Omega_l, \tau_l, \phi_l)\}_{l=1}^{n} \cong \mathbf{\Gamma}_n$. Here we only provide a summary of the relevant computational quantities, leaving the major derivations and full explanation to Appendix A. We start by writing

$$F_z(\omega) := \left[\mathbf{R}^{(z)}(\omega)\right]^* \left[\mathbf{R}^{(z)}(\omega)\right]^T \quad \textit{(Dephasing Filter-Transfer Function)}, \tag{13}$$

$$F_\Omega(\omega) := \left[\mathbf{R}^{(\Omega)}(\omega)\right]^* \left[\mathbf{R}^{(\Omega)}(\omega)\right]^T \quad \textit{(Amplitude Filter-Transfer Function)}, \tag{14}$$

where the row vectors $\mathbf{R}^{(z)}(\omega), \mathbf{R}^{(\Omega)}(\omega) \in \mathbb{R}^3$ are obtained by Fourier transforming relevant time-domain functions associated with the control evolution dynamics. In Appendix B we derive the explicit computational forms

$$\mathbf{R}^{(z)}(\omega) = \sum_{l=1}^{n} e^{i\omega t_{l-1}} \mathbf{R}_z^{P_l}(\omega) \mathbf{\Lambda}^{(l-1)} \quad \textit{(Dephasing Control Vector)}, \tag{15}$$

$$\mathbf{R}^{(\Omega)}(\omega) = \sum_{l=1}^{n} \left[e^{i\omega t_{l-1}} - e^{i\omega t_l}\right] \vec{\mathbb{T}}^{(l)} \mathbf{\Lambda}^{(l-1)} \quad \textit{(Amplitude Control Vector)}. \tag{16}$$

The row vector $\mathbf{R}_z^{P_l}(\omega) \in \mathbb{R}^3$ captures the spectral response in the dephasing noise quadrature contributed during the lth unitary control segment. This takes the form (see Appendix B)

$$\mathbf{R}_z^{P_l}(\omega) = \frac{\omega}{\omega^2 - \Omega_l^2} \begin{pmatrix} \sin(\phi_l)[i\Omega_l e^{i\omega\tau_l}\cos(\Omega_l\tau_l) + \omega e^{i\omega\tau_l}\sin(\Omega_l\tau_l) - i\Omega_l] \\ -\cos(\phi_l)[i\Omega_l e^{i\omega\tau_l}\cos(\Omega_l\tau_l) + \omega e^{i\omega\tau_l}\sin(\Omega_l\tau_l) - i\Omega_l] \\ i\Omega_l e^{i\omega\tau_l}\sin(\Omega_l\tau_l) - \omega e^{i\omega\tau_l}\cos(\Omega_l\tau_l) + \omega \end{pmatrix}^T. \tag{17}$$

We also define the *lth-Segment Projection Vector*

$$\vec{\mathbb{T}}^{(l)} := \frac{\Omega_l}{2}\left(\cos(\phi_l), \sin(\phi_l), 0\right) \tag{18}$$

to compactly express the control variables - namely Rabi rate and the rotation-axis vector, projected onto the xy plane of the Bloch sphere - associated with evolution during the lth unitary. In fact, inspection of Eqs. (15) and (16) reveals that $\vec{\mathbb{T}}^{(l)}$ is the computational analogue of $\mathbf{R}_z^{P_l}$ for the amplitude noise quadrature. The simpler dependence of $\vec{\mathbb{T}}^{(l)}$ on the control variables, however, reflects the fact that amplitude noise in our model is always coaxial, and hence commutes with, the control.

On the other hand, the 3 × 3 *Control History Matrix* $\mathbf{\Lambda}^{(l-1)}$, defined by

$$\Lambda_{ij}^{(l-1)} = \frac{1}{2}\operatorname{Tr}\left[Q_{l-1}^{\dagger}\hat{\sigma}_i Q_{l-1}\hat{\sigma}_j\right], \tag{19}$$

is the result of expanding the operator $Q_{l-1}^{\dagger}\hat{\sigma}_i Q_{l-1} \equiv \sum_j \Lambda_{ij}^{(l-1)}\hat{\sigma}_j$, with $i,j \in \{x,y,z\}$, in the Pauli basis, and identifying the coefficients $\Lambda_{ij}^{(l-1)}$ as the matrix elements. $\mathbf{\Lambda}^{(l-1)}$ thereby captures the accumulated effect of the previous $l-1$ *completed* unitaries, implemented via the cumulative operator Q_{l-1}.

4 Characteristics of noise filters

The power of the noise filtering formalism lies in the simple interpretation of the filter-transfer functions $F_i(\omega)$, which may be characterized in a standard engineering approach, considering passbands, stopbands, and filter order [42, 48, 49, 60]. In particular, error suppression corresponds to minimizing $F_i(\omega)$, $i \in \{z,\Omega\}$ in the spectral region where the corresponding PSDs are non-negligible. This can, in principle, be achieved by judicious construction of the control sequence since the filter-transfer functions are completely parametrized in variables describing the time-domain control applied to the qubit.

We are now in a position to examine the characteristics of the filter-transfer functions for an arbitrary control sequence $\mathbf{\Gamma}_n$, formally indicating the functional dependence of the filter-transfer functions on the control variables by writing $F_i(\omega\tau) = F_i(\omega\tau;\mathbf{\Gamma}_n)$, $i \in \{z,\Omega\}$. Inversely, we may commence a study of filter design based on constructing control sequences satisfying some *desired* filter property - our main goal. We now advance the main mathematical framework used in this paper to study filter design, pulling together the ideas introduced in the previous sections.

4.1 The filter cost function

A definition of the cost function associated with filter performance - captured through the *filter order* - leads us naturally to the imposition of constraints on the available space of controls. This cost function therefore lies at the heart of our attempts to craft control protocols appropriate for a given noise environment.

From the spectral overlap in Eq. (4), minimizing the infidelity contributed by the noise process $S_i(\omega)$ corresponds to minimizing the area under $F_i(\omega\tau;\mathbf{\Gamma}_n)$ in the spectral region of interest. We therefore define a cost function over a user-defined frequency band taking the form

$$A_i(\mathbf{\Gamma}_n) := \int_{\omega_L}^{\omega_c} d\omega F_i(\omega\tau;\mathbf{\Gamma}_n), \quad i \in \{z,\Omega\} \tag{20}$$

to diagnose the filtering effectiveness achieved by the control sequence $\mathbf{\Gamma}_n$. The smaller the integral $A_i(\mathbf{\Gamma}_n)$, the more effective the noise filtering over this band, in this noise quadrature. Since $\mathbf{\Gamma}_n$ is defined continuously over $\mathfrak{C}_n$ for a given n, we may in principle construct a variational procedure over this control space to derive minimizing 'values' of $\mathbf{\Gamma}_n$ satisfying a given cost function. In effect, the problem involves solving for paths in the control space over which the functional $A_i(\mathbf{\Gamma}_n)$ is minimized (up to some order).

Typically one would define the band $[\omega_L,\omega_c]$ over which the cost function is defined to fall within the *stopband* of $F_i(\omega\tau)$, below which filtering generally takes place. In general the band $[\omega_L, \omega_c]$ may be tailored to target specific spectral regions in the noise PSD. Doing so may produce highly effective filtering over this narrow spectral region, though out-of-band behaviour can be quite poor if not specifically optimized.[d]

4.2 The filter order

Again, following concepts from filtering in classical control engineering, we may define a *filter order* which will play a central role in efficiently realizing effective noise filters. We will mainly consider high-pass filters for low-frequency noise, setting $\omega_L = 0$ such that filtering takes place in the stopband up to the cutoff ω_c. In this case it is useful to perform the Taylor expansion of the filter-transfer function about $\omega = 0$, written

$$F_i(\omega\tau; \mathbf{\Gamma}_n) = \sum_{k=1}^{\infty} C_{2k}^{(i)}(\mathbf{\Gamma}_n)(\omega\tau)^{2k}, \tag{21}$$

where the dependence of the expansion coefficients $C_{2k}^{(i)}$ on $\mathbf{\Gamma}_n$ has been made explicit, and we include only even powers of $\omega\tau$ due to the evenness of $F_i(\omega\tau)$. Assuming sufficiently low-frequency noise ($\omega_c < 1/\tau$), the approximation $F(\omega\tau) \propto (\omega\tau)^{2p}$ holds for some p associated with the most significant power law expansion term. This defines a high-pass filter with *filter order* (determined by p) visualized as the slope in the stopband on a log-log plot.[e]

Using this notation, and working in the low-frequency limit, we then say the control sequence $\mathbf{\Gamma}_n \in \mathfrak{C}_n$ filters $\beta_i(t)$ noise to order $(p-1)$ over the band $[0, \omega_c]$ if $\mathbf{\Gamma}_n$ is a *concurrent zero* of the first $(p-1)$ Taylor coefficients. That is, if $C_2^{(i)}(\mathbf{\Gamma}_n) = C_4^{(i)}(\mathbf{\Gamma}_n) = \cdots = C_{2(p-1)}^{(i)}(\mathbf{\Gamma}_n) = 0$. In this case we approximate $F_i(\omega\tau; \mathbf{\Gamma}_n) \approx C_{2p}^{(i)}(\mathbf{\Gamma}_n)(\omega\tau)^{2p}$ and consequently $A_i(\mathbf{\Gamma}_n) \approx C_{2p}^{(i)}(\mathbf{\Gamma}_n)\frac{(\omega_c\tau)^{2p+1}}{2p+1}$. Thus $\mathbf{\Gamma}_n$ is a $(p-1)$-order (high-pass) filter in the ith noise quadrature if the following equivalent conditions are satisfied

$$\frac{A_i(\mathbf{\Gamma}_n)}{C_{2p}^{(i)}(\mathbf{\Gamma}_n)} = \mathcal{O}\left(\frac{(\omega_c\tau)^{2p+1}}{2p+1}\right) \quad \Longleftrightarrow \quad C_2^{(i)}(\mathbf{\Gamma}_n) = \cdots = C_{2(p-1)}^{(i)}(\mathbf{\Gamma}_n) = 0. \tag{22}$$

This metric will play a central role in the analyses that follow.

It is important to disambiguate the asymptotic filter order $(p-1)$, introduced above for characterizing the behaviour near zero frequency, from a more general metric capable of describing filter performance over an arbitrary spectral band. For this we introduce the *local filter order* $(p^* - 1)$ by the property that, over the band $[\omega_L, \omega_c]$ the filter-transfer function is well approximated by $F_i \propto (\omega\tau)^{2p^*}$. One may take the limit that $\omega_L \to \omega_c \to \omega^*$ and thereby obtain the *instantaneous filter order*, effectively measuring the power-law behaviour at ω^*. Both local and instantaneous filter order reduce to the asymptotic filter order over the stopband if over this region F_i is well-characterized by its the zero-frequency behaviour.

4.3 Time-domain filter order vs. Magnus order

Both the asymptotic and instantaneous filter orders defined above for time-domain noise must be distinguished from the *Magnus order* of error cancellation. The latter is familiar from work in NMR in which quasi-static errors can be cancelled by suitable composite pulse sequence design. The regime of quasistatic errors coincides with the DC limit for the time-dependent noise fields introduced in Section 2. That is, the time-dependent noise fields reduce to scalar constants $\beta_z(\beta_\Omega)$. The Magnus expansion terms in (70), now denoted $\Phi_\mu^{(DC)}$, are then evaluated strictly as time integrals over *ideal* control operations, scaled by factors $\beta_z^\mu(\beta_\Omega^\mu)$ specifying the power law dependence on the magnitude of these static offsets errors. A pulse sequence for which $\Phi_1^{(DC)} = \cdots = \Phi_{\mu-1}^{(DC)} = 0$ is then said to compensate offset errors to Magnus order $(\mu - 1)$. In this case the total error operator satisfies $\Phi^{(DC)}(\tau) = \mathcal{O}(\Phi_\mu^{(DC)})$ and is dominated by the residual error proportional to the μth power in the magnitude of the error.

This is quite distinct from time-dependent noise where the error expansion used to calculate the fidelity contains terms of various Magnus order but equivalent time-dependent error norm in the ensemble average (see, *e.g.* Eq. 1 in Ref. [48]). The net result is the observation that *high-order error suppression in the Magnus expansion does not imply high-order time-domain noise filtering.* This has been validated using experiments on trapped ions [37], and formalized rigorously in Ref. [38], where it has been shown that $p \leq \mu$, but p^* over a user-defined band is unrelated to μ. Our focus throughout this work will be on crafting efficient noise filters rather than high-order error suppressing gates.

5 Filter design by Walsh synthesis

Even with the general insights into the appropriate modulation protocols outlined above, it is desirable to bound the dimensionality of the control space, and hence the complexity of the filter-design task, by imposing physically motivated constraints on the form of $\mathbf{\Gamma}_n$. In practice the achievable filter order is typically limited by the number of unitary operations in the control sequence; one may increase $(p - 1)$ at the cost of increasing n. From an experimental standpoint, faced with the physical limitation set by a maximum achievable Rabi rate, this cost manifests as a longer total sequence duration $\tau = \hat{\sigma}_l^n \tau_l$. This may offset the proposed benefit of the higher-order filter due to a longer noise interaction time. From a theoretical standpoint the cost is in the greater complexity of the variational search; the number of (free) variational parameters in $\mathbf{\Gamma}_n$ grow as $3n$ and the number of matrix products in Eqs. (15) and (16) grows as n.

We are able to effectively bound the synthesis space while still achieving highly effective gates by *synthesizing* relevant time-domain control fields in the basis set of Walsh functions - square wave analogues of the sines and cosines [51, 61] - using the concept of functional analysis. Walsh functions are defined in a uniform piecewise-constant fashion (Figure 2), building intrinsic compatibility with discrete clocking [54] and classical digital logic. Since their formulation in the first half of the twentieth century [62] they have played an important role in scientific and engineering applications. Their development and utilization has been strongly influenced by parallel developments in digital electronics and computer science since the 1960s, with Walsh-type transforms replacing Fourier transforms in a range of engineering applications such as communication, signal processing, pattern recognition, noise filtering and so forth [61, 63].

More recently the Walsh functions have been identified as an attractive resource in quantum information, with applications in time-resolved magnetometry using nitrogen-vacancy centres in diamond [64] and in DD for digital-efficient pulse sequencing [51]. Notably, in the latter scheme the decoupling performance was found to be determined by the distinct symmetry and spectral properties of the Walsh basis. These properties enable us to establish *analytic design rules* (see Section 5.3) to further streamline Walsh-synthesized filter construction.

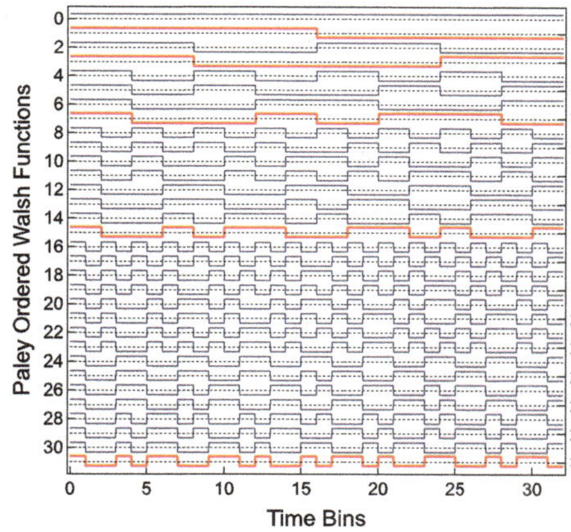

Figure 2 Walsh functions. The first 32 Paley-ordered Walsh functions PAL_k, $k \in \{0, \ldots, 31\}$. Functions with maximum Hamming weight $r(k)$ (hence maximum number of Rademacher functions) for given $m(k)$, corresponding to Paley orders $2^{m(k)} - 1$, are highlighted in red.

We begin by reviewing the relevant mathematical details of the Walsh basis. Two equivalent representations are introduced, *Paley ordering* and the *Hadamard representation*, which shall be used throughout this paper.

5.1 The Paley and Hadamard representations

The set of Walsh functions $w_k : [0,1] \to \{\pm 1\}, k \in \mathbb{N}$ form an orthonormal-complete family of binary-valued square waves defined on the unit interval. They are aperiodic and hence do not admit to a unique ordering, in contrast with the Fourier basis in which sinusoids are ordered by increasing *frequency*. A number of different orderings exist [61, 65, 66] due to the different ways in which the basis elements may be defined. We employ the *Paley ordering* [67] in which basis functions are generated from products of *Rademacher functions* [68], defined by

$$R_j(x) := \mathrm{sgn}\left[\sin(2^j \pi x)\right], \quad x \in [0,1], j \geq 0. \tag{23}$$

The jth Rademacher function $R_j(x)$ is thus a periodic square wave switching 2^{j-1} times between ± 1 over the interval $[0,1]$. The Walsh function of Paley order k, here denoted $\mathrm{PAL}_k(x)$, is then defined by

$$\mathrm{PAL}_k(x) = \prod_{j=1}^{m} R_j(x)^{b_j}, \tag{24}$$

where $(b_m, b_{m-1}, \ldots, b_1)_2$ is the binary representation of k. That is, $k = b_m 2^{m-1} + b_{m-1} 2^{m-2} + \cdots + b_1 2^0$, where $m(k)$ indexes the most significant binary digit, having defined $b_m \equiv 1$. Consequently, $\mathrm{PAL}_k(x)$ has factors $R_j(x)$ whenever b_j is a nonzero binary digit of k; the total number of Rademacher functions in the construction of $\mathrm{PAL}_k(x)$ is thus given by the number of nonzero b_j's in k - namely, the *Hamming weight* denoted $r(k)$. For a given value of $m(k)$, the maximum number Rademacher functions therefore occur for $\mathrm{PAL}_{2^{m(k)}-1}(x)$. For example, setting $m(k) = 3$, a maximum of three Rademacher functions are used to construct $\mathrm{PAL}_7(x) = R_3(x)R_2(x)R_1(x)$, corresponding to the three nonzero digits $b_{1,2,3}$ in the binary expansion $k = 7 = (1,1,1)_2$. For illustration, the first 32 Walsh functions in the Paley ordering are shown in Figure 2.

The discrete-timestep properties of these basis functions produce, under linear superposition, piecewise-constant waveforms with digitized segment lengths. In our framework these segments are used to specify the a modulation of the control field, ultimately defining a piecewise-constant sequence of unitaries. We therefore require a straightforward expression for the envelope of an arbitrary synthesis $\sum_{k=0}^{N} X_k \, \mathrm{PAL}_k(x)$. Due to the aperiodicity of the Walsh functions, however, a general expression in Paley ordering is difficult. To overcome this it is convenient to use the Hadamard representation.

The unique sign-switching envelope of $\mathrm{PAL}_k(x)$ is determined by the sign-switching of the constituent Rademacher functions. Since any $R_j(x)$ switches sign uniformly 2^j times over the interval $[0,1]$, the fastest sign-modulation rate in $\mathrm{PAL}_k(x)$ derives from the highest order Rademacher function $R_{m(k)}(x)$, which switches sign $2^{m(k)}$ times over $[0,1]$. Provided $m(k) \leq n$, we may therefore partition $[0,1]$ into 2^n equal time bins such that $\mathrm{PAL}_k(x)$ is constant valued on each bin. Any basis function $\mathrm{PAL}_k(x)$ then projects completely onto a digital vector in $\mathbb{R}^{2^n}$ with the jth element, $P_j^{(k)} \in \{\pm 1\}$, defined by the value of $\mathrm{PAL}_k(x)$ in the jth bin. That is, $\mathrm{PAL}_k(x)$ is isomorphic to the discrete digital vector written

$$\mathbf{P}_{2^n}^{(k)} \equiv \left[P_1^{(k)}, \quad P_2^{(k)}, \quad \ldots, \quad P_{2^n}^{(k)}\right]. \tag{25}$$

This projection is possible for all $k \in \{0, 1, \ldots, 2^n - 1\}$ for which the condition $m(k) \leq n$ is true, resulting in a set of 2^n vectors. Since these vectors inherit the orthogonality of the $\mathrm{PAL}_k(x)$, moreover, they form a *discrete* Walsh basis spanning $\mathbb{R}^{2^n}$.

In the Hadamard representation, these vectors occur as columns (rows) of the Hadamard matrix of dimension 2^n. Using so-called *Sylvester construction* [69] the 2^n-dimensional Hadamard matrix H_{2^n} is generated recursively by

$$H_{2^n} = \begin{bmatrix} H_{2^{n-1}} & H_{2^{n-1}} \\ H_{2^{n-1}} & -H_{2^{n-1}} \end{bmatrix} = S^{\otimes n}, \qquad S = \begin{bmatrix} 1 & 1 \\ 1 & -1 \end{bmatrix}, \qquad H_1 = 1, \tag{26}$$

where S is the Sylvester matrix, and $\otimes n$ denotes $n \geq 1$ applications of the Kronecker product. In this construction $\mathbf{P}_{2^n}^{(k)}$ defines the $i(k) = 1 + \sum_{j=1}^{m(k)} b_j 2^{n-j}$ column (row) of H_{2^n}. The orthogonality of the Walsh basis is thereby reflected in the familiar property that $H_{2^n} H_{2^n}^T = 2^n I$, implying the orthogonality of the Hadamard matrices.

The Hadamard representation of the Walsh functions has the distinct advantage of naturally specifying the piecewise-constant structure of time domain sequences constructed via linear combinations of Walsh functions. Any function synthesized in the Paley-ordered Walsh basis, $f(x) = \sum_{k=0}^{2^n-1} q_k \, \mathrm{PAL}_k(x)$, has a vector representation in the column space of H^{2^n}. In this section we will use this observation to efficiently construct Walsh-synthesized filters, whose properties map compactly onto the Walsh spectrum.

5.2 Walsh-synthesized filters

The basic physics of commutation relations between Pauli operators immediately suggests an immediate constraint on the available modulation, broadly involving structuring of the Rabi rate or control phase in the time domain

$$\mathbf{\Gamma}_n^{\mathrm{AMF}} \cong \left\{(\tau_l, \Omega_l)\right\}_{l=1}^{n}, \quad \phi_l = \phi_0 \; \forall l \in \{1, \ldots, n\}, \tag{27}$$

$$\mathbf{\Gamma}_n^{\mathrm{PMF}} \cong \left\{(\tau_l, \phi_l)\right\}_{l=1}^{n}, \quad \Omega_l = \Omega_0 \; \forall l \in \{1, \ldots, n\} \tag{28}$$

referred to as *single-axis amplitude-modulated filters* (AMFs) and *constant-amplitude phase-modulated filters* (PMFs). These constrained forms may be used to design filters for dephasing and amplitude noise *separately* using minimal control resources. For $\hat{\sigma}_z$ (dephasing) noise it is sufficient to employ rotations about a single (orthogonal) axis in the xy plane and therefore restrict attention to AMFs. On the other hand, unless implementing the trivial identity gate such that the total gate rotation angle $\Theta \equiv \sum_{l=1}^{\mathcal{M}} \theta_l = 0$, strict amplitude modulation is insufficient for filtering amplitude noise.[f] For nontrivial gates, amplitude noise filters generally require control over the rotation axis, and for this purpose we employ PMFs.

In the Walsh synthesis framework, the modulated structures $\mathbf{\Gamma}_n^{\mathrm{AMF}}$ and $\mathbf{\Gamma}_n^{\mathrm{PMF}}$ are further constrained by synthesizing the time-domain Rabi rate $\Omega(t)$ or phase $\phi(t)$ as linear superpositions of Walsh functions

$$\Omega(t) = \sum_{k=0}^{N} X_k \,\mathrm{PAL}_k(t/\tau), \qquad \phi(t) = \sum_{k=0}^{N} Y_k \,\mathrm{PAL}_k(t/\tau), \quad t \in [0, \tau]. \tag{29}$$

Here the synthesis spectra are denoted in terms of X_k (Y_k) to distinguish Walsh modulation in the amplitude (phase) quadratures. We refer to the resulting sequences as *Walsh amplitude-* (WAMF) and *phase-* (WPMF) *modulated filters*. To compactly express these modulated control forms as sequences of unitaries we now employ the Hadamard representation.

Consider an arbitrary function $f(x) = \sum_{k=0}^{N} q_k \,\mathrm{PAL}_k(x)$ synthesized in the Walsh basis up to Paley order N. From Section 5.1 all basis functions in this synthesis projected onto a Hadamard matrix of (minimum) dimension $\mathcal{M}(N) \equiv 2^{m(N)}$. A discrete representation of the function $f(x)$ therefore exists as a projection onto the column space of $H_{\mathcal{M}}$ by writing

$$\mathbf{f} = H_{\mathcal{M}}\tilde{\mathbf{q}}, \quad \tilde{\mathbf{q}} = (\tilde{q}_1, \tilde{q}_2, \ldots, \tilde{q}_{\mathcal{M}})^T, \tilde{q}_{i(k)} = \begin{cases} q_k & \text{for } 0 \le k \le N, \\ 0 & \text{for } N < k < \mathcal{M}, \end{cases} \tag{30}$$

where the column vector $\tilde{\mathbf{q}}$ consists of the reordered Paley spectral amplitudes q_k according to the change of basis map specified by $i(k) = 1 + \sum_{j=1}^{m(k)} b_j 2^{n-j}$. Thus, in the Hadamard representation, the piecewise-constant structure of $f(x)$ is extracted from the vector $\mathbf{f} = (f_1, f_2, \ldots, f_{\mathcal{M}})^T$, with $\tilde{\mathbf{q}}$ representing the synthesis spectrum. In this case Eq. (29) implies the forms

$$\mathbf{\Gamma}_{\mathcal{M}}^{(\mathrm{WAMF})} = \begin{matrix} \\ p_1 \\ p_2 \\ \vdots \\ p_{\mathcal{M}} \end{matrix} \begin{matrix} \Omega_l & \tau_l & \theta_l & \phi_l \\ \left[\begin{matrix} \\ \\ \vec{\mathbf{\Omega}} \\ \\ \\ \end{matrix} \right. & \begin{matrix} \tau_{\mathcal{M}} \\ \tau_{\mathcal{M}} \\ \vdots \\ \tau_{\mathcal{M}} \end{matrix} & \begin{matrix} \tau_{\mathcal{M}}\Omega_1 \\ \tau_{\mathcal{M}}\Omega_2 \\ \vdots \\ \tau_{\mathcal{M}}\Omega_{\mathcal{M}} \end{matrix} & \left. \begin{matrix} \phi_0 \\ \phi_0 \\ \vdots \\ \phi_0 \end{matrix} \right] \end{matrix}, \quad \vec{\mathbf{\Omega}} = H_{\mathcal{M}}\tilde{\mathbf{X}}, \tag{31}$$

$$\Gamma_{\mathcal{M}}^{(\text{WPMF})} = \begin{matrix} p_1 \\ p_2 \\ \vdots \\ p_{\mathcal{M}} \end{matrix} \overset{\begin{matrix} \Omega_l & \tau_l & \theta_l & \phi_l \end{matrix}}{\left[\begin{array}{ccc|c} \Omega_0 & \tau_{\mathcal{M}} & \tau_{\mathcal{M}}\Omega_0 & \\ \Omega_0 & \tau_{\mathcal{M}} & \tau_{\mathcal{M}}\Omega_0 & \vec{\boldsymbol{\phi}} \\ \vdots & \vdots & \vdots & \\ \Omega_0 & \tau_{\mathcal{M}} & \tau_{\mathcal{M}}\Omega_0 & \end{array}\right]}, \quad \vec{\boldsymbol{\phi}} = H_{\mathcal{M}}\tilde{\mathbf{Y}} \tag{32}$$

with $\mathcal{M} \equiv 2^{m(N)}$ and $\tau_{\mathcal{M}} \equiv \tau/\mathcal{M}$. The Rabi rate- or phase-modulation is thus defined by the vectors $\vec{\boldsymbol{\Omega}} = (\Omega_1, \Omega_2, \ldots, \Omega_{\mathcal{M}})^T$ and $\vec{\boldsymbol{\phi}} = (\phi_1, \phi_2, \ldots, \phi_{\mathcal{M}})^T$ whose components

$$\Omega_l \equiv \Omega_l(X_0, X_1, \ldots, X_N), \qquad \phi_l \equiv \phi_l(Y_0, Y_1, \ldots, Y_N), \quad l \in \{1, \ldots, \mathcal{M}\} \tag{33}$$

specify the control variables during lth timestep. In this case τ_l takes a fixed discrete value, though consecutive segments with the same values of Ω_l and ϕ_l may be combined sequentially to form effective operations of longer duration. The remaining degrees of freedom reside in the functional dependencies of $\Omega_l(\mathbf{X})$ and $\phi_l(\mathbf{Y})$ on the Walsh spectra,[g] the explicit forms of which are determined by the Hadamard matrix equations above.

The reduced control space, now compactly parameterized by the Walsh spectra, thus consists of bounded-strength unitary sequences inheriting the discrete timing properties of the Walsh basis. This contrasts with similar composite pulse methods in NMR and quantum information [20, 27, 70] which generally rely on structures defined in continuous time.[h] In the next section we identify useful properties of the Walsh basis which capture filter performance and hence inform effective filter design.

5.3 Analytic filter-design rules

From Eqs. (31) and (32) the WAMF (WPMF) constructs are completely parameterized by the Walsh spectra $\mathbf{X}^{(i)}$, $i \in \{z, \Omega\}$. Here, for compactness, we denote $\mathbf{X}^{(z)}(\mathbf{X}^{(\Omega)}) = \mathbf{X}(\mathbf{Y})$. Filter properties and gate characteristics consequently map onto the basis functions in the synthesis.

To target these properties it is convenient to partition the Walsh spectrum $\mathbf{X}^{(i)} \equiv (\mathbf{X}_\nu^{(i)}, \mathbf{X}_\rho^{(i)})$ into spectral-amplitude classes to be treated as *variational* ($\mathbf{X}_\nu^{(i)}$) and *fixed parameters* ($\mathbf{X}_\rho^{(i)}$). Making the formal substitution $\mathbf{\Gamma}_{\mathcal{M}} \to \mathbf{X}^{(i)}$, the cost function in Section 4 is consequently re-expressed

$$A_i\big(\mathbf{X}_\nu^{(i)}; \mathbf{X}_\rho^{(i)}\big) := \int_{\omega_L}^{\omega_c} d\omega F_i\big(\omega\tau; \mathbf{X}_\nu^{(i)}; \mathbf{X}_\rho^{(i)}\big), \quad i \in \{z, \Omega\}, \tag{34}$$

where it is understood that A_i is minimized over the space spanned by $\mathbf{X}_\nu^{(i)}$ with $\mathbf{X}_\rho^{(i)}$ held constant. Similarly, high-pass $(p-1)$-order filters satisfy the conditions

$$\frac{A_i(\mathbf{X}^{(i)})}{C_{2p}(\mathbf{X}^{(i)})} = \mathcal{O}\left(\frac{(\omega\tau_c)^{2p+1}}{2p+1}\right) \iff C_2^{(i)}\big(\mathbf{X}^{(i)}\big) = \cdots = C_{2(p-1)}^{(i)}\big(\mathbf{X}^{(i)}\big) = 0. \tag{35}$$

We are now in a position to establish a range of analytic filter-design rules to refine our search space and streamline Walsh synthesis leveraging approaches similar to electrical or digital signal filter construction. In particular, the well defined spectral properties and symmetries of the Walsh functions may be used to inform effective filter construction with improved performance. These include:

(1) Alternate modulation quadratures for dephasing or amplitude noise.
(2) Restricting Walsh synthesis by symmetry considerations.
(3) Constraining Walsh spectra for target gate angle.
(4) Achievable filtering characteristics determined by $m(k)$ and $r(k)$.

We address each of these in turn.

(i) *Alternate modulation quadratures for dephasing or amplitude noise* - As the most basic element of design, we first reiterate the statements made above establishing WAMFs (WPMFs) as useful for filtering dephasing (amplitude) noise *separately*. In Section 9, however, we derive universal noise filters by concatenating these two filter constructs.

(ii) *Restricting Walsh synthesis by symmetry considerations* - As with the cosines (sines) constituting the Fourier basis, the Walsh basis separates into so-called CAL (SAL) functions with even (odd) parity. Restricting the synthesis to the CAL subset ensures the modulated waveform has time-reversal symmetry about the sequence midpoint $\tau/2$. This can be a convenient and effective method in filter design, in line with the observation in dynamic decoupling literature [13, 71] that sequence performance is often improved using time-symmetric over -asymmetric building blocks.[i]

(iii) *Constraining Walsh spectra for target gate angle* - Imposing desired physical properties on a candidate control sequence may generally be achieved by holding some subset $\mathbf{X}_p^{(i)}$ of the Walsh-spectral amplitudes constant. For example, we may fix the total rotation angle of the Bloch vector in order to implement a target logic operation. For WAMFs this involves a very straightforward constraint on the Walsh spectrum: the total rotation angle depends only on X_0. This can be seen as follows. First observe for Paley orders $k \geq 1$ the Walsh functions are balanced in the sense that $\int_0^1 \mathrm{PAL}_k(x)\,dx = \delta_{0k}$, where δ_{ij} denotes the Kronecker delta. For WAMFs the total gate angle $\Theta \equiv \int_0^\tau dt\Omega(t)$ therefore takes the form

$$\Theta = \int_0^\tau \sum_{k=1}^{N} X_k\,\mathrm{PAL}_k(t/\tau)\,dt = \tau \sum_{k=1}^{N} X_k \int_0^1 \mathrm{PAL}_k(x)\,dx = \tau \sum_{k=1}^{N} X_k \delta_{0k} = X_0 \tau.$$

The effective gate rotation, $\theta = \Theta \bmod 2\pi$, is consequently given by

$$\theta = X_0 \tau \bmod 2\pi \tag{36}$$

implying the necessary constraint on X_0 for a desired θ.

(iv) *Achievable filtering characteristics determined by m(k) and r(k)* - The achievable filter order over the entire stopband is essentially limited by the number of constituent control operations: one may achieve higher p at the cost of higher n. For the Walsh-synthesized filters in Eqs. (31) and (32), with N the highest-order basis function, $n \equiv 2^{m(N)}$. Hence higher-order Walsh functions generally produce higher-order filters.

For high-pass filters further insight is gained by examining the low-frequency spectral properties of the $\mathrm{PAL}_k(t/\tau)$. This reflects the fact that the filter-transfer functions are closely related to Fourier transforms of relevant time-domain control functions. In particular, the Fourier transform of $\mathrm{PAL}_k(t/\tau)$, near zero frequency, has a power-law expansion [51]

$$\mathcal{F}_x\left[\mathrm{PAL}_k(x)\right] \propto (\omega\tau)^{r(k)}, \tag{37}$$

where $r(k)$ is the Hamming weight. Here $x \equiv t/\tau$ is a non-dimensional time-domain variable and $\mathcal{F}_x[\text{PAL}_k(x)]$ denotes the forward Fourier transform of $\text{PAL}_k(x)$ from x to the (non-dimensional) angular frequency variable $\omega\tau$ in Fourier space. Increasing the low-frequency *roll-off* is therefore associated with maximizing $r(k)$ for a given number of control operations $n = 2^{m(k)}$. This corresponds to maximizing the number of Rademacher functions in the construction[j] and immediately identifies Paley orders $k = 2^{\alpha} - 1$, $\alpha \in \mathbb{N}$, (see Figure 2) as key design resources.

6 Walsh Amplitude Modulated Filters (WAMFs)

Having introduced the basic physical picture and mathematical basis for Walsh filter synthesis, we move on to demonstrate explicit realizations of WAMFs for dephasing noise. Both first and second-order filters with high-pass filter characteristics are constructed.

6.1 First-order WAMFs

We begin by considering first-order filters for dephasing noise implementing target single-qubit rotations. Construction begins by considering the design rules (i)-(iv) outlined in Section 5.3. For filtering noise in this quadrature (i) implies we should employ the WAMF construction (Eq. (31)). In this case, invoking (iii), the requirement of implementing non-trivial gates dictates we include Paley order $k = 0$ in the synthesis. The average Rabi rate (and hence rotation angle) is then determined by X_0, the spectral amplitude of $\text{PAL}_0(t/\tau)$. The remaining synthesis choices include basis functions of Paley order $k > 0$ and are in principle unbounded.

As a first application, we pursue the construction minimizing the number $n = 2^{m(N)}$ of unitary operations in the synthesized sequence such that error suppression is still attainable. In line with design rules (ii) and (iv), time-reversal symmetry is ensured and the number of Rademacher functions is maximized by reducing the remaining synthesis choices to the single basis function $\text{PAL}_3(t/\tau)$ (Figure 2). Hence, in this simple example, $N = 3$, and $\mathcal{M}(N) = 2^{m(N)} = 2^2$, yielding 4-segment gates with segment lengths $\tau_{\mathcal{M}} = \tau/4$. These represent the lowest-order constructions with error suppression capabilites.

The Rabi rate is consequently written $\Omega(t) = X_0\,\text{PAL}_0(t/\tau) + X_3\,\text{PAL}_3(t/\tau)$. Physically, X_3 specifies the modulation depth of the resulting Rabi rate envelope (see inset to Figure 3(c)) while X_0 determines the average value as described above. Accordingly, for a particular target rotation, we treat X_0 as a *fixed* parameter (see Eq. (36)) while X_3 is treated as a *variational* parameter by which to optimize the (dephasing) cost function (Eq. (34)). Thus, values of X_3 for which $A_z(X_3; X_0)$ is minimized specify the optimum modulation depths for an effective filter.

Using Eq. (30) the Walsh synthesis spectrum is represented $\tilde{\mathbf{X}} \equiv (\tilde{X}_1, \tilde{X}_2, \tilde{X}_3, \tilde{X}_4)^T = (X_0, 0, 0, X_3)^T$, yielding the Hadamard representation of the modulated Rabi rate

$$\vec{\mathbf{\Omega}} = \begin{bmatrix} 1 & 1 & 1 & 1 \\ 1 & -1 & 1 & -1 \\ 1 & 1 & -1 & -1 \\ 1 & -1 & -1 & 1 \end{bmatrix} \begin{bmatrix} X_0 \\ 0 \\ 0 \\ X_3 \end{bmatrix} = \begin{bmatrix} X_0 + X_3 \\ X_0 - X_3 \\ X_0 - X_3 \\ X_0 + X_3 \end{bmatrix}. \tag{38}$$

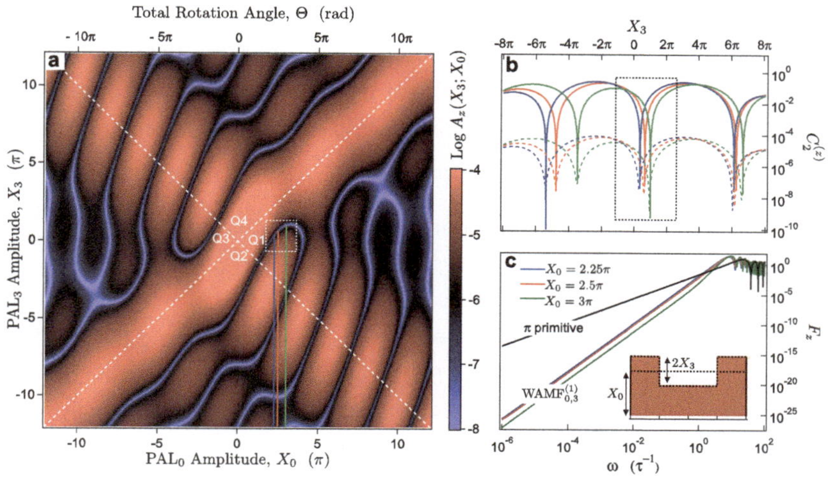

Figure 3 First-order WAMFs. Construction of $\text{WAMF}_{0,3}^{(1)}$ for dephasing noise filtering. (**a**) Log-scale color plot of the cost function $A_z(X_3;X_0)$ integrated over $\omega \in [10^{-9},10^{-1}]\tau^{-1}$. Total gate angle $\Theta = X_0$ ($\tau \equiv 1$). Blue regions indicate minima in $A_z(X_3;X_0)$, implying optimized filter synthesis. Coloured lines (blue, red, green) at $X_0 = (2\frac{1}{4}, 2\frac{1}{2}, 3)\pi$ correspond to rotation angles $\theta = (\frac{\pi}{4}, \frac{\pi}{2}, \pi)$. These lines terminate at values $X_3 = (0.36\ldots, 0.65\ldots, 1)\pi$ on a blue contour (boxed) and indicate representative points in the X_0X_3 plane for which first-order filtering is achieved. In this plot, for $|X_3| > |X_0|$ the Rabi rates $X_\pm$ have opposite sign, implying a π-phase shift in addition to amplitude modulation (e.g. see Eq. **(145)**). We therefore distinguish quadrants Q1 and Q3 in the X_0X_3 plane in which $|X_3| \leq |X_0|$ (strict amplitude modulation) and Q2 and Q4 in which $|X_3| > |X_0|$ (sign-switching amplitude modulation). (**b**) *Solid lines*: first order Taylor coefficient $C_2^{(z)}(X_3;X_0)$ as a function of X_3 with $X_0 = (2\frac{1}{4}, 2\frac{1}{2}, 3)\pi$; zeros appear as dips on log-scale. *Dotted lines*: one-dimensional slices of $A_z(X_3;X_0)$ for same fixed values of X_0. Boxed dips correspond to boxed points in a) where the colored lines intersect with the blue contour. (**c**) Filter-transfer functions for the spectrally optimized $\text{WAMF}_{0,3}^{(1)}$ gates identified by the boxed features in (a) and (b).

The resulting WAMF construction, denoted $\text{WAMF}_{0,3}^{(1)}$, is thus represented

$$\text{WAMF}_{0,3}^{(1)} = \begin{array}{c} \\ P_1 \\ P_2 \\ P_3 \\ P_4 \end{array} \begin{array}{c} \begin{array}{cccc} \Omega_l & \tau_l & \theta_l & \phi_l \end{array} \\ \left[\begin{array}{cccc} \frac{X_+}{\tau} & \frac{\tau}{4} & \frac{X_+}{4} & 0 \\ \frac{X_-}{\tau} & \frac{\tau}{4} & \frac{X_-}{4} & 0 \\ \frac{X_-}{\tau} & \frac{\tau}{4} & \frac{X_-}{4} & 0 \\ \frac{X_+}{\tau} & \frac{\tau}{4} & \frac{X_+}{4} & 0 \end{array}\right] \end{array}, \tag{39}$$

where $X_\pm := X_0 \pm X_3$, and the superscript in this notation denotes first-order filter capabilities. Hence these gates inhabit the two-dimensional control space spanned by the X_0X_3 plane (see Figure 3(a)). The dephasing filter-transfer function $F_z(\omega;X_3;X_0)$ for an arbitrary $\text{WAMF}_{0,3}^{(1)}$ gate is derived by substituting Eq. (39) into Eq. (13). The cost function $A_z(X_3;X_0)$ may then be numerically integrated.

Figure 3(a) shows a two-dimensional representation of $A_z(X_3;X_0)$ integrated over the stopband $\omega \in [10^{-9},10^{-1}]\tau^{-1}$. The value of $\text{Log}_{10}[A_z(X_3;X_0)]$ is indicated by the color scale. Total sequence length is normalized to $\tau = 1$ in this data, so the total gate rotation angle $\Theta \equiv X_0$ is given directly by the X_0-axis. As can be seen, for any fixed X_0 there exist quasi-periodic tunings of X_3 which minimize the cost function. In other words, we have a prescription for synthesizing *spectrally-optimized dephasing filters which implement arbitrary rotation angles*. Interestingly, the point $(X_0,X_3) \equiv (3\pi,\pi)$ reproduces the previously derived first-order DCG NOT construction [19].

Blue regions, where the cost function has been minimized, represent *first-order* filters for low-frequency noise due to the restrictions placed on the synthesis space.[k] To demonstrate that these optimized $\mathrm{WAMF}^{(1)}_{0,3}$ gates perform as first-order filters we Taylor expand $F_z(\omega; X_3; X_0)$ as in Eq. (35), and derive an easy analytic expression for the first order coefficient

$$C_2^{(z)}(X_3; X_0) = 4\left[\frac{(X_0 - X_3)\sin(\frac{X_0}{2}) + 2X_3 \sin(\frac{X_0 - X_3}{4})}{(X_0^2 - X_3^2)}\right]^2. \quad (40)$$

In principle we may now solve $C_2^{(z)}(X_0; X_3) = 0$ to find values of X_3 giving first-order filters for a given X_0. In Figure 3(b) we plot $C_2^{(z)}(X_0; X_3)$ (solid lines) as a function of X_3 for the choices $X_0 = (2\frac{1}{4}, 2\frac{1}{2}, 3)\pi$ as above. Zeros of $C_2^{(z)}(X_0, X_3)$, appearing as dips on the log scale, occur quasi-periodically in X_3 and match with points in Figure 3(a) where corresponding lines of constant X_0 intersect with the blue contours. To demonstrate this we plot one-dimensional slices of $A_z(X_3; X_0)$ for fixed values $X_0 = (2\frac{1}{4}, 2\frac{1}{2}, 3)\pi$ (dotted lines). We find the minima in $A_z(X_3; X_0)$ align with zeros of $C_2^{(z)}(X_0; X_3)$, implying the blue contours in the X_0X_3 plane do indeed produce first-order filters, with $(p - 1) = 1$. The boxed zeros near $X_3 = (0.36\ldots, 0.65\ldots, 1)\pi$ correspond to the termination points of the colored lines in Figure 3(a) (also boxed). These indicate representative points in the X_0X_3 plane producing first-order filters with nontrivial rotations. In particular, these filters implement $\theta = (\frac{\pi}{4}, \frac{\pi}{2}, \pi)$ rotations.

The corresponding dephasing filter-transfer functions for these three optimized gates are shown in Figure 3(c). As expected from Eq. (21), with $C_2^{(z)} = 0$, these approximately satisfy $F_z \propto (\omega\tau)^4$ in the stopband, producing first-order filters with $(p - 1) = 1$. For comparison we include the dephasing filter-transfer function for a primitive π rotation where $F_z \propto (\omega\tau)^2$, implying $(p - 1) = 0$. The steeper slopes, or *roll-offs*, for the optimized $\mathrm{WAMF}^{(1)}_{0,3}$ gates captures this difference. This filter design method, and the performance of the $\mathrm{WAMF}^{(1)}_{0,3}$ filters, has recently been experimentally validated by our group [37].

6.2 Second-order WAMFs

We now consider higher-order dephasing filters by increasing the number n of segments in the sequence. In particular we consider 8-segment gates. Construction again begins by considering the design rules (i)-(iv) outlined in Section 5.3.

Using (i) and (iii) we employ the WAMF construction and include Paley order $k = 0$ to ensure nontrivial rotation angles. Extending to 8-segments, however, increases the accessible range of Walsh functions in the synthesis as identified in design rule (iv). Specifically we extend the synthesis to Paley orders $k \leq 7$ corresponding to the complexity class $m(k) \leq 3$, implying a $2^3 = 8$ segments construction in the Hadamard representation. We denote these constructions by $\mathrm{WAMF}^{(2)}_{0:7}$ where the superscript indicates second-order filtering capabilities, as will be shown. Imposing time-reversal symmetry about $\tau/2$ further restricts the synthesis space to $k \in \{3, 5, 6\}$, corresponding to CAL functions referenced in design rule (ii). We therefore study synthesized filters with spectral amplitudes partitioned into fixed $\mathbf{X}_\rho = X_0$ and variational $\mathbf{X}_\nu = (X_3, X_5, X_6)$ classes.

As a representative example we set $X_0 = 3\pi$ and restrict attention to filters implementing a net π rotation ($\tau \equiv 1$). Our cost function consequently takes the form $A_z(X_3, X_5, X_6; 3\pi) = \int_{\omega_L}^{\omega_c} F_z(\omega\tau; X_3, X_5, X_6; 3\pi)\, d\omega$, implying a three-dimensional *varia-*

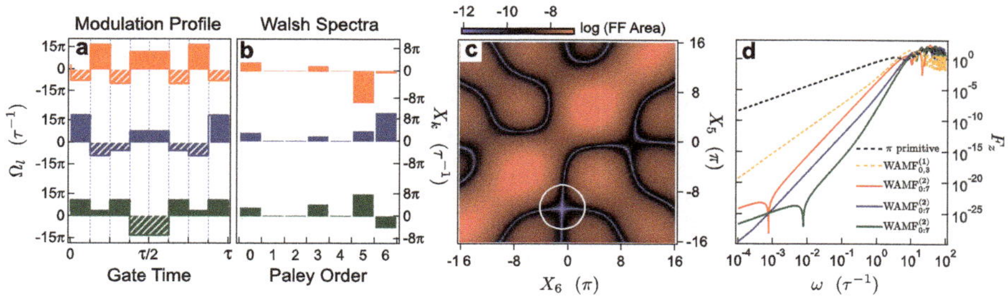

Figure 4 Higher-order WAMFs. Construction of $\mathrm{WAMF}^{(2)}_{0:7}$ for dephasing noise filtering. (**a**) Representative amplitude-modulated profiles for spectrally-optimized 8-segment $\mathrm{WAMF}^{(2)}_{0:7}$ gates. Vertical axes indicates Rabi rate values Ω_l in units of $1/\tau$ for the 8-segments. (**b**) Corresponding (Paley ordered) Walsh spectra. Vertical axes indicate values of the Walsh spectral amplitudes X_k in units of $1/\tau$. Optimized spectra obtained via Nelder-Mead search. (**c**) Log-scale color plot of the cost function $A_z(X_5,X_6)$ (integrated over $\omega \in [10^{-9},10^{-1}]\tau^{-1}$) defined on representative two-dimensional cross section of $\mathbf{X}_\nu$-domain. Blue regions indicate minima in $A_z(X_5,X_6)$, implying *second-order* optimized filter synthesis. 'Cross-region' (circled) indicates robustness region with respect to errors in $X_{5,6}$. (**d**) Dephasing filter-transfer functions for the optimized $\mathrm{WAMF}^{(2)}_{0:7}$ gates in (a), compared against primitive π_x rotation and optimized π_x $\mathrm{WAMF}^{(1)}_{0,3}$ gate. For the blue, red and green traces the cost function $A_z(X_3,X_5,X_6)$ was defined over the band $[\omega_L,\omega_c]$ with $\omega_c=\tau^{-1}$ and $\omega_L=(10^{-4},10^{-3},10^{-2})\tau^{-1}$.

tional control space over which to derive spectrally-optimized filters. We accomplish this using a Nelder-Mead search to minimize $A_z(\mathbf{X}_\nu;3\pi)$ over the $\mathbf{X}_\nu$-domain.

Representative examples of spectrally-optimized $\mathrm{WAMF}^{(2)}_{0:7}$ constructions are shown in Figure 4. The 8-segment time-domain amplitude-modulated profiles are represented in Figure 4(a), with corresponding Walsh spectra shown in Figure 4(b). The blue, red and green spectra were obtained using a Nelder-Mead optimization of the cost function $A_z(X_3,X_5,X_6)$ defined over $[\omega_L,\omega_c]$ with $\omega_c=\tau^{-1}$ and $\omega_L=(10^{-4},10^{-3},10^{-2})\tau^{-1}$ respectively. The corresponding dephasing filter-transfer functions $F_z(\omega)$ are plotted as solid blue, red and green traces in Figure 4(d). Within the respective cost function bands these satisfy $F_z\propto(\omega\tau)^{2p^*}$, with the *instantaneous filter order* ranging between $2<(p^*-1)<3.8$ at various points.[1] For comparison we also plot the dephasing filter-transfer function for a primitive π rotation (black dashed trace) and an optimized $\mathrm{WAMF}^{(1)}_{0,3}$ gate (yellow dashed trace). These respectively satisfy $F_z\propto(\omega\tau)^{2,4}$ over the whole stopband and are well characterized by the (asymptotic) filter orders $(p-1)=0,1$.

Figure 4(a) shows a two-dimensional representation of $A_z(X_5,X_6)$ defined on a two-dimensional cross-section of the $\mathbf{X}_\nu$-domain, holding X_3 fixed, and integrated over the stopband $\omega\in[10^{-9},10^{-1}]\tau^{-1}$. The value of $\mathrm{Log}_{10}[A_z(X_5,X_6)]$ is indicated by the color scale. Areas in blue indicate minima in $A_z(X_5,X_6)$, indicating optimized paths in X_5X_6 plane the over which effective filters may be found. Notably, it is possible to find 'cross-regions' (circled) in which the spectral amplitudes X_5 and X_6 may independently be varied substantially without the cost function moving off a local minima. This potentially indicates the existence of classes of WAMFs which may be robust to errors in the Walsh spectrum itself.

7 Walsh Phase Modulated Filters (WPMFs)

We now turn to filters for amplitude-damping noise constructed via phase modulation using the WPMF construction set out in Eq. (32). Following the same procedure described above for WAMFs, one can implement a Nelder-Mead search to derive spectrally-optimized WPMFs which implement nontrivial rotations. For these constructions, however, the target rotation angle is dependent on both the Rabi rate and the sequence of phase modulations. Consequently it is less straightforward to impose a constraint during the optimization procedure to ensure a particular target rotation. Although we do not pursue the general problem in detail in this paper, we demonstrate the approach in this and the following sections, deriving a family of WPMFs in which the synthesis space is limited to a variety of simple combinations of Walsh function.

In the remainder of this section we study a variation on the strict WPMF structure which resolves the difficulty of imposing a target rotation and enables us to make some useful connections with existing composite pulse sequences in NMR. This variation involves partitioning the control modulation into a target rotation $P(\theta, 0)$ followed by a sequence of phase-modulated *identity* operations $\prod_{l=1}^{\mathcal{M}} P(2\pi, \phi_l)$ with the ϕ_l chosen such that amplitude noise is filtered to some order. Here the operator $P(\theta, \phi)$ denotes the rotation through angle θ about $\hat{\sigma}_\phi$. By insisting these $\mathcal{M}$ 'correction' segments are all identity operations, the phase modulations do not produce complicated rotation paths and the net rotation is determined simply by the initial target pulse.

We assume a constant Rabi rate Ω_0 so that each correction segment has equal duration $\tau_{2\pi} = 2\pi/\Omega_0$. Provided $\mathcal{M}$ is a power of 2, the phase modulation describing the correction sequence may therefore be constructed as a Walsh-synthesized waveform consisting of Paley orders $k \leq \mathcal{M} - 1$. The simplest such 'synthesis' derives from a single Walsh function $\mathrm{PAL}_k(x)$ with spectral weight Y_k, yielding the sequence

$$\mathrm{WPMF}_k^{(c)}(\theta) \equiv \begin{matrix} \\ p_1 \\ p_2 \\ \vdots \\ p_{\mathcal{M}+1} \end{matrix} \begin{matrix} \begin{matrix} \Omega_l & \tau_l & \theta_l & \phi_l \end{matrix} \\ \left[\begin{array}{ccc|c} \Omega_0 & \tau_\theta & \theta & 0 \\ \hline \Omega_0 & \tau_{2\pi} & 2\pi & \\ \vdots & \vdots & \vdots & \vec{\phi}_c \\ \Omega_0 & \tau_{2\pi} & 2\pi & \end{array}\right] \end{matrix}, \quad \vec{\phi}_c = Y_k \mathbf{P}_{\mathcal{M}}^{(k)}, \mathcal{M}(k) = 2^{m(k)}, \tag{41}$$

where, as in Eq. (25), the column vector $\mathbf{P}_{\mathcal{M}}^{(k)}$ specifies the sequence of values taken by $\mathrm{PAL}_k(\frac{t-\tau_\theta}{\tau-\tau_\theta})$ over the interval $[\tau_\theta, \tau]$ partitioned into the minimum $\mathcal{M}(k)$ equal time bins. We include the superscript (c) and write the vectorized phase $\vec{\phi}_c$ to indicate Walsh modulation during the 'correction stage' of the sequence, disambiguating this from the strict WPMF structure. For a given θ we may now treat Y_k as a tuning parameter which may be optimized by minimizing the cost function $A_\Omega(Y_k;\theta) \equiv \int_0^{\omega_c} d\omega F_\Omega(\omega\tau; Y_k; \theta)$. The optimized Y_k are thereby defined as an implicit function of θ.

In fact we may analytically show this construction yields first-order filters for amplitude noise by Taylor expanding $F_\Omega(\omega)$ and solving the first-order filter condition $C_2^{(\Omega)}(Y_k;\theta) = 0$. We compute $C_2^{(\Omega)}(Y_k;\Omega) = [\theta + 2\pi\mathcal{M}(k)\cos(Y_k)]^2/4$, implying the optimized Walsh spectral amplitude

$$Y_k(\theta) = \cos^{-1}\left(-\frac{\theta}{2\pi\mathcal{M}(k)}\right). \tag{42}$$

On the other hand, computing the second-order Taylor coefficient and substituting in the first-order-optimized spectral value in Eq. (42), one finds $C_4^{(\Omega)} \neq 0$. Hence the $\mathrm{WPMF}_k^{(c)}$ sequence has $(p-1) = 1$ filtering properties.

We make the interesting observation that Eqs. (41) and (42) produce the first-order Solovay-Kitaev SK1(θ) sequence [72, 73] and the second-order Wimperis passband P2(θ)

Table 1 Filter characteristics of $\mathrm{WPMF}_k^{(c)}$ constructions corresponding to well-known NMR sequences, SK1 and P2, originally designed to compensate for *static* amplitude errors to first and second *Magnus order* respectively (Section 4.3)

	$\mathrm{WPMF}_k^{(c)}$ construction				Amplitude errors	
	k	$\mathcal{M}(k)$	$\vec{\phi}$	$Y_k(\theta)$	$(\mu-1)$	$(p-1)$
SK1(θ)	1	2	$Y_1(\theta)\,\mathrm{PAL}_1$	$\cos^{-1}(-\theta/4\pi)$	1	1
P2(θ)	3	4	$Y_3(\theta)\,\mathrm{PAL}_3$	$\cos^{-1}(-\theta/8\pi)$	2	1

sequence [27] by setting $k = 1, 3$ respectively. Hence these well-known NMR sequences, originally designed to compensate for *static* amplitude errors to first and second *Magnus order* respectively (see Section 4.3), appear in the Walsh filter space as phase-modulated filters for non-Markovian amplitude noise. Table 1 summarizes this.

Another remarkable result is found using the synthesis $\vec{\phi}_c = H_{\mathcal{M}}\tilde{\mathbf{Y}}$ over the two Walsh functions PAL_0 and PAL_3, setting $\tilde{\mathbf{Y}} \equiv (Y_0, 0, 0, Y_3)^T$ in analogy with the Walsh spectrum defining amplitude modulation in the $\mathrm{WAMF}_{0,3}^{(1)}$ construction. The first-order filtering condition $C_2^{(\Omega)}(Y_0, Y_3, \Omega_0; \theta) = 0$ then implies solutions

$$\tilde{\mathbf{Y}} = (2\phi_{\mathrm{BB1}}, 0, 0, -\phi_{\mathrm{BB1}})^T, \qquad \phi_{\mathrm{BB1}}(\theta) \equiv \cos^{-1}\left(-\frac{\theta}{4\pi}\right), \qquad \Omega_0 = \frac{4\pi + \theta}{\tau} \tag{43}$$

yielding the Wimperis broadband BB1 sequence [74]

$$\mathrm{BB1}(\theta) = \begin{array}{c} \\ p_1 \\ p_2 \\ p_3 \\ p_4 \\ p_5 \end{array}\!\!\begin{array}{c} \begin{array}{cccc} \Omega_l & \tau_l & \theta_l & \phi_l \end{array} \\ \left[\begin{array}{cccc} \Omega_0 & \tau_\theta & \theta & 0 \\ \Omega_0 & \frac{\tau-\tau_\theta}{4} & \pi & Y_+ \\ \Omega_0 & \frac{\tau-\tau_\theta}{4} & \pi & Y_- \\ \Omega_0 & \frac{\tau-\tau_\theta}{4} & \pi & Y_- \\ \Omega_0 & \frac{\tau-\tau_\theta}{4} & \pi & Y_+ \end{array}\right] \end{array} \cong \begin{array}{c} \\ p_1 \\ p_2 \\ p_3 \\ p_4 \end{array}\!\!\begin{array}{c} \begin{array}{cccc} \Omega_l & \tau_l & \theta_l & \phi_l \end{array} \\ \left[\begin{array}{cccc} \Omega_0 & \tau_\theta & \theta & 0 \\ \Omega_0 & \frac{\tau-\tau_\theta}{4} & \pi & \phi_{\mathrm{BB1}} \\ \Omega_0 & \frac{\tau-\tau_\theta}{2} & 2\pi & 3\phi_{\mathrm{BB1}} \\ \Omega_0 & \frac{\tau-\tau_\theta}{4} & \pi & \phi_{\mathrm{BB1}} \end{array}\right] \end{array}. \tag{44}$$

Here $Y_\pm := Y_0 \pm Y_3$ and in the last equality we have collapsed the array to show the BB1 construction explicitly. Computing the second-order Taylor coefficient and in substituting the first-order-optimized spectral values, however, we find $C_4^{(\Omega)} \neq 0$. Thus, although BB1 was originally derived to compensate *static* amplitude errors to second *Magnus order*, it only provides *first-order* noise filtering.

8 Walsh Rotary Spin Echo (WRSE)

In this section we treat a sub-class of Walsh modulated filters which may be described either in terms of phase- or amplitude-modulation. The phase-modulation consists of applying a sequence of π-phase shifts, relative to some offset ϕ_0, on the driving field with a constant amplitude Ω_0. This construction generalizes the rotary spin echo (RSE) sequence from NMR, analogous to the Hahn-echo sequence for driven systems, consisting of a single π-phase shift applied at the sequence midpoint $\tau/2$. In quantum information RSE has been employed, for example, in relaxation noise spectroscopy [75] and in mitigating low-frequency off-resonant noise [76] in driven superconducting flux qubits. In contrast with previous approaches, our generalization permits higher-order filter performance in both amplitude and dephasing quadratures and may prove of significant use.

In our construction the temporal profile of the phase is expressed $\phi(t) = \phi_0 + \frac{\pi}{2}(1 - y(t))$, where $y(t) \in \{\pm 1\}$ is a binary-valued *switching function* defined to change sign at the

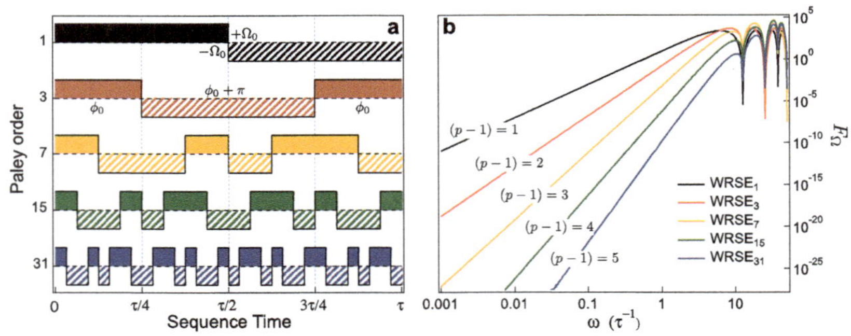

Figure 5 WRSEs as amplitude-noise filters. WRSE_k amplitude noise filter characteristics, $k \in \{1, 3, 7, 15, 31\}$. (**a**) Modulation profiles of WRSE_k sequences. Amplitude modulation involves switching the sign $\pm\Omega_0$; this is equivalent to holding Ω_0 constant and instead shifting the phase ϕ_0 by π, indicated by the hatching. (**b**) Corresponding amplitude filter-transfer functions $F_\Omega(\omega)$, showing filter order increases with Hamming weights $r(k) \in \{1, 2, 3, 4, 5\}$.

application of each π-shift. Specifically, we consider sequences based on Walsh functions by defining the switching function

$$y(t) := \mathrm{PAL}_k(t/\tau), \quad t \in \{0, \tau\}. \tag{45}$$

Consequently the phase has Walsh synthesis $\phi(t) = Y_0\,\mathrm{PAL}_0(t/\tau) + Y_k\,\mathrm{PAL}_k(t/\tau)$ where $Y_0 = \phi_0 + \frac{\pi}{2}$ and $Y_k = -\frac{\pi}{2}$. These gates only perform the identity operation with $\Theta \equiv 0$ owing to the property that $\hat{\sigma}_{\phi_0+\pi} = -\hat{\sigma}_{\phi_0}$ (see Eq. (6)). This implies the direction of unitary rotation is reversed by the application of each π shift, and since any Walsh function of Paley order $k > 0$ is equally distributed between values ± 1 over the domain these rotations perfectly cancel, yielding zero total rotation. This is formally equivalent to modulating the *sign* of the Rabi rate and holding the phase ϕ_0 constant (see Appendix C), as schematically illustrated in Figure 5(a). These sequences, referred to as *Walsh rotary spin echo* order k (WRSE_k), thus take the form

$$\mathrm{WRSE}_k \equiv \prod_{l=1}^{\mathcal{M}(k)} \exp\left(i\frac{\Omega_0}{2}\tau_{\mathcal{M}}\hat{\sigma}_{\phi_l}\right), \qquad \vec{\boldsymbol{\phi}} = (\phi_1, \phi_2, \ldots, \phi_{\mathcal{M}})^T = H_{\mathcal{M}}\tilde{\mathbf{Y}}, \tag{46}$$

where, referring to Eqs. (32) and (30), $\mathcal{M}(k) \equiv 2^{m(k)}$ and the Walsh spectrum $\tilde{\mathbf{Y}}$ has only two nonzero elements: $\tilde{Y}_1 = \phi_0 + \frac{\pi}{2}$ and $\tilde{Y}_{i(k)} = -\frac{\pi}{2}$. This naming convention reflects the fact that the WRSE family generalize the traditional RSE sequence: in particular, $\mathrm{WRSE}_1 \equiv$ RSE. We now proceed to derive the filtering properties of the WRSE family, starting with the amplitude quadrature then moving to the dephasing quadrature.

8.1 WRSE as amplitude filters

The amplitude filter-transfer function, in the stopband, is determined by the Hamming weight of the chosen Walsh function by

$$F_\Omega(\omega\tau) \propto (\omega\tau)^{2(r(k)+1)}. \tag{47}$$

Comparing this with the low-frequency approximation $F_\Omega(\omega\tau; k) \propto (\omega\tau)^{2p}$ from Eq. (22), we conclude the time-domain filter order is given by $(p-1) = r(k)$. That is, high-pass filter performance is completely determined by the Hamming weight. Figure 5(c) demonstrates this by plotting $F_\Omega(\omega)$ for the WRSE_k sequences with $k \in \{1, 3, 7, 15, 31\}$, corresponding to the Hamming weights $r(k) \in \{1, 2, 3, 4, 5\}$. The corresponding filter order increase is clear from the steepening roll-off.

This result follows from deriving $F_\Omega(\omega)$ for Eq. (46) which, owing to the fact that the noise Hamiltonian in this quadrature always commutes with control operations, takes the relatively simple analytical form (see Appendix C)

$$F_\Omega(\omega) = \frac{\Omega_0^2}{4}\left|\sum_{l=1}^{\mathcal{M}} P_l^{(k)}\left[e^{i\omega t_{l-1}} - e^{i\omega t_l}\right]\right|^2. \tag{48}$$

Our key insight now is to observe the sum inside the modulus square above satisfies $\sum_{l=1}^{\mathcal{M}} P_l^{(k)}[e^{i\omega t_l} - e^{i\omega t_{l-1}}] = i\omega\tau\mathcal{F}_x[\text{PAL}_k(x)]$, where $x \equiv t/\tau$ is a non-dimensional time-domain variable. We now invoking Eq. (37) in design rule (iii), namely $\mathcal{F}_x[\text{PAL}_k(x)] \propto (\omega\tau)^{r(k)}$, to map the low-frequency spectral properties onto the Hamming weight $r(k)$ of the chosen Walsh function. Substituting this into Eq. (48) then yields Eq. (47).

8.2 WRSE as dephasing filters

The dephasing filter performance for WRSE_k is more complicated to study as noise terms in this quadrature do not commute with our control, obfuscating a compact expression for $F_z(\omega)$. It is convenient instead to study the zeros of the Taylor coefficients $C_{2j}^{(z)}$ as in Eq. (22). Since the Rabi rate is the only free variable in the WRSE_k, for a given k, it follows the Taylor coefficients are functions only of Ω_0. Filtering to order $(p-1)$ then corresponds to the condition

$$C_2^{(z)}(\eta; k) = C_4^{(k)}(\eta; k) = \cdots = C_{2(p-1)}^{(z)}(\eta; k) = 0, \tag{49}$$

where η denotes some value of Ω_0 for which the above coefficients are concurrently zero. Here we have included the Paley order as a parameter of the coefficients. Analysis shows, however, concurrent zeros exist only for $j \in \{1, 2\}$. We effectively obtain the following 'no-go theorem': WRSE_k *sequences perform as (high-pass) dephasing noise filters up to (but not beyond) second order.* This result may be of use to characterize the relevant quadrature of an unknown noise source, by probing with higher-order WRSE_k sequences and determining the resulting filtering properties.

The general insight supporting this statement is developed in Appendix C, by explicitly studying the representative case for WRSE_3. The Taylor coefficients $C_{2,4,6}^{(z)}$ for this representative case are plotted in Figure 6(a) as functions of Ω_0. Zeros of $C_2^{(z)}$ occur at multiples of 4π, and concurrent zeros of $C_{2,4}^{(z)}$ occur at multiples of 8π. However, since $C_6^{(z)}$ is never zero at multiples of 8π (see inset to Figure 6(a)), it follows we can never achieve higher than second-order filtering. We verify this by examining the slope of $F_z(\omega; \Omega_0)$ for the values $\Omega_0 = 2\pi q$, $q \in \{1, \ldots, 8\}$ (see Figure 6(b)). Similar considerations for other values of k generalize the result.

9 Universal Walsh Modulated Filters (UWMFs)

In the previous sections we have considered WAMF and WPMF gates which implement target qubit rotations while filtering, to some order, *either* dephasing or amplitude noise

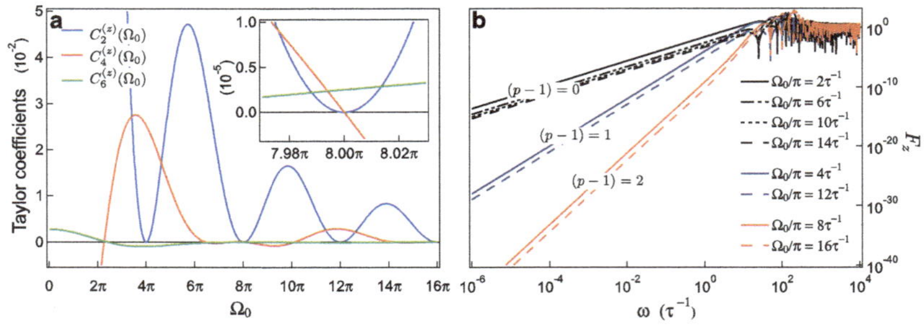

Figure 6 WRSEs as dephasing-noise filters. WRSE_3 dephasing noise filter characteristics. (**a**) Taylor expansion coefficients $C^{(z)}_{2,4,6}(\Omega_0;3)$ for $F_z(\omega)$. The inset shows typical behaviour: $\Omega_0 = 8\pi$ is a concurrent zero of $C^{(z)}_{2,4}(\Omega_0;3)$, but not of $C^{(z)}_{6}(\Omega_0;3)$. Hence WRSE_3 can only filter up to second-order. (**b**) Dephasing filter-transfer functions for WRSE_3 corresponding to $\Omega_0 = 2\pi q$, $q \in \{1,\ldots,8\}$. When Ω_0 is a multiple of 8π we achieve second-order filters, that is $(p-1) = 2$.

respectively. In this section we derive filters for universal noise by concatenating both filter types into a composite structure that filters both noise quadratures simultaneously, while still implementing a target qubit rotation. We refer to such constructions as *universal Walsh modulated filters* (UWMFs).

The general approach is conveniently illustrated through a particular concatenated structure obtained by embedding the $\mathrm{WPMF}^{(c)}_1(\theta) \equiv \mathrm{SK1}(\theta)$ gate (Section 7) within the various segments of the $\mathrm{WAMF}^{(1)}_{0,3}$ gate (Section 6.1). The former is explicitly written

$$\mathrm{WPMF}^{(c)}_1(\theta) \equiv \begin{matrix} \\ P_1 \\ P_2 \\ P_3 \end{matrix} \begin{matrix} \begin{matrix} \Omega_l & \tau_l & \theta_l & \phi_l \end{matrix} \\ \begin{bmatrix} \Omega_0 & \tau_\theta & \theta & 0 \\ \Omega_0 & \tau_{2\pi} & 2\pi & \phi_{\mathrm{SK1}} \\ \Omega_0 & \tau_{2\pi} & 2\pi & -\phi_{\mathrm{SK1}} \end{bmatrix} \end{matrix}, \tag{50}$$

$$\Omega_0 = \frac{\theta + 4\pi}{\tau_{\mathrm{SK1}}}, \qquad \tau_\theta = \frac{\theta}{\Omega_0}, \qquad \phi_{\mathrm{SK1}}(\theta) \equiv Y_1(\theta) = \cos^{-1}\left(-\frac{\theta}{4\pi}\right). \tag{51}$$

The basic concatenation procedure is now to replace each constant-amplitude pulse in $\mathrm{WAMF}^{(1)}_{0,3}$ with the constant-amplitude phase-modulated sequence implementing the equivalent rotation. Doing so effectively distributes the dephasing filter across the composite sequence, each subsequence of which filters amplitude noise. In our example, this takes place via the operator substitutions

$$P_1(X_+/4, 0) \to \mathrm{SK1}^{(1)}(X_+/4), \qquad \tau^{(1)}_{\mathrm{SK1}} = \tau/4, \tag{52}$$

$$P_2(X_-/2, 0) \to \mathrm{SK1}^{(2)}(X_-/2), \qquad \tau^{(2)}_{\mathrm{SK1}} = \tau/2, \tag{53}$$

$$P_3(X_+/4, 0) \to \mathrm{SK1}^{(3)}(X_+/4), \qquad \tau^{(3)}_{\mathrm{SK1}} = \tau/4, \tag{54}$$

where, as indicated, each SK1 is equal in duration to the original $\mathrm{WAMF}^{(1)}_{0,3}$ pulse being replaced. We denote this concatenated structure by $\mathrm{UWMF}_{1,\mathrm{SK1}}$. The temporal profile is shown in Figure 7(a). The SK1 phase flips $\phi = \pm\phi_{\mathrm{SK1}}$ are indicated by the oppositely oriented hatching within each constant-amplitude segment of the $\mathrm{WAMF}^{(1)}_{0,3}$ envelope; $\phi = 0$ is indicated by white fill. Once again the total gate rotation is determined by Eq. (36) and, as in Figure 3, X_3 may be treated as an independent tuning parameter to achieve first-order

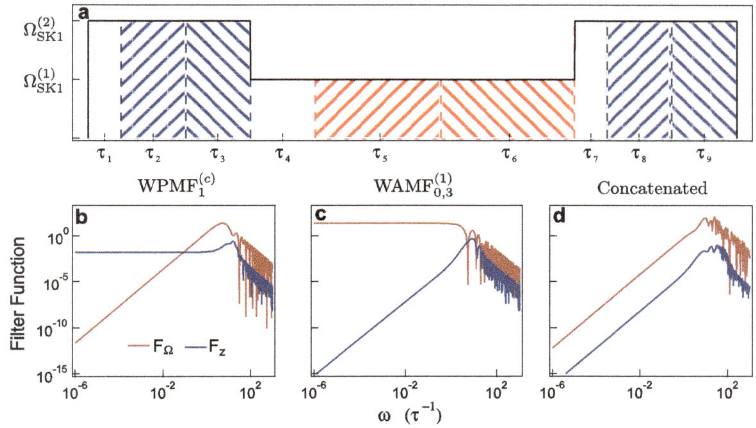

Figure 7 UWMFs. Concatenation scheme for universal noise suppression. (**a**) Concatenation of $\mathrm{WPMF}_1^{(c)} \equiv \mathrm{SK1}$ within $\mathrm{WAMF}_{0,3}^{(1)}$ sequence yielding $\mathrm{UWMF}_{1,SK1}$. White fill indicates rotations enacted with $\phi = 0$; orientation of hatching denotes SK1 phase flips $\phi = \pm\phi_{\mathrm{SK1}}$. (**b**) Filter-transfer functions for $\mathrm{WPMF}_1^{(c)} \equiv$ SK1 sequence. (**c**) Filter-transfer functions for four-segment $\mathrm{WAMF}_{0,3}^{(1)}$ sequence. (**d**) Filter-transfer functions for concatenated sequence.

filtering against dephasing noise. The dephasing and amplitude filter-transfer functions for the concatenated and tuned sequence (in this case for a net π rotation) are shown in Figure 7(d), indicating effective filtering of both amplitude and dephasing noise. Below we detail two alternative constructions for realizing the UWMF structure.

9.1 Concatenation Method 1: Constrain Sequencing of WAMF Envelope

The amplitude-modulated envelope of the $\mathrm{WAMF}_{0,3}^{(1)}$ construction, as defined by Eq. (39), may be viewed as a sequence of 3 piecewise-constant rotations $\vec{\theta} = (\frac{X_+}{4}, \frac{X_-}{2}, \frac{X_+}{4})$, implemented over durations $(\frac{\tau}{4}, \frac{\tau}{2}, \frac{\tau}{4})$ with Rabi rates $\vec{\Omega} = (\frac{X_+}{\tau}, \frac{X_-}{\tau}, \frac{X_+}{\tau})$. In our first concatenation method each rotation θ_l, $l \in \{1,2,3\}$, is replaced by a constant-amplitude phase-modulated operation $\mathrm{SK1}(\theta_l)$, implemented over a duration $\tau_{\mathrm{SK1}}^{(l)}$ equal to that of the original rotation. That is, we constrain $(\tau_{\mathrm{SK1}}^{(1)}, \tau_{\mathrm{SK1}}^{(2)}, \tau_{\mathrm{SK1}}^{(3)}) = (\frac{\tau}{4}, \frac{\tau}{2}, \frac{\tau}{4})$. The constant Rabi rate $\Omega_{\mathrm{SK1}}^{(l)} = \theta_l/\tau_{\mathrm{SK1}}^{(l)}$ driving each SK1 sequence is thus given by

$$\Omega_{\mathrm{SK1}}^{(1)} = \Omega_{\mathrm{SK1}}^{(3)} = \frac{1}{\tau}(X_+ + 16\pi), \qquad \Omega_{\mathrm{SK1}}^{(2)} = \frac{1}{\tau}(X_- + 8\pi). \tag{55}$$

The composite structure may then be written

$$\mathrm{UWMF}_{\mathrm{SK1},1} \equiv \begin{matrix} & \\ P_1 \\ P_2 \\ P_3 \\ P_4 \\ P_5 \\ P_6 \\ P_7 \\ P_8 \\ P_9 \end{matrix} \begin{matrix} \Omega_l & \tau_l & \theta_l & \phi_l \\ \left[\begin{matrix} \Omega_{\mathrm{SK1}}^{(1)} \\ \Omega_{\mathrm{SK1}}^{(1)} \\ \Omega_{\mathrm{SK1}}^{(1)} \\ \Omega_{\mathrm{SK1}}^{(2)} \\ \Omega_{\mathrm{SK1}}^{(2)} \\ \Omega_{\mathrm{SK1}}^{(2)} \\ \Omega_{\mathrm{SK1}}^{(3)} \\ \Omega_{\mathrm{SK1}}^{(3)} \\ \Omega_{\mathrm{SK1}}^{(3)} \end{matrix}\right. & \begin{matrix} \tau_1 \\ \tau_2 \\ \tau_3 \\ \tau_4 \\ \tau_5 \\ \tau_6 \\ \tau_7 \\ \tau_8 \\ \tau_9 \end{matrix} & \begin{matrix} \frac{X_+}{4} \\ 2\pi \\ 2\pi \\ \frac{X_-}{2} \\ 2\pi \\ 2\pi \\ \frac{X_+}{4} \\ 2\pi \\ 2\pi \end{matrix} & \left.\begin{matrix} 0 \\ \phi_{\mathrm{SK1}}^{(1)} \\ -\phi_{\mathrm{SK1}}^{(1)} \\ 0 \\ \phi_{\mathrm{SK1}}^{(2)} \\ -\phi_{\mathrm{SK1}}^{(2)} \\ 0 \\ \phi_{\mathrm{SK1}}^{(3)} \\ -\phi_{\mathrm{SK1}}^{(3)} \end{matrix}\right] \end{matrix}, \tag{56}$$

$$\tau_{1,7} = \frac{\tau}{4}\left(\frac{X_+}{X_+ + 16\pi}\right), \qquad \tau_{2,3,8,9} = \tau\left(\frac{2\pi}{X_+ + 16\pi}\right), \tag{57}$$

$$\tau_4 = \frac{\tau}{2}\left(\frac{X_-}{X_- + 8\pi}\right), \qquad \tau_{5,6} = \tau\left(\frac{2\pi}{X_- + 8\pi}\right), \tag{58}$$

$$\phi^{(1)}_{\mathrm{SK1}} = \phi^{(3)}_{\mathrm{SK1}} = \cos^{-1}\left(-\frac{X_+}{16\pi}\right), \qquad \phi^{(2)}_{\mathrm{SK1}} = \cos^{-1}\left(-\frac{X_-}{8\pi}\right). \tag{59}$$

Using this method one finds tunings of X_0 and X_3 such that both dephasing and amplitude noise are filtered to first-order, as in Figure 7. These do *not*, however, correspond directly to the optimum Walsh coefficients found for simple $\mathrm{WAMF}^{(1)}_{0,3}$ construction shown in Figure 3(a). Rather, an equivalent tuning plot may be generated over the X_0X_3 domain, essentially identical to Figure 3(a) but with minima shifted by a constant factor. The second method, detailed below, involves a slightly different construction in which we recover the original $\mathrm{WAMF}^{(1)}_{0,3}$ tuning plot.

9.2 Concatenation Method 2: Constrain Sequencing of Target SK1 Rotations

In the second construction we impose the constraint that $\tau_1 : \tau_4 : \tau_7 = 1 : 2 : 1$. That is, the *target rotations* within the three successive SK1 sequences follow the same timing sequence as the three constant-amplitude pulses being replaced, previously constituting the amplitude-modulated WAMG_1 envelope. Thus we write $(\tau_1, \tau_4, \tau_7) = \nu(1, 2, 1)$ where ν is some fraction of the total duration τ of the composite structure to be determined.

Now, we know pulses (1, 4, 7) execute the rotations $(\frac{X_+}{4}, \frac{X_-}{2}, \frac{X_+}{4})$, given by the third row of Eq. (39). Hence the Rabi rates for these pulses in the composite structure must take the form $(\Omega_1, \Omega_4, \Omega_7) = (X_+, X_-, X_+)/4\nu$. However each SK1 sequence has constant Rabi rate and we therefore conclude

$$\Omega^{(1)}_{\mathrm{SK1}} = \Omega^{(3)}_{\mathrm{SK1}} = \frac{1}{4\nu}X_+, \qquad \Omega^{(2)}_{\mathrm{SK1}} = \frac{1}{4\nu}X_-. \tag{60}$$

Now, the duration of the composite sequence must satisfy

$$\tau = \sum_{l=1}^{9} \frac{\theta_l}{\Omega_1} \tag{61}$$

$$= 2\left(\frac{2\pi}{\Omega^{(1)}_{\mathrm{SK1}}}\right) + 2\left(\frac{2\pi}{\Omega^{(2)}_{\mathrm{SK1}}}\right) + 2\left(\frac{2\pi}{\Omega^{(3)}_{\mathrm{SK1}}}\right) + 4\nu \tag{62}$$

$$= 4\pi\left(\frac{1}{\Omega^{(1)}_{\mathrm{SK1}}} + \frac{1}{\Omega^{(2)}_{\mathrm{SK1}}} + \frac{1}{\Omega^{(3)}_{\mathrm{SK1}}}\right) + 4\nu. \tag{63}$$

Substituting in the results from Eq. (60) we therefore find $\tau = 16\pi\nu(2/X_+ + 1/X_-) + 4\nu$. Or, solving for ν,

$$\nu = \frac{1}{4}\tau(1 + \pi\kappa)^{-1}, \quad \kappa := 4\left(\frac{2}{X_+} + \frac{1}{X_-}\right), \tag{64}$$

concluding the construction. The composite structure may then be written

$$\mathrm{UWMF}_{\mathrm{SK1},1} \equiv \begin{array}{c} \\ p_1 \\ p_2 \\ p_3 \\ p_4 \\ p_5 \\ p_6 \\ p_7 \\ p_8 \\ p_9 \end{array} \begin{array}{c} \begin{array}{cccc} \Omega_l & \tau_l & \theta_l & \phi_l \end{array} \\ \left[\begin{array}{cccc} \frac{X_+}{4\nu} & \nu & \frac{X_+}{4} & 0 \\ \frac{X_+}{4\nu} & \frac{8\pi\nu}{X_+} & 2\pi & \phi^{(1)}_{\mathrm{SK1}} \\ \frac{X_+}{4\nu} & \frac{8\pi\nu}{X_+} & 2\pi & -\phi^{(1)}_{\mathrm{SK1}} \\ \frac{X_-}{4\nu} & 2\nu & \frac{X_-}{2} & 0 \\ \frac{X_-}{4\nu} & \frac{8\pi\nu}{X_-} & 2\pi & \phi^{(2)}_{\mathrm{SK1}} \\ \frac{X_-}{4\nu} & \frac{8\pi\nu}{X_-} & 2\pi & -\phi^{(2)}_{\mathrm{SK1}} \\ \frac{X_+}{4\nu} & \nu & \frac{X_+}{4} & 0 \\ \frac{X_+}{4\nu} & \frac{8\pi\nu}{X_+} & 2\pi & \phi^{(3)}_{\mathrm{SK1}} \\ \frac{X_+}{4\nu} & \frac{8\pi\nu}{X_+} & 2\pi & -\phi^{(3)}_{\mathrm{SK1}} \end{array}\right] \end{array}, \tag{65}$$

$$\phi^{(1)}_{\mathrm{SK1}} = \phi^{(3)}_{\mathrm{SK1}} = \cos^{-1}\left(-\frac{X_+}{16\pi}\right), \qquad \phi^{(2)}_{\mathrm{SK1}} = \cos^{-1}\left(-\frac{X_-}{8\pi}\right). \tag{66}$$

Minima of the cost function in the X_0X_3 domain for this construction are found in identical regions to those shown in Figure 3(a) for the simple $\mathrm{WAMF}^{(1)}_{0,3}$ construction. Substituting these minimizing values of $X_{0,3}$ into Eq. (65) thus optimizes the *concatenated* structure, yielding a desired net rotation (dictated by X_0) which filters both amplitude and dephasing noise to first order simultaneously.

10 Effect of bandwidth limits on Walsh filters

In the preceding sections of this paper filter design is based on optimizing the Walsh spectrum from which the relevant control structures are synthesized. This necessarily assumes perfectly square waveforms. Real control hardware, however, may suffer from bandwidth limitations which 'smooth out' the squareness of the pulse on the timescale of the application, leading to reduced filter performance. Here we show the assumption of perfect square pulses may be readily relaxed, with useful filter construction a simple matter of re-optimization in the Walsh-synthesis framework.

To illustrate the general procedure we consider the $\mathcal{M}$-segment WAMF. Each segment implements a rotation through angle $\theta_l = \Omega_l\tau_l$ over duration $\tau_l = \tau/\mathcal{M}$ and with constant Rabi rate Ω_l, $l \in \{1,\ldots,\mathcal{M}\}$. The squareness of the resulting amplitude-modulated waveform may be relaxed by replacing the constant value Ω_l with an arbitrarily varying function of time in each segment.

In order to achieve this we consider Walsh synthesis over the *rotation angle* implemented in a single segment rather than the Rabi rate. That is, we write $\theta_l = \theta_l(X_0, X_1, \ldots, X_N)$ with the dependence on the Walsh spectra defined by the Hadamard-matrix equation

$$\vec{\boldsymbol{\theta}} = (\theta_1, \theta_2, \ldots, \theta_{\mathcal{M}})^T = (\tau/\mathcal{M})H_{\mathcal{M}}\tilde{\mathbf{X}}. \tag{67}$$

Defined in this way, the $\mathcal{M}$-segment arbitrary-pulse sequence shares with the WAMF construction the property that the total gate rotation angle $\Theta = \sum_{l=1}^{n}\theta_l = X_0\tau$ is completely determined by the spectral amplitude of PAL_0. The symmetry-based design rules similarly carry over and filter optimization proceeds in the same manner as for ordinary Walsh-modulated control by minimizing the filter cost function with respect to the Walsh spectrum.

As a natural example of this approach we assume a Gaussian profile $G_l(t;\mu_l,\sigma_l)$ defined on $t \in [t_{l-1}, t_l]$ with mean μ_l and standard deviation σ_l. Specifically, we construct

$$G_l(t;\mu_l,\sigma_l) := \frac{\theta_l}{C_l\sigma_l\sqrt{2\pi}}\exp\left[-\frac{(t-\mu_l)^2}{2\sigma_l^2}\right], \quad \mu_l = \frac{t_{l-1}+t_l}{2}, \sigma_l = g\tau/\mathcal{M} \tag{68}$$

with μ_l the segment midpoint and σ_l expressed as a multiple g of the segment duration. The normalizing factor

$$C_l := \int_{t_{l-1}}^{t_l} \frac{1}{\sigma_l\sqrt{2\pi}}\exp\left[-\frac{(t-\mu_l)^2}{2\sigma_l^2}\right]dt \tag{69}$$

is included to ensure the total rotation implemented by the Gaussian pulse in the lth segment is given by $\int_{t_l}^{t_{l-1}} G_l(t;\mu_l,\sigma_l)\,dt = \theta_l$. We now impose the same structure on the segment rotations θ_l as done for WAMFs in the Walsh-synthesis framework, such that the smooth-pulse sequence remains strictly parametrized in the Walsh spectrum $\mathbf{X}$.

For concreteness, we examine the Gaussian-pulse variation on the 4-segment $\text{WAMF}_{0,3}^{(1)}$ filter described by Eq. (39) for two different Gaussian profiles, illustrated in Figure 8(a), (c). The cost function $A_z(X_3;X_0) = \int_{\omega_L}^{\omega_c} d\omega F_z(\omega\tau;X_3;X_0)$ may be computed by partitioning the time domain into a large number N_s of subintervals on which the continuous Gaussian envelope is treated as approximately constant. Figure 8(b), (d) show a two-dimensional representation of $A_z(X_3;X_0)$ integrated over the interval $\omega \in [10^{-9}, 10^{-6}]\tau^{-1}$. The value of $\text{Log}_{10}[A_z(X_3;X_0)]$ is indicated by the color scale. Total sequence length is normalized to $\tau = 1$ in this data, so the total gate rotation angle $\Theta \equiv X_0$ is given directly by the X_0-axis. Regions in blue represent effective (first-order) filter constructions, where the cost function is minimized.

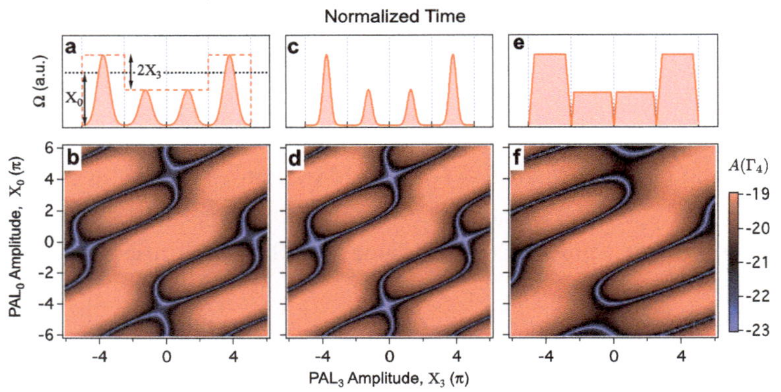

Figure 8 Shaped-pulse filter constructions. Construction of the first-order Walsh amplitude modulated dephasing-suppressing filter using shaped pulse segments. (**a**), (**c**), (**e**) Schematic representation of Walsh synthesis for a four segment gate of discrete Gaussian or trapezoidal segments. Walsh synthesis determines the overall amplitude of individual pulses with fixed duration and standard deviation, setting the effective pulse area in each segment. The metric g takes value 1/6 in panel (a), and 1/12 in panel (c). (e) Trapezoidal pulses are characterized by a constant slope such that all angles are a fraction of a square waveform defined as $F\frac{\pi}{2}$. Here $F = 0.992$. In all pulse constructions the pulse profile is computed over 100 discrete time steps, permitting calculation of relevant filter-transfer functions. (**b**), (**d**), (**f**) Two-dimensional representation of the integral metric defining our target cost function, $A(\mathbf{\Gamma}_4)$ integrated over the stopband $\omega \in [10^{-9}, 10^{-6}]\tau^{-1}$ for the corresponding pulse forms above. Areas in blue minimize $A(\mathbf{\Gamma}_4)$, representing effective filter constructions. The X_0 determines the net rotation enacted in a gate while X_3 determines the modulation depth, as represented in (a).

Comparing with Figure 3 we conclude useful filter construction using Gaussian pulses is a simple matter of re-optimization in the Walsh-synthesis framework. This is readily achieved using a Nelder-Mead optimization of $A_z(X_3; X_0)$ for any particular choice of g, ω_L, ω_c, X_0 or N_s. Minor changes in the filter performance and optimal constructions arise with changes in Gaussian pulse parameters such as g. Comparison with pulses constructed using a trapezoidal form (Figure 8(e)) we find a different optimization outcome that more closely approximates standard square pulses. Nonetheless, these results show that, irrespective the specific pulse form, re-optimization over the Walsh coefficients remains a direct method to construct useful filters. In cases where unknown waveform distortion is likely in hardware, it is possible to implement automated feedback mechanisms, as has previously been demonstrated in dynamical decoupling experiments [15].

We may also explore the impact of finite modulation bandwidth on the application of square pulses if re-optimization of the waveform is, for some reason, not possible. In order to explore these effects we systematically relax the infinite-modulation-bandwidth assumption underlying any square-pulse approximation by processing the ideal time-domain profile through a bandlimited digital filter with a user-defined cutoff. This results in an imperfect (bandlimited) profile envelope, effectively due to a reconstruction based on a truncated Fourier series. These profile distortions, manifesting as implementation errors, reduce filter performance, quantified by an increase in the area under the corresponding filter-transfer function.

In this example we again consider the amplitude-modulated profile associated with the $\text{WAMF}^{(1)}_{0,3}$ gate ($X_0 = 3\pi$, $X_3 = \pi$), and impose band limitations using a digital first-order Butterworth filter. Here the time domain is partitioned into $1/f_s$ subintervals, where $f_s = 1/2^{11}$ is the sampling rate of the digital filter. The bandlimited envelope of reconstructed waveform is then a function of the Butterworth cutoff frequency f_c. As the cutoff approaches the Nyquist frequency $f_s/2$ (the maximum possible value), the bandlimited effects are reduced and the reconstructed waveform approaches the ideal square-pulse envelope. Figure 9(a) illustrates this as we increase f_c/f_s. The corresponding filter characteristics are shown in Figure 9(c) where we plot $F_z(\omega)$ as a function of f_c/f_s. To compute these filter-transfer functions we treat the reconstructed waveforms as amplitude-modulated sequences consisting of $n = 1/f_s$ segments whose Rabi-rates are determined by the bandlimited envelope (see Figure 9(b)). The solid black traces in Figure 9(a), (b), (c) respectively show the ideal profile and corresponding filter-transfer function, against which the filter performance of the bandlimited gates are benchmarked. As we decrease the cut-

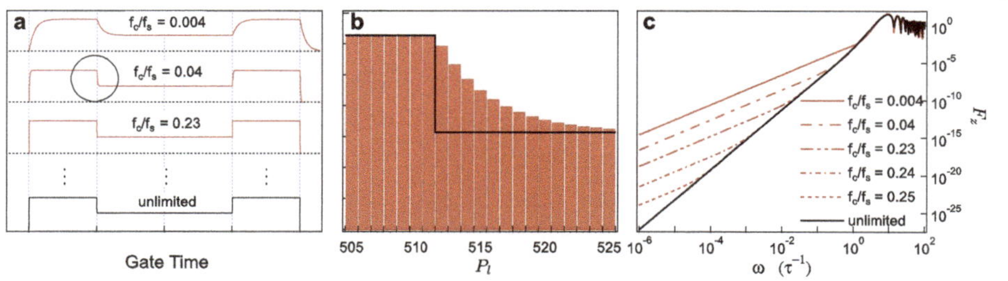

Figure 9 Bandlimited filters. Effect of bandwidth limits on Walsh filters. (**a**) Bandlimited reconstructions of $\text{WAMF}^{(1)}_{0,3}$ modulation envelope as a function of first-order Butterworth filter parameter f_c/f_s. (**b**) *Red*: Closeup of circled region in (a); *Black* ideal square profile. (**c**) Filter-transfer functions for profiles in (a).

off the integrated area under $F_z(\omega)$ gradually increases. However as this manifests first in the low-frequency region, even these bandlimited gates can provide useful filtering, given sufficient hardware precision. Similarly motivated studies in dynamic decoupling show analogous low-frequency filter performance decay with pulse-timing errors.

11 Conclusion

As the size and complexity of quantum information processing technologies increase, resource-efficiency will play a vital role in selecting methods designed to reduce errors in quantum coherent hardware systems. The pressure to minimize quantum-hardware overhead is likely to make *open-loop control* protocols a key element in the design of error-robust quantum information systems [14]. For these to be practically useful, however, it is important to move toward realistic control and noise models.

Decoherence in real driven systems is predominantly due to low-frequency *correlated* noise environments. This strongly motivates our study of bounded-strength control as a noise filtering problem using time-dependent, non-Markovian error models. Moreover, in contrast with traditional DD schemes, the added complications of treating bounded-strength control - due to the continual presence of noise interactions during control operations and the resulting nonlinear dynamics - necessitates a streamlined approach to the design of noise-filtering control. The generalized filter-transfer function framework we employ takes as input experimentally measurable characteristics of the environment - namely noise power spectra - and provides a simple framework for both control construction and the calculation of predicted operational fidelities. It also efficiently captures the control nonlinearities implicit in situations where control and noise Hamiltonians do not commute. We have exploited these strengths to pursue a simple variational procedure for filter design by minimizing a cost function over the relevant control space.

A key strength of the method we have presented is derived from our use of functional analysis for the crafting of effective noise-filtering control protocols. In particular, employing the Walsh basis brings an intuitive set of analytic design rules for filter construction that further constrain the possible filter-construction space [51]. For instance, a user-imposed limit on the acceptable number of pulse segments in a filter construction impose additional constraints due to Walsh-function symmetry, spectral properties, and the level of recursiveness of the Walsh functions (measured by the Hamming weight of the Walsh Paley-ordered index).

In addition to efficiency of synthesis is the intrinsic compatibility with hardware controllers that comes with the selection of Walsh functions as our basis set. This is particularly important in the layered architecture for quantum information systems mentioned throughout this paper. In such a setting, it rapidly becomes undesirable to mandate a significant amount of communication between the physical qubit layer and hardware at the highest levels of system abstraction. This suggests that controllers implementing dynamic error suppression protocols (here producing noise filters for arbitrary driven operations) should be reasonably simple to implement in standard digital hardware and should require only limited communication bandwidth to higher levels of the system.

These considerations are explicitly met in crafting control solutions from the Walsh basis. First, the Walsh functions are defined using integer multiples of a fundamental clock period, meaning that limitations of finite timing precision in the definition of a control protocol are automatically inbuilt. Further, given a particular Walsh-modulated control protocol is entirely defined by its Walsh spectrum, programming of the controller can in principle be reduced to a simple vector of numbers representing the Walsh spectrum and minimum timestep. All other information *e.g.* the total time, total number of timesteps, etc., is carried implicitly in the spectrum. Moreover, the actual Walsh-function *generation* is compatible with simple hardware systems (adding of various harmonics of a fundamental square-wave clock) and when Walsh synthesis is performed at the level of the controller hardware, this may provide a path to on-the-fly synthesis of the required modulation waveform. Such capabilities also reduce the complexity of running automated hardware-driven optimization procedures for finding relevant control waveforms [15], by allowing efficient generation of many trial waveforms without the need for large memory stores at the local controller.

Synthesizing all of these considerations the Walsh-modulated noise filters we have developed in this work provide one of the first solutions for error suppression at the physical-qubit level simultaneously meeting the physical and engineering requirements we outline above for scalable control solutions. Using this framework we have derived a range of novel filters, chiefly WAMFs for dephasing noise and WPMFs for amplitude noise. Both are capable of spectral optimization subject to physically motivated constraints such as implementing target qubit rotations. These design forms are also compatible with concatenation for filtering universal noise. Interestingly, our approach unifies a number of existing composite pulse sequencing schemes; we have revealed how Walsh-modulated filter construction naturally incorporates familiar sequences (e.g., DCG, SK1, P2, BB1) in a non-Markovian time-dependent noise context. This potential to incorporate other approaches may prove useful in building a consistent picture of the scope and applicability of the many and varied schemes that continue to be developed by the quantum control community. These considerations make the Walsh basis an attractive design platform and we believe this simple framework will provide a straightforward path for the development of improved quantum control techniques.

Appendix A: Detailed filter-transfer function derivation

In this appendix we derive the computational form of the first-order infidelity $\langle a_1^2 \rangle = \langle \mathbf{a}_1 \mathbf{a}_1^T \rangle$ expressed in Eq. (4). Recall, we write the total evolution operator $U(t) = U_c(t)\tilde{U}(t)$ where the *error propagator* $\tilde{U}(t)$ satisfies the Schrodinger equation $i\dot{\tilde{U}}(t) = \tilde{H}_0(t)\tilde{U}(t)$ in a frame co-rotating with the control, defined by the *toggling frame Hamiltonian* $\tilde{H}_0(t) := U_c^\dagger(t)H_0(t)U_c(t)$. We may obtain an arbitrarily accurate, unitary estimate of the error propagator by performing a Magnus expansion, whereby $\tilde{U}(\tau) = \exp[-i\Phi(\tau)]$ and the effective error operator $\Phi(\tau) = \sum_{\mu=1}^{\infty} \Phi_\mu(\tau)$ at the end of the interaction has Magnus expansion terms

$$
\begin{aligned}
\Phi_1(\tau) &= \int_0^\tau dt \tilde{H}_0(t), \\
\Phi_2(\tau) &= -\frac{i}{2}\int_0^\tau dt_1 \int_0^{t_1} dt_2 \big[\tilde{H}_0(t_1), \tilde{H}_0(t_2)\big], \\
\Phi_3(\tau) &= \frac{1}{6}\int_0^\tau dt_1 \int_0^{t_1} dt_2 \int_0^{t_2} dt_3 \big\{\big[\tilde{H}_0(t_1), \big[\tilde{H}_0(t_2), \tilde{H}_0(t_3)\big]\big] \\
&\quad + \big[\tilde{H}_0(t_3), \big[\tilde{H}_0(t_2), \tilde{H}_0(t_1)\big]\big]\big\}, \\
&\dots
\end{aligned} \tag{70}
$$

These generally take the form of time-ordered integrals over nested commutators in $\tilde{H}_0(t)$.

We define the *error vector* $\mathbf{a}(\tau)$ by re-expressing the operator $\Phi(\tau) \equiv \mathbf{a}(\tau) \cdot \boldsymbol{\sigma}$ in the basis of Pauli operators.[m] Assuming unitary processes, one may then employ vector identities to expand $\mathbf{a}(\tau) = \sum_{\mu=1}^{\infty} \mathbf{a}_{\mu}(\tau)$ in an infinite power series such that $\Phi_{\mu}(\tau) = \mathbf{a}_{\mu}(\tau) \cdot \boldsymbol{\sigma}, \forall \mu \in \mathbb{N}$ [49]. The control propagator is therefore written $\tilde{U}(\tau) = \exp[-i \sum_{\mu=1}^{\infty} \mathbf{a}_{\mu}(\tau) \cdot \boldsymbol{\sigma}]$ and may be approximated, with arbitrary accuracy, as a unitary operator in simple exponential form. In this paper we consider the first-order approximation $\mathbf{a}(\tau) \approx \mathbf{a}_1(\tau)$. Hence we restrict attention to deriving the form of $\mathbf{a}_1(\tau)$ which, using Eq. (70), satisfies

$$\mathbf{a}_1(\tau) \cdot \boldsymbol{\sigma} = \Phi_1(\tau) = \int_0^{\tau} dt \tilde{H}_0(t). \tag{71}$$

Our first task is therefore to derive a computationally useful forms for the toggling frame Hamiltonian $\tilde{H}_0(t)$. This is done in the following section.

Toggling frame Hamiltonian $\tilde{H}_0(t)$: computational form

The noise Hamiltonian $H_0(t) = H_0^{(z)}(t) + H_0^{(\Omega)}(t)$, presented in Section 2 is linear in the dephasing and amplitude contributions. Since the $\tilde{H}_0(t)$ is linear in $H_0(t)$ we may write

$$\tilde{H}_0(t) = \tilde{H}_0^{(z)}(t) + \tilde{H}_0^{(\Omega)}(t), \tag{72}$$

where we have defined the dephasing and amplitude toggling frame Hamiltonians by

$$\tilde{H}_0^{(z)}(t) := U_c^{\dagger}(t) H_0^{(z)}(t) U_c(t), \tag{73}$$

$$\tilde{H}_0^{(\Omega)}(t) := U_c^{\dagger}(t) H_0^{(\Omega)}(t) U_c(t). \tag{74}$$

It is convenient to employ the definitions of the scalar functions $R_{ij}(t)$, $R_{ij}^{P_l}(t - t_{l-1})$ and $\Lambda_{ij}^{(l-1)}$, $i, j \in \{x, y, z\}$ introduced by Green *et. al* [49]. These are defined as the Cartesian expansion coefficients, in the basis of Pauli matrices, of the following operators

$$U_c^{\dagger}(t) \hat{\sigma}_i U_c(t) = \sum_{j=x,y,z} R_{ij}(t) \hat{\sigma}_j, \tag{75}$$

$$U_c^{\dagger}(t, t_{l-1}) \hat{\sigma}_i U_c(t, t_{l-1}) = \sum_{j=x,y,z} R_{ij}^{P_l}(t - t_{l-1}) \hat{\sigma}_j, \tag{76}$$

$$Q_{l-1}^{\dagger} \hat{\sigma}_i Q_{l-1} = \sum_{j=x,y,z} \Lambda_{ij}^{(l-1)} \hat{\sigma}_j. \tag{77}$$

These functions then serve as matrix elements, defining the computational matrices

$$\textit{Total Control Matrix}: \quad [\mathbf{R}(t)]_{ij} := \frac{1}{2} \mathrm{Tr}\big[U_c^{\dagger}(t) \hat{\sigma}_i U_c(t) \hat{\sigma}_j\big], \tag{78}$$

$$\textit{Local Control Matrix}: \quad [\mathbf{R}^{P_l}(t - t_{l-1})]_{ij} := \frac{1}{2} \mathrm{Tr}\big[U_c^{\dagger}(t, t_{l-1}) \hat{\sigma}_i U_c(t, t_{l-1}) \hat{\sigma}_j\big], \tag{79}$$

$$\textit{Control History Matrix}: \quad [\boldsymbol{\Lambda}^{(l-1)}]_{ij} := \frac{1}{2} \mathrm{Tr}\big[Q_{l-1}^{\dagger} \hat{\sigma}_i Q_{l-1} \hat{\sigma}_j\big], \tag{80}$$

where the above expressions follow from post-multiplying Eqs. (75), (76) and (77) by $\hat{\sigma}_j$, taking the trace and using the linearity of the trace operation. The matrix $\mathbf{R}(t)$ captures the qubit dynamics, in the time-domain, due to the control Hamiltonian *at any time*; $\mathbf{R}^{Pl}(t - t_{l-1})$ captures essentially the same information, but restricted to the time interval $t \in I_l$. That is, during the lth pulse. The 3×3 *Control History Matrix* $\mathbf{\Lambda}^{(l-1)}$, on the other hand, captures the accumulated effect of the previous $l-1$ completed pulses, implemented via the cumulative operator Q_{l-1}.

Dephasing toggling Hamiltonian $\tilde{H}_0^{(z)}(t)$

Substituting Eq. (2) into Eq. (73) the dephasing noise component of $\tilde{H}_0(t)$ takes the form

$$\tilde{H}_0^{(z)}(t) = \beta_z(t) U_c^\dagger(t) \hat{\sigma}_z U_c(t). \tag{81}$$

Using Eq. (75), we may express $U_c^\dagger(t)\hat{\sigma}_z U_c(t) = \sum_j R_{zj}(t)\hat{\sigma}_j$, $j \in \{x, y, z\}$, yielding

$$\tilde{H}_0^{(z)}(t) = \beta_z(t) \mathbf{R}^{(z)}(t) \boldsymbol{\sigma}, \tag{82}$$

where the time-domain *Dephasing Control Vector* $\mathbf{R}^{(z)}(t)$ is defined as the third row of $\mathbf{R}(t)$. Here, and in the following derivations, for notational simplicity $\boldsymbol{\sigma}$ is understood to be the column vector

$$\boldsymbol{\sigma} \equiv (\hat{\sigma}_x, \hat{\sigma}_y, \hat{\sigma}_z)^T. \tag{83}$$

The computational form of $\mathbf{R}^{(z)}(t)$, by inspection of the more general computational form for $\mathbf{R}(t)$ derived in Appendix B, is

$$\mathbf{R}^{(z)}(t) = \sum_{l=1}^{n} G^{(l)}(t) \mathbf{R}_z^{P_l}(t - t_{l-1}) \mathbf{\Lambda}^{(l-1)}. \tag{84}$$

Amplitude toggling Hamiltonian $\tilde{H}_0^{(\Omega)}(t)$

Similarly, substituting Eq. (3) into Eq. (74) we find

$$\tilde{H}_0^{(\Omega)}(t) = \beta_\Omega(t) \sum_{l=1}^{n} G^{(l)}(t) \frac{\Omega_l}{2} \left\{ U_c^\dagger(t) \hat{\sigma}_{\phi_l} U_c(t) \right\} \tag{85}$$

$$= \beta_\Omega(t) \sum_{l=1}^{n} G^{(l)}(t) \frac{\Omega_l}{2} Q_{l-1}^\dagger U_c^\dagger(t, t_{l-1}) \hat{\sigma}_{\phi_l} U_c(t, t_{l-1}) Q_{l-1}, \tag{86}$$

where in the second line we have substituted $U_c(t) = \sum_{l=1}^{n} G^{(l)}(t) U_c(t, t_{l-1}) Q_{l-1}$ using Eq. (10). From Eq. (8) the control operator $U_c(t, t_{l-1}) = \exp[-i\frac{\Omega_l}{2}\hat{\sigma}_{\phi_l}(t - t_{l-1})]$ commutes with $\hat{\sigma}_{\phi_l}$, $\forall l \in \{1, \ldots, n\}$. That is, coaxial amplitude noise always 'tracks' the direction of control. Hence Eq. (86) reduces to

$$\tilde{H}_0^{(\Omega)}(t) = \beta_\Omega(t) \sum_{l=1}^{n} G^{(l)}(t) \frac{\Omega_l}{2} Q_{l-1}^\dagger \hat{\sigma}_{\phi_l} Q_{l-1}. \tag{87}$$

Now, from Eq. (77), we know $Q_{l-1}^\dagger \hat{\sigma}_i Q_{l-1} = \mathbf{\Lambda}_i^{(l-1)} \boldsymbol{\sigma}$, where $\mathbf{\Lambda}_i^{(l-1)}$ denotes the ith row of $\mathbf{\Lambda}^{(l-1)}$. We may therefore write

$$\frac{\Omega_l}{2} Q_{l-1}^\dagger, \hat{\sigma}_{\phi_l} Q_{l-1} = \frac{\Omega_l}{2} \left[\cos(\phi_l) \mathbf{\Lambda}_x^{(l-1)} + \sin(\phi_l) \mathbf{\Lambda}_y^{(l-1)} \right] = \mathbb{T}^{(l)} \mathbf{\Lambda}^{(l-1)} \boldsymbol{\sigma}, \tag{88}$$

where $\vec{\mathbb{T}}^{(l)} \equiv \frac{\Omega_l}{2}(\cos(\phi_l), \sin(\phi_l), 0)$ is the *Projection Vector* defined in Eq. (18). Consequently, substituting into Eq. (87), the amplitude toggling Hamiltonian is re-expressed

$$\tilde{H}_0^{(\Omega)}(t) = \beta_\Omega(t)\mathbf{R}^{(\Omega)}(t)\boldsymbol{\sigma}, \tag{89}$$

where, for compactness, we have defined the time-domain *Amplitude Control Vector*

$$\mathbf{R}^{(\Omega)}(t) := \sum_{l=1}^{n} G^{(l)}(t)\vec{\mathbb{T}}^{(l)}\boldsymbol{\Lambda}^{(l-1)}. \tag{90}$$

First-order error vector

To summarize of the previous sections, the error propagator is now written

$$\begin{aligned}
\tilde{U}(\tau) &= \exp\left[-i\Phi(\tau)\right] &(91)\\
&\approx \exp\left[-i\mathbf{a}_1(\tau)\boldsymbol{\sigma}\right] &(92)\\
&= \exp\left[-i\int_0^\tau dt\tilde{H}_0(t)\right] &(93)\\
&= \exp\left[-i\left(\int_0^\tau dt\tilde{H}_0^{(z)}(t) + \int_0^\tau dt\tilde{H}_0^{(\Omega)}(t)\right)\right] &(94)\\
&= \exp\left[-i\left(\int_0^\tau dt\beta_z(t)\mathbf{R}^{(z)}(t) + \int_0^\tau dt\beta_\Omega(t)\mathbf{R}^{(\Omega)}(t)\right)\boldsymbol{\sigma}\right]. &(95)
\end{aligned}$$

The first-order error vector consequently takes the form

$$\begin{aligned}
\mathbf{a}_1(\tau) &= \int_0^\tau dt\beta_z(t)\mathbf{R}^{(z)}(t) + \int_0^\tau dt\beta_\Omega(t)\mathbf{R}^{(\Omega)}(t) &(96)\\
&= \mathbf{a}_1^{(z)} + \mathbf{a}_1^{(\Omega)}, &(97)
\end{aligned}$$

where we define components

$$\mathbf{a}_1^{(z)}(\tau) := \int_0^\tau dt\beta_z(t)\mathbf{R}^{(z)}(t) \quad \textit{(Dephasing Error Vector)}, \tag{98}$$

$$\mathbf{a}_1^{(\Omega)}(\tau) := \int_0^\tau dt\beta_\Omega(t)\mathbf{R}^{(\Omega)}(t) \quad \textit{(Amplitude Error Vector)}. \tag{99}$$

First-order infidelity

The time-domain representation of the first-order infidelity $\langle a_1^2\rangle = \langle \mathbf{a}_1\mathbf{a}_1^T\rangle$ now follows directly from Eq. (96). For filtering time-dependent noise, however, it is more useful to transform to a spectral representation in which case $\langle a_1^2\rangle$ separates into dephasing and amplitude noise terms, each appearing as an overlap integral between the noise PSD and a frequency-domain filter-transfer function. In this section we summarize this derivation, and define the filter-transfer functions. In the following section we present the final forms for these filter-transfer functions.

The modulus square of the first-order error vector, using Eq. (96), is given by

$$\begin{aligned}
\mathbf{a}_1\mathbf{a}_1^T &= \left(\mathbf{a}_1^{(z)} + \mathbf{a}_1^{(\Omega)}\right)\left(\mathbf{a}_1^{(z)} + \mathbf{a}_1^{(\Omega)}\right)^T &(100)\\
&= \mathbf{a}_1^{(z)}\mathbf{a}_1^{(z)T} + \mathbf{a}_1^{(z)}\mathbf{a}_1^{(\Omega)T} + \mathbf{a}_1^{(\Omega)}\mathbf{a}_1^{(z)T} + \mathbf{a}_1^{(\Omega)}\mathbf{a}_1^{(\Omega)T}. &(101)
\end{aligned}$$

Substituting Eqs. (98) and (99) and taking the ensemble average over time, the first-order infidelity $\langle a_1^2\rangle = \langle \mathbf{a}_1\mathbf{a}_1^T\rangle$ takes the form

$$\langle a_1^2\rangle = \int_0^\tau dt_1 \int_0^\tau dt_2 \langle \beta_\Omega(t_1)\beta_\Omega(t_2)\rangle \mathbf{R}^{(\Omega)}(t_1)\left[\mathbf{R}^{(\Omega)}(t_2)\right]^T \tag{102}$$

$$+ \int_0^\tau dt_1 \int_0^\tau dt_4 \langle \beta_\Omega(t_1)\beta_z(t_4)\rangle \mathbf{R}^{(\Omega)}(t_1)\left[\mathbf{R}^{(z)}(t_4)\right]^T \tag{103}$$

$$+ \int_0^\tau dt_3 \int_0^\tau dt_2 \langle \beta_z(t_3)\beta_\Omega(t_2)\rangle \mathbf{R}^{(z)}(t_3)\left[\mathbf{R}^{(\Omega)}(t_2)\right]^T \tag{104}$$

$$+ \int_0^\tau dt_3 \int_0^\tau dt_4 \langle \beta_z(t_3)\beta_z(t_4)\rangle \mathbf{R}^{(z)}(t_3)\left[\mathbf{R}^{(z)}(t_4)\right]^T. \tag{105}$$

Here the time average only operates on the noise fields, not on the control vectors since these are deterministic. Assuming, as in Section 2, $\beta_\Omega(t)$ and $\beta_z(t)$ are uncorrelated, classical random variables with zero mean, the two-point cross-correlation functions $\langle\beta_\Omega(t_i)\beta_z(t_j)\rangle = \langle\beta_z(t_k)\beta_\Omega(t_l)\rangle = 0$. Hence the infidelity reduces to

$$\langle a_1^2\rangle = \int_0^\tau dt_1 \int_0^\tau dt_2 \langle \beta_\Omega(t_1)\beta_\Omega(t_2)\rangle \mathbf{R}^{(\Omega)}(t_1)\left[\mathbf{R}^{(\Omega)}(t_2)\right]^T \tag{106}$$

$$+ \int_0^\tau dt_3 \int_0^\tau dt_4 \langle \beta_z(t_3)\beta_z(t_4)\rangle \mathbf{R}^{(z)}(t_3)\left[\mathbf{R}^{(z)}(t_4)\right]^T. \tag{107}$$

We further assuming *wide-sense-stationary*, so the remaining two-point correlation functions depend only on the time difference, and therefore reduce to *auto-correlation functions*. In this case we may invoke the Wiener-Khinchin theorem [77],

$$\langle \beta(t_1)\beta(t_2)\rangle = \frac{1}{2\pi}\int_{-\infty}^{\infty} S_\beta(\omega)e^{i\omega(t_2-t_1)}\,d\omega \tag{108}$$

which relates the autocorrelation function of a signal $\beta(t)$ to the Fourier transform of its PSD $S_\beta(\omega)$. Denoting the dephasing and amplitude noise PSDs by $S_z(\omega)$ and $S_\Omega(\omega)$, we may therefore re-express $\langle a_1^2\rangle$ in terms of the noise spectral properties, yielding

$$\langle a_1^2\rangle = \frac{1}{2\pi}\int_{-\infty}^{\infty} d\omega S_\Omega(\omega)\left[\int_0^\tau dt_1 e^{-i\omega t_1}\mathbf{R}^{(\Omega)}(t_1)\int_0^\tau dt_2 e^{i\omega t_2}\left[\mathbf{R}^{(\Omega)}(t_2)\right]^T\right] \tag{109}$$

$$+ \frac{1}{2\pi}\int_{-\infty}^{\infty} d\omega' S_z(\omega')\left[\int_0^\tau dt_3 e^{-i\omega' t_3}\mathbf{R}^{(z)}(t_3)\int_0^\tau dt_4 e^{i\omega' t_4}\left[\mathbf{R}^{(z)}(t_4)\right]^T\right]. \tag{110}$$

Defining the frequency-domain control vectors via the integral transforms

$$\mathbf{R}^{(\Omega)}(\omega) = -i\omega\int_0^\tau dt e^{i\omega t}\mathbf{R}^{(\Omega)}(t), \qquad \mathbf{R}^{(z)}(\omega) = -i\omega\int_0^\tau dt e^{i\omega t}\mathbf{R}^{(z)}(t) \tag{111}$$

it is then straightforward to further re-express the infidelity in terms the spectral properties of the *control*, yielding

$$\langle a_1^2\rangle = \frac{1}{2\pi}\int_{-\infty}^{\infty}\frac{d\omega}{\omega^2} S_\Omega(\omega)\left[\mathbf{R}^{(\Omega)}(\omega)\right]^*\left[\mathbf{R}^{(\Omega)}(\omega)\right]^T \tag{112}$$

$$+ \frac{1}{2\pi}\int_{-\infty}^{\infty}\frac{d\omega'}{\omega'^2} S_z(\omega')\left[\mathbf{R}^{(z)}(\omega')\right]^*\left[\mathbf{R}^{(z)}(\omega')\right]^T. \tag{113}$$

Defining the frequency-domain filter-transfer functions,

$$F_z(\omega) := \left[\mathbf{R}^{(z)}(\omega)\right]^* \left[\mathbf{R}^{(z)}(\omega)\right]^T \quad (\textit{Dephasing Filter-Transfer Function}), \tag{114}$$

$$F_\Omega(\omega) := \left[\mathbf{R}^{(\Omega)}(\omega)\right]^* \left[\mathbf{R}^{(\Omega)}(\omega)\right]^T \quad (\textit{Amplitude Filter-Transfer Function}). \tag{115}$$

we therefore recover Eq. (4)

$$\langle a_1^2 \rangle = \frac{1}{2\pi}\int_{-\infty}^{\infty} \frac{d\omega}{\omega^2} S_z(\omega) F_z(\omega) + \frac{1}{2\pi}\int_{-\infty}^{\infty} \frac{d\omega'}{\omega'^2} S_\Omega(\omega') F_\Omega(\omega'). \tag{116}$$

Appendix B: Control vectors: computational forms

The dephasing and amplitude filter-transfer functions $F_z(\omega)$ and $F_\Omega(\omega)$ are obtained by taking the modulus square respectively of the frequency-domain dephasing and amplitude control vectors $\mathbf{R}^{(z)}(\omega)$ and $\mathbf{R}^{(\Omega)}(\omega)$, defined by Eq. (111) in terms of a Fourier-type transform. In this section we derive the computationally useful forms of $\mathbf{R}^{(z)}(\omega)$ and $\mathbf{R}^{(\Omega)}(\omega)$.

Total Control Matrix R(*t*) computational form

The time-domain *Total Control Matrix* $\mathbf{R}(t)$ is defined by Eq. (78) with elements $R_{ij}(t) := \frac{1}{2}\mathrm{Tr}[U_c^\dagger(t)\hat{\sigma}_i U_c(t)\hat{\sigma}_j]$. Substituting in Eq. (10) we then obtain

$$R_{ij}(t) = \frac{1}{2}\sum_{l=1}^{n} G^{(l)}(t)\,\mathrm{Tr}\left[Q_{l-1}^\dagger\{U_c^\dagger(t,t_{l-1})\hat{\sigma}_i U_c(t,t_{l-1})\}Q_{l-1}\hat{\sigma}_j\right] \tag{117}$$

$$= \frac{1}{2}\sum_{l=1}^{n} G^{(l)}(t)\,\mathrm{Tr}\left[Q_{l-1}^\dagger\left\{\sum_{k=x,y,z} R_{ik}^{P_l}(t-t_{l-1})\hat{\sigma}_k\right\}Q_{l-1}\hat{\sigma}_j\right], \tag{118}$$

where we have used Eq. (76) to re-express $U_c^\dagger(t,t_{l-1})\hat{\sigma}_i U_c(t,t_{l-1})$ in terms of the $R_{ij}^{P_l}$. Using the linearity of the trace operation and recalling the definition of the of the *Control History Matrix* elements Λ_{ij}^{l-1} from Eq. (80), we then obtain

$$R_{ij}(t) = \frac{1}{2}\sum_{k=x,y,z}\sum_{l=1}^{n} G^{(l)}(t)\,\mathrm{Tr}\left[R_{ik}^{P_l}(t-t_{l-1})Q_{l-1}^\dagger\hat{\sigma}_k Q_{l-1}\hat{\sigma}_j\right] \tag{119}$$

$$= \sum_{k=x,y,z}\sum_{l=1}^{n} G^{(l)}(t) R_{ik}^{P_l}(t-t_{l-1})\left\{\frac{1}{2}\mathrm{Tr}\left[Q_{l-1}^\dagger\hat{\sigma}_k Q_{l-1}\hat{\sigma}_j\right]\right\} \tag{120}$$

$$\equiv \sum_{l=1}^{n} G^{(l)}(t)\sum_{k=x,y,z}\{R_{ik}^{P_l}(t-t_{l-1})\}\{\Lambda_{kj}^{(l-1)}\}. \tag{121}$$

Hence, by definition of matrix multiplication, the time domain total control matrix takes the form

$$\mathbf{R}(t) = \sum_{l=1}^{n} G^{(l)}(t)\mathbf{R}^{P_l}(t-t_{l-1})\mathbf{\Lambda}^{(l-1)}. \tag{122}$$

We move to the frequency domain by performing the integral transform on $\mathbf{R}(t)$ defined by

$$\mathbf{R}(\omega) := -i\omega \int_0^{\tau} dt e^{i\omega t}\mathbf{R}(t) \tag{123}$$

$$= -i\omega \sum_{l=1}^{n} \left\{ \int_0^{\tau} dt e^{i\omega t} G^{(l)}(t)\mathbf{R}^{P_l}(t - t_{l-1}) \right\} \mathbf{\Lambda}^{(l-1)} \tag{124}$$

$$= -i\omega \sum_{l=1}^{n} \left\{ \int_{t_{l-1}}^{t_l} dt e^{i\omega t}\mathbf{R}^{P_l}(t - t_{l-1}) \right\} \mathbf{\Lambda}^{(l-1)} \tag{125}$$

$$= \sum_{l=1}^{n} e^{i\omega t_{l-1}} \left\{ -i\omega \int_0^{\tau_l} dt' e^{i\omega t'} \mathbf{R}^{P_l}(t') \right\} \mathbf{\Lambda}^{(l-1)}. \tag{126}$$

In the last line we have performed a change of variables using $t' = t - t_{l-1}$, $\forall l \in \{1, \ldots, n\}$. The frequency-domain *Total Control Matrix* thus takes the computational form

$$\mathbf{R}(\omega) = \sum_{l=1}^{n} e^{i\omega t_{l-1}} \mathbf{R}^{P_l}(\omega)\mathbf{\Lambda}^{(l-1)}, \tag{127}$$

where we have defined the frequency-domain *Local Control Matrix* by

$$\mathbf{R}^{P_l}(\omega) := -i\omega \int_0^{\tau_l} dt' e^{i\omega t'} \mathbf{R}^{P_l}(t'). \tag{128}$$

The matrix elements of $\mathbf{R}^{P_l}(\omega)$ are derived as functions of the control sequence in the section bellow.

Local Control Matrix elements

As defined in Eq. (79), the matrix elements of the time-domain *Local Control Matrix* are given by $R_{ij}^{P_l}(t - t_{l-1}) = \frac{1}{2}\,\mathrm{Tr}[U_c^{\dagger}(t, t_{l-1})\hat{\sigma}_i U_c(t, t_{l-1})\hat{\sigma}_j]$, where $U_c(t, t_{l-1}) \equiv \exp[-i\frac{\Omega_l}{2}\hat{\sigma}_{\phi_l}(t - t_{l-1})]$. The frequency domain representation $\mathbf{R}^{P_l}(\omega)$ then follows from the integral transform defined by Eq. (128), with matrix elements expressed as functions of the control parameters $\{\Omega_l, \tau_l, \phi_l\}$. These matrix elements take the form

$$R_{xx}^{P_l}(\omega) = \cos^2(\phi_l)\left[1 - e^{i\omega\tau_l}\right] + \frac{\omega \sin^2(\phi_l)}{\omega^2 - \Omega_l^2} V_l(\omega), \tag{129}$$

$$R_{xy}^{P_l}(\omega) = \frac{1}{2}\sin(2\phi_l)\left[1 - e^{i\omega\tau_l}\right] - \frac{\omega \sin(2\phi_l)}{2(\omega^2 - \Omega_l^2)} V_l(\omega), \tag{130}$$

$$R_{xz}^{P_l}(\omega) = -\frac{\omega}{\omega^2 - \Omega_l^2}\sin(\phi_l)B_l(\omega), \tag{131}$$

$$R_{yx}^{P_l}(\omega) = \frac{1}{2}\sin(2\phi_l)\left[1 - e^{i\omega\tau_l}\right] - \frac{\omega \sin(2\phi_l)}{2(\omega^2 - \Omega_l^2)} V_l(\omega), \tag{132}$$

$$R_{yy}^{P_l}(\omega) = \sin^2(\phi_l)\left[1 - e^{i\omega\tau_l}\right] + \frac{\omega \cos^2(\phi_l)}{\omega^2 - \Omega_l^2} V_l(\omega), \tag{133}$$

$$R_{yz}^{P_l}(\omega) = \frac{\omega}{\omega^2 - \Omega_l^2}\cos(\phi_l)B_l(\omega), \tag{134}$$

$$R_{zx}^{P_l}(\omega) = \frac{\omega}{\omega^2 - \Omega_l^2}\sin(\phi_l)B_l(\omega), \tag{135}$$

$$R_{zy}^{P_l}(\omega) = -\frac{\omega}{\omega^2 - \Omega_l^2}\cos(\phi_l)B_l(\omega), \tag{136}$$

$$R_{zz}^{P_l}(\omega) = \frac{\omega}{\omega^2 - \Omega_l^2}V_l(\omega), \tag{137}$$

where we have defined

$$V_l(\omega) := \left[i\Omega_l e^{i\omega\tau_l}\sin(\Omega_l\tau_l) - \omega e^{i\omega\tau_l}\cos(\Omega_l\tau_l) + \omega\right], \tag{138}$$

$$B_l(\omega) := \left[i\Omega_l e^{i\omega\tau_l}\cos(\Omega_l\tau_l) + \omega e^{i\omega\tau_l}\sin(\Omega_l\tau_l) - i\Omega_l\right]. \tag{139}$$

Amplitude Control Vector

The computational form for the frequency-domain *Amplitude Control Vector* $\mathbf{R}^{(\Omega)}(\omega)$ follows from substituting Eq. (90) into Eq. (111), yielding

$$\mathbf{R}^{(\Omega)}(\omega) := -i\omega\int_0^\tau dt e^{i\omega t}\mathbf{R}^{(\Omega)}(t) \tag{140}$$

$$= -i\omega\left\{\int_0^\tau dt e^{i\omega t}\sum_{l=1}^{n} G^{(l)}(t)\right\}\vec{\mathbb{T}}^{(l)}\mathbf{\Lambda}^{(l-1)} \tag{141}$$

$$= -i\omega\sum_{l=1}^{n}\left\{\int_{t_{l-1}}^{t_l} dt e^{i\omega t}\right\}\vec{\mathbb{T}}^{(l)}\mathbf{\Lambda}^{(l-1)} \tag{142}$$

$$= -i\omega\sum_{l=1}^{n}\frac{1}{i\omega}\left[e^{i\omega t_l} - e^{i\omega t_{l-1}}\right]\vec{\mathbb{T}}^{(l)}\mathbf{\Lambda}^{(l-1)} \tag{143}$$

$$= \sum_{l=1}^{n}\left[e^{i\omega t_{l-1}} - e^{i\omega t_l}\right]\vec{\mathbb{T}}^{(l)}\mathbf{\Lambda}^{(l-1)}. \tag{144}$$

Appendix C: Walsh Rotary Spin Echo derivations

The WRSE$_k$ sequence is defined by the phase modulation $\phi(t) = \phi_0 + \frac{\pi}{2}(1 - y(t))$, where $y(t) \in \{\pm 1\}$. Referring to Eq. (6), however, the spin operator $\hat{\sigma}_{\phi(t)}$ satisfies

$$\hat{\sigma}_{\phi_0+\pi} = -\hat{\sigma}_{\phi_0} \iff \hat{\sigma}_{\phi_0+\frac{\pi}{2}(1-y(t))} = y(t)\hat{\sigma}_{\phi_0}. \tag{145}$$

Consequently the sign-inversion may be absorbed into a modulated Rabi-rate defined by $\Omega_y(t) := y(t)\Omega_0$. The sequence is then conveniently recast as *amplitude* modulation with constant phase ϕ_0. Defining $y(t) := \mathrm{PAL}_k(t/\tau)$ we therefore obtain the Walsh synthesis $\Omega_k(t) = \Omega_0\,\mathrm{PAL}_k(t/\tau)$ consisting of a single Walsh function. Referring to Eq. (31) the associated amplitude modulation is thus given by

$$\vec{\mathbf{\Omega}} = H_{\mathcal{M}}\tilde{\mathbf{X}} = \Omega_0\mathbf{P}_{\mathcal{M}}^{(k)}, \quad \mathcal{M} = 2^{m(k)}, \tag{146}$$

where $\tilde{X}_{i(k)} = \Omega_0$ is the only nonzero element of $\tilde{\mathbf{X}}$ and, as in Section 5.1, $\mathbf{P}_{\mathcal{M}}^{(k)} := (P_1^{(k)}, P_2^{(k)}, \ldots, P_{\mathcal{M}}^{(k)})^T$ defines the $i(k)$th column of $H_{\mathcal{M}}$. Eq. (46) is then more conveniently re-expressed

$$\mathrm{WRSE}_k \equiv \prod_{l=1}^{\mathcal{M}(k)} \exp\left(i\frac{P_l^{(k)}\Omega_0}{2}\tau_{\mathcal{M}}\hat{\sigma}_{\phi_0}\right). \tag{147}$$

In the next section we use this form to analyze the amplitude noise filtering properties. The subsequent section treats the filtering properties in the dephasing quadrature.

WRSE$_k$ as amplitude noise filters

Referring to Eq. (147), the rotation operator $\hat{\sigma}_{\phi_0}$ for WRSE$_k$ is treated as fixed across all pulses. Thus it commutes with P_l and, consequently, with $Q_l = P_l P_{l-1} \cdots P_0 \ \forall l \in \{1, \ldots, \mathcal{M}(k)\}$. It follows $Q_{l-1}^\dagger \hat{\sigma}_{\phi_0} Q_{l-1} = \hat{\sigma}_{\phi_0}$ which, post-multiplying by $\hat{\sigma}_k$ and taking the trace of both sides, yields the identity

$$\frac{1}{2}\mathrm{Tr}[\hat{\sigma}_{\phi_0}\hat{\sigma}_k] = \frac{1}{2}\mathrm{Tr}\left[Q_{l-1}^\dagger \hat{\sigma}_{\phi_0} Q_{l-1}\hat{\sigma}_k\right], \quad k \in \{x, y, z\}. \tag{148}$$

The LHS expands to $\delta_{xk}\cos(\phi_0) + \delta_{yk}\sin(\phi_0)$ (where δ_{lk} is the Kronecker delta), and the RHS expands to $\cos(\phi_0)\Lambda_{xk}^{(l-1)} + \sin(\phi_0)\Lambda_{yk}^{(l-1)}$ (using the definition of the *Pulse History Matrix* in Eq. (80)). We thus obtain the following three identities $\forall \phi_0$:

$$k = x: \quad \cos(\phi_0)\Lambda_{xx}^{(l-1)} + \sin(\phi_0)\Lambda_{yx}^{(l-1)} = \cos(\phi_0), \tag{149}$$

$$k = y: \quad \cos(\phi_0)\Lambda_{xy}^{(l-1)} + \sin(\phi_0)\Lambda_{yy}^{(l-1)} = \sin(\phi_0), \tag{150}$$

$$k = z: \quad \cos(\phi_0)\Lambda_{xz}^{(l-1)} + \sin(\phi_0)\Lambda_{yz}^{(l-1)} = 0. \tag{151}$$

Now, setting $\Omega_l \equiv P_l^{(k)}\Omega_0$ the *Projection Vector* defined in Eq. (18) becomes $\vec{\mathbb{T}}^{(l)} = (\Omega_0 P_l^{(k)}/2)[\cos(\phi_0), \sin(\phi_0), 0]$, in which case

$$\vec{\mathbb{T}}^{(l)}\mathbf{\Lambda}^{(l-1)} = \frac{\Omega_0}{2}P_l^{(k)}\left\{\cos(\phi_0)\mathbf{\Lambda}_x^{(l-1)} + \sin(\phi_0)\mathbf{\Lambda}_y^{(l-1)}\right\} \tag{152}$$

$$= \frac{\Omega_0}{2}P_l^{(k)}\begin{bmatrix}\cos(\phi_0)\Lambda_{xx}^{(l-1)} + \sin(\phi_0)\Lambda_{yx}^{(l-1)} \\ \cos(\phi_0)\Lambda_{xy}^{(l-1)} + \sin(\phi_0)\Lambda_{yy}^{(l-1)} \\ \cos(\phi_0)\Lambda_{xz}^{(l-1)} + \sin(\phi_0)\Lambda_{yz}^{(l-1)}\end{bmatrix}^T \tag{153}$$

$$= \frac{\Omega_0}{2}P_l^{(k)}\left[\cos(\phi_0), \ \ \sin(\phi_0), \ \ 0\right], \tag{154}$$

where in the last equality we have used the identities derived above in Eqs. (149), (150) and (151). Using Eq. (16) we therefore obtain

$$\mathbf{R}^{(\Omega)}(\omega) = \frac{\Omega_0}{2}\sum_{l=1}^{\mathcal{M}} P_l^{(k)}\left[e^{i\omega t_{l-1}} - e^{i\omega t_l}\right]\begin{bmatrix}\cos(\phi_0) \\ \sin(\phi_0) \\ 0\end{bmatrix}^T, \tag{155}$$

where $t_l = l\tau/\mathcal{M}(k)$. From Eq. (14) the amplitude filter-transfer function therefore becomes

$$F_\Omega(\omega) = \frac{\Omega^2}{4}\left|\sum_{l=1}^{\mathcal{M}} P_l^{(k)}\left[e^{i\omega t_{l-1}} - e^{i\omega t_l}\right]\right|^2, \tag{156}$$

where, on taking the modulus square, the ϕ_0 dependence amounts to $\cos(\phi_0)^2 + \sin(\phi_0)^2 = 1$ and hence vanishes.

WRSE_k as dephasing noise filters

For a general WRSE_k sequence, one can show the first-order Taylor coefficient for $F_z(\omega)$ takes the analytic form

$$C_2^{(z)}(\Omega_0;k) = \mathrm{sinc}^2\left(\frac{\Omega_0}{2^{\kappa(k)}}\right), \quad \kappa(k) = \begin{cases} m(k) & \text{if } r(k) \neq 1, \\ m(k)+1 & \text{if } r(k) = 1 \end{cases} \tag{157}$$

yielding the family of zeros $Z_2^{(k)} = \{2^{\kappa(k)}\pi q | q \in \mathbb{N}\}$. Hence it is always possible to produce a first-order filter with $(p-1) = 1$ by setting $\Omega_0 \in Z_2^{(k)}$. Higher-order filters for dephasing noise - that is, such that $(p-1) > 1$ - then correspond to some η satisfying Eq. (49) such that $\eta \in Z_2^{(k)}$. Although a general analytical form for these higher-order coefficients is not easy to express[n] we may still make progress, however, by simply substituting in the candidate values $\Omega_0 = 2^{\kappa(k)}\pi q$ and determining which $q \in \mathbb{N}$ produce concurrent zeros of the $C_{2j}^{(z)}(\Omega_0)$. As a representative example we study the particular case for WRSE_3, deriving the coefficients

$$C_4^{(z)}(\Omega_0;3) = \frac{1}{\Omega_0^4}\left[(\Omega_0^2 - 16)\cos\frac{\Omega_0}{2} - 2\Omega_0^2\cos\frac{\Omega_0}{4} - 8\Omega_0\sin\frac{\Omega_0}{4} + (\Omega_0^2+16)\right], \tag{158}$$

$$C_6^{(z)}(\Omega_0;3) = \frac{1}{48\Omega_0^6}\left[(\Omega_0^4 - 96\Omega_0^2 + 1{,}152)\cos\frac{\Omega_0}{2} - (5\Omega_0^4 - 192\Omega_0^2)\cos\frac{\Omega_0}{4} - (28\Omega_0^3 - 768\Omega_0)\sin\frac{\Omega_0}{4} + 4(\Omega_0^4 - 36\Omega_0^2 - 288)\right]. \tag{159}$$

From above, the choice $k = 3$ implies $\kappa(k) = 2$ and consequently the candidate zeros take the form $\Omega_0 = 4\pi q$. Substituting into the above expressions yields

$$C_4^{(z)}(4\pi q;3) = \frac{1-(-1)^q}{8q^2\pi^2}, \quad q \in \mathbb{N}, \tag{160}$$

$$C_6^{(z)}(4\pi q;3) = \begin{cases} \frac{1}{(8\pi q)^4} & \text{if } q \text{ even}, \\ \frac{27-10\pi^2(1+2q)^2}{7\pi^4(1+2q)^4} & \text{if } q \text{ odd}. \end{cases} \tag{161}$$

Thus q must be even to ensure $C_4^{(z)} = 0$. However this choice implies $C_6^{(z)} > 0$ (in fact $C_6^{(z)} > 0$ for any choice of q) and it follows WRSE_3 is at maximum a second-order filter. In Figure 6(a) we plot $C_{2j}^{(z)}(\Omega_0;3)$, $j \in \{1,2,3\}$ showing the existence of concurrent zeros only for $j \in \{1,2\}$. The inset shows, as a representative case, the behaviour near $\Omega_0 = 8\pi$. In Figure 6(b) we plot $F_z(\omega)$ setting $\Omega_0 = 2\pi q$, $q \in \{1,\ldots,8\}$, showing values for which first and second-order filtering is achieved. Repeating this general procedure for other values of k we find similar results and conclude the WRSE_k family are capable of up to second-order filtering against dephasing noise.

Competing interests

The authors declare that they have no competing interests.

Authors' contributions

All authors contributed equally to the writing of this paper. All authors read and approved the final manuscript.

Acknowledgements

We thank K Brown, JT Merrill and L Viola for useful discussions. This work partially supported by the US Army Research Office under Contract Number W911NF-11-1-0068, and the Australian Research Council Centre of Excellence for Engineered Quantum Systems CE110001013.

Endnotes

[a] The assumption of independence is reasonable, for instance, in the case of a driving field where random fluctuations in frequency and amplitude arise from different physical processes. A general model including correlations between noise processes is possible, however, following the approach outlined by *Green et al.* [49].

[b] The first-order approximation has recently been experimentally tested and demonstrated to produce good agreement in the weak noise limit [37]. For the noise field $\beta(t)$, this regime is sufficiently characterized by requiring $\xi^2 \ll 1$, where the *smallness parameter* is defined by $\xi^2 \equiv \langle \beta^2(t) \rangle \tau^2 \equiv \tau^2 \int_{-\infty}^{+\infty} d\omega S_\beta(\omega)$ [48]. The condition $\xi^2 < 1$ is also required for the Magnus series to formally converge.

[c] For a resonantly driven qubit ϕ_l is the phase of the driving field and Ω_l is linearly proportional to the driving amplitude.

[d] This effect is captured by the multiple slopes in Figure 4(h) which clearly show the difference between the asymptotic zero-frequency *roll-off* and the local slope over targeted regions $[\omega_L, \omega_c]$ in the stopband.

[e] All stopbands 'turn on' with a finite response, the functional form of which determines the filter order and the effectiveness of noise suppression. In the stopband this is quantified by the slope, or *roll off* in the language of filter design.

[f] This can be shown by Taylor expanding the amplitude noise filter function $F_\Omega(\omega\tau; \mathbf{\Gamma}_n^{\mathrm{AMF}})$ and deriving the result that $C_2^{(\Omega)}(\mathbf{\Gamma}_n^{\mathrm{AMF}}) = \frac{1}{4}(\sum_{l=1}^{n} \theta_l)^2$.

[g] We use the vectors $\mathbf{X} \equiv (X_0, X_1, \ldots, X_N)$ and $\mathbf{Y} \equiv (Y_0, Y_1, \ldots, Y_N)$ to compactly write the Paley ordered Walsh spectra implied by Eq. (29) in synthesizing $\Omega(t)$ and $\phi(t)$.

[h] Pulse periods taking non-integer multiples values of $\tau_{\mathbf{min}}$ then have intrinsic conflict with implementation in discretized time via digital control, giving rise to residual errors.

[i] Our studies have not produced proof that this symmetry is strictly necessary. In fact for WPMFs it is not required. However WAMF constructions possessing time-reversal symmetry do appear to yield results more readily, and all WAMFs we have discovered have this property.

[j] Maximizing the number of Rademacher functions does *not* correspond to maximizing the switching rate of $\mathrm{PAL}_k(x)$. In fact, for a given $m(k)$ the maximum switching rate for $\mathrm{PAL}_k(x)$ corresponds to $k = 2^{m(k)-1}$, which consists of the single Rademacher function $R_{m(k)-1}(x)$.

[k] These points may also be derived using Nelder-Mead optimization of $A_z(X_0; X_3)$ over the two-dimensional domain. This method is useful for more complex constructions (see Section 6.2) where spectral optimization becomes a more multi-dimensional task.

[l] Since the $\mathrm{WAMF}_{0:7}^{(2)}$ gates in Figure 4 were derived by optimizing the cost function over local regions in the stopband, the asymptotic filter order $(p-1)$ associated with Taylor expanding $F_z(\omega)$ about $\omega = 0$ is not a meaningful descriptor of these filters. Hence we do not expect $C_{2,4}^{(z)} = 0$ and do not pursue such a calculation. Instead the instantaneous filter order is used.

[m] This is valid since $\Phi(\tau)$ belongs to the Lie algebra of $SU(2)$, inheriting this property from the toggling frame Hamiltonians from which it is derived.

[n] The higher-order $C_{2j}^{(k)}$ involve terms oscillating at multiple frequencies and have nontrivial dependencies on Ω_0. Their zeros must in general be determined numerically.

References

1. Van Harlingen DJ, Plourde BLT, Robertson TL, Reichardt PA, Clarke J. Decoherence in flux qubits due to 1/*f* noise in Josephson junctions. In: Leggett A, Ruggiero B, Silvestini P, editors. Quantum computing and quantum bits in mesoscopic systems. New York: Kluwer Academic; 2004. p. 171-84.
2. Faoro L, Viola L. Dynamical suppression of 1/*f* noise processes in qubit systems. Phys Rev Lett. 2004;92:117905.
3. Bylander J, Gustavsson S, Yan F, Yoshihara F, Harrabi K, Fitch G, Cory DG, Nakamura Y, Tsai J-S, Oliver WD. Noise spectroscopy through dynamical decoupling with a superconducting flux qubit. Nat Phys. 2011;7:565-70.
4. Zhang W, Dobrovitski VV, Santos LF, Viola L, Harmon BN. Dynamical control of electron spin coherence in a quantum dot: a theoretical study. Phys Rev B. 2007;75:201302.
5. Biercuk MJ, Uys H, VanDevender AP, Shiga N, Itano WM, Bollinger JJ. High-fidelity quantum control using ion crystals in a Penning trap. Quantum Inf Comput. 2009;9:920-49.
6. Rutman J. Characterization of phase and frequency instabilities in precision frequency sources: fifteen years of progress. Proc IEEE. 1978;66:1048.
7. Tarn TJ, Huang G, Clark JW. Modelling of quantum mechanical control systems. Math Model. 1980;1:109.
8. Clark JW, Lucarelli DG, Tarn TJ. Control of quantum systems. Int J Mod Phys B. 2003;17:5397.
9. Bouten L, Van Handel R, James MR. An introduction to quantum filtering. SIAM J Control Optim. 2007;46:2199.
10. Nurdin HI, James MR, Petersen IR. Coherent quantum LGQ control. Automatica. 2009;45:1837.
11. Viola L, Lloyd S, Knill E. Universal control of decoupled quantum systems. Phys Rev Lett. 1999;83:4888.
12. Ticozzi F, Nishio K, Altafini C. Stabilization of stochastic quantum dynamics via open and closed loop control. IEEE Trans Autom Control. 2013;58:74-85.

13. Lidar DA, Brun TA. Quantum error correction. New York: Cambridge University Press; 2013.
14. Jones NC, Van Meter R, Fowler AG, McMahon PL, Kim J, Ladd TD, Yamamoto Y. Layered architecture for quantum computing. Phys Rev X. 2012;2:031007.
15. Biercuk MJ, Uys H, VanDevender AP, Shiga N, Itano WM, Bollinger JJ. Optimized dynamical decoupling in a model quantum memory. Nature. 2009;458:996.
16. Khodjasteh K, Sastrawan J, Hayes D, Green TJ, Biercuk MJ, Viola L. Designing a practical high-fidelity long-time quantum memory. Nat Commun. 2013;4:2045.
17. Du J, Rong X, Zhao N, Wang Y, Yang J, Liu RB. Preserving electron spin coherence in solids by optimal dynamical decoupling. Nature. 2009;461:1265.
18. Sagi Y, Almog I, Davidson N. Process tomography of dynamical decoupling in a dense cold atomic ensemble. Phys Rev Lett. 2010;105:053201.
19. Khodjasteh K, Viola L. Dynamically error-corrected gates for universal quantum computation. Phys Rev Lett. 2009;102:080501.
20. Khodjasteh K, Lidar DA, Viola L. Arbitrarily accurate dynamical control in open quantum systems. Phys Rev Lett. 2010;104:090501.
21. Wang X, Bishop LS, Kestner JP, Barnes E, Sun K, Das Sarma S. Composite pulses for robust universal control of singlet-triplet qubits. Nat Commun. 2012;3:997.
22. Souza AM, Álvarez GA, Suter D. Experimental protection of quantum gates against decoherence and control errors. Phys Rev A. 2012;86:050301.
23. van der Sar T, Wang ZH, Blok MS, Bernien H, Taminiau TH, Toyli DM, Lidar DA, Awschalom DD, Hanson R, Dobrovitski VV. Decoherence-protected quantum gates for a hybrid solid-state spin register. Nature. 2012;484:82-6.
24. Liu G-Q, Po HC, Du J, Liu RB, Pan X-Y. Noise-resilient quantum evolution steered by dynamical decoupling. Nat Commun. 2013;4:2254.
25. Fauseweh B, Pasini S, Uhrig GS. Frequency-modulated pulses for quantum bits coupled to time-dependent baths. Phys Rev A. 2012;85:022310.
26. Vandersypen LMK, Chuang IL. NMR techniques for quantum control and computation. Rev Mod Phys. 2005;76:1037.
27. Merrill JT, Brown KR. Progress in compensating pulse sequences for quantum computation. arXiv:1203.6392 (2012).
28. Kabytayev C, Green TJ, Khodjasteh K, Biercuk MJ, Viola L, Brown KR. Robustness of composite pulses to time-dependent control noise. Phys Rev A. 2014;90:012316.
29. Viola L, Knill E. Robust dynamical decoupling of quantum systems with bounded controls. Phys Rev Lett. 2003;90:037901.
30. Jones NC, Ladd TD, Fong BH. Dynamical decoupling of a qubit with always-on control fields. New J Phys. 2012;14:093045.
31. Cai J-M, Naydenov B, Pfeiffer R, McGuinness LP, Jahnke KD, Jelezko F, Plenio MB, Retzker A. Robust dynamical decoupling with concatenated continuous driving. New J Phys. 2012;14:113023.
32. Fanchini FF, Hornos JEM, Napolitano RdJ. Continuously decoupling single-qubit operations from a perturbing thermal bath of scalar bosons. Phys Rev A. 2007;75:022329.
33. Xu X, Wang Z, Duan C, Huang P, Wang P, Wang Y, Xu N, Kong X, Shi F, Rong X, Du J. Coherence-protected quantum gate by continuous dynamical decoupling in diamond. Phys Rev Lett. 2012;109:070502.
34. Bermudez A, Schmidt PO, Plenio MB, Retzker A. Robust trapped-ion quantum logic gates by continuous dynamical decoupling. Phys Rev A. 2012;85:040302.
35. Chaudhry AZ, Gong J. Decoherence control: universal protection of two-qubit states and two-qubit gates using continuous driving fields. Phys Rev A. 2012;85:012315.
36. Lemmer A, Bermudez A, Plenio MB. Driven geometric phase gates with trapped ions. New J Phys. 2013;15:083001.
37. Soare A, Ball H, Hayes D, Sastrawan J, Jarratt MC, McLoughlin JJ, Zhen X, Green TJ, Biercuk MJ. Experimental noise filtering by quantum control. Nat Phys. 2014;10:825-9.
38. Silva GP, Viola L. A general transfer-function approach to noise filtering in open-loop quantum control. arXiv:1408.3836 (2014).
39. Kofman AG, Kurizki G. Universal dynamical control of quantum mechanical decay: modulation of the coupling to the continuum. Phys Rev Lett. 2001;87:270405.
40. Kofman AG, Kurizki G. Unified theory of dynamically suppressed qubit decoherence in thermal baths. Phys Rev Lett. 2004;93:130406.
41. Girod B, Rabenstein R, Stenger A. Signals and systems. New York: Wiley; 2001.
42. Biercuk MJ, Doherty AC, Uys H. Dynamical decoupling sequence construction as a filter-design problem. J Phys B. 2011;44:154002.
43. Martinis JM, Nam S, Aumentado J, Lang KM, Urbina C. Decoherence of a superconducting qubit due to bias noise. Phys Rev B. 2003;67:094510.
44. Kuopanportti P, Mottonen M, Bergholm V, Saira O-P, Zhang J, Whaley KB. Suppression of $1/f^{\alpha}$ noise in one-qubit systems. Phys Rev A. 2008;77:032334.
45. Uys H, Biercuk MJ, Bollinger JJ. Optimized noise filtration through dynamical decoupling. Phys Rev Lett. 2009;103:040501.
46. Uhrig GS. Keeping a quantum bit alive by optimized pi-pulse sequences. Phys Rev Lett. 2007;98:100504.
47. Cywinski L, Lutchyn RM, Nave CP, Sarma SD. How to enhance dephasing time in superconducting qubits. Phys Rev B. 2008;77:174509.
48. Green TJ, Uys H, Biercuk MJ. High-order noise filtering in nontrivial quantum logic gates. Phys Rev Lett. 2012;109:020501.
49. Green TJ, Sastrawan J, Uys H, Biercuk MJ. Arbitrary quantum control of qubits in the presence of universal noise. New J Phys. 2013;15:095004.
50. Hayes D, Clark SM, Debnath S, Hucul D, Inlek IV, Lee KW, Quraishi Q, Monroe C. Coherent error suppression in multiqubit entangling gates. Phys Rev Lett. 2012;109:020503.
51. Hayes D, Khodjasteh K, Viola L, Biercuk MJ. Reducing sequencing complexity in dynamical quantum error suppression by Walsh modulation. Phys Rev A. 2011;84:062323.

52. Green TJ, Biercuk MJ. Phase-modulated decoupling and error suppression in qubit-oscillator systems. arXiv:1408.2749 (2014).
53. Owrutsky P, Khaneja N. Control of inhomogeneous ensembles on the Bloch sphere. Phys Rev A. 2012;86:022315.
54. Hodgson TE, Viola L, D'Amico I. Towards optimized suppression of dephasing in systems subject to pulse timing constraints. Phys Rev A. 2010;81:062321.
55. Schumacher B. Sending entanglement through noisy quantum channels. Phys Rev A. 1996;54:2614.
56. Blanes S, Cases F, Oteo JA, Ros J. The Magnus expansion and some of its applications. Phys Rep. 2009;470:151.
57. Magnus W. On the exponential solution of differential equations for a linear operator. Commun Pure Appl Math. 1954;7:649.
58. Haeberlen U, Waugh JS. Coherent averaging effects in magnetic resonance. Phys Rev. 1968;175:453.
59. Ernst RR, Bodenhausen GB, Wokaun A. Principles of nuclear magnetic resonance in one and two dimensions. New York: Oxford University Press; 1987.
60. Ajoy A, Álvarez GA, Suter D. Optimal pulse spacing for dynamical decoupling in the presence of a purely-dephasing spin-bath. Phys Rev A. 2011;83:032303.
61. Beauchamp KG. Walsh functions and their applications. London: Academic Press; 1975.
62. Walsh JL. A closed set of normal orthogonal functions. Am J Math. 1923;45:5.
63. Tzafestas SG. Walsh functions in signal and systems analysis and design. New York: Van Nostrand Reinhold; 1985.
64. Cooper A, Magesan E, Yum HN, Cappellaro P. Time-resolved magnetic sensing with electronic spins in diamond. Nat Commun. 2014;5:3141.
65. Harmuth HF. Transmission of information by orthogonal functions. New York: Springer; 1969.
66. Harmuth HF. Applications of Walsh functions in communications. IEEE Spectr. 1969;6:82.
67. Paley R. A remarkable series of orthogonal functions (I). Proc Lond Math Soc. 1932;2:241.
68. Rademacher H. Einige Sätze über Reihen von allgemeinen Orthogonalfunktionen. Math Ann. 1922;87:241.
69. Horadam KJ. Hadamard matrices and their applications. Princeton: Princeton University Press; 2007.
70. Khodjasteh K, Viola L. Dynamical quantum error correction of unitary operations with bounded controls. Phys Rev A. 2009;80:032314.
71. Souza AM, Álvarez GA, Suter D. Effects of time-reversal symmetry in dynamical decoupling. Phys Rev A. 2012;85:032306.
72. Brown KR, Harrow AW, Chuang IL. Arbitrarily accurate composite pulse sequences. Phys Rev A. 2004;70:052318.
73. Brown KR, Harrow AW, Chuang IL. Erratum: Arbitrarily accurate composite pulse sequences [Phys. Rev. A 70, 052318 (2004)]. Phys Rev A. 2005;72:039905.
74. Wimperis S. Broadband, narrowband, and passband composite pulses for use in advanced {NMR} experiments. J Magn Reson, Ser A. 1994;109:221.
75. Yan F, Gustavsson S, Bylander J, Jin X, Yoshihara F, Cory DG, Nakamura Y, Orlando TP, Oliver WD. Rotating-frame relaxation as a noise spectrum analyser of a superconducting qubit undergoing driven evolution. Nat Commun. 2013;4:2337.
76. Gustavsson S, Bylander J, Yan F, Forn-Díaz P, Bolkhovsky V, Braje D, Fitch G, Harrabi K, Lennon D, Miloshi J, Murphy P, Slattery R, Spector S, Turek B, Weir T, Welander PB, Yoshihara F, Cory DG, Nakamura Y, Orlando TP, Oliver WD. Driven dynamics and rotary echo of a qubit tunably coupled to a harmonic oscillator. Phys Rev Lett. 2012;108:170503.
77. Miller SL, Childers D. Probability and random processes with applications to signal processing and communications. Boston: Academic Press; 2012.

16

One-dimensional waveguide coupled to multiple qubits: photon-photon correlations

Yao-Lung L Fang, Huaixiu Zheng and Harold U Baranger*

*Correspondence: baranger@phy.duke.edu
Department of Physics, Duke University, P.O. Box 90305, Durham, NC 27708, USA

Abstract

For a one-dimensional (1D) waveguide coupled to two or three qubits, we show that the photon-photon correlations have a wide variety of behavior, with structure that depends sensitively on the frequency and on the qubit-qubit separation L. We study the correlations by calculating the second-order correlation function $g_2(t)$ in which the interference among the photons multiply scattered from the qubits causes rich structure. In one case, for example, transmitted and reflected photons are both bunched initially, but then become strongly anti-bunched for a long time interval. We first calculate the correlation function $g_2(t)$ including non-Markovian effects and then show that a much simpler Markovian treatment, which can be solved analytically, is accurate for small qubit separation. As a result, the non-classical properties of microwaves in a 1D waveguide coupled to many superconducting qubits with experimentally accessible separation L could be readily explored with our approach.

Keywords: quantum electrodynamics; quantum photonics; circuit QED; non-classical light

1 Introduction

One-dimensional (1D) waveguide-QED systems are currently generating increasing interest - systems in which photons confined in one-dimension interact with one or several two-level systems (qubits). Part of the motivation comes from the striking quantum optics effects that can be seen in these strongly coupled systems [1–19]. Another motivating factor is the promise of waveguide-QED systems for quantum information processing [20–27]. Finally, a key driver of the interest in waveguide-QED systems is the tremendous experimental progress that has been made recently in a number of systems [28–41]. Perhaps the leading system for waveguide-QED investigations and applications is an open microwave transmission line coupled to superconducting qubits [34–37, 42, 43]. While much of the work to date has focused on systems in which there is a single qubit, and there is a growing literature on the case of two qubits [10, 17, 19, 41, 44–49], an important future direction for both fundamental effects and possible applications is to study a waveguide coupled to multiple (or many) qubits. As a step in this direction, here we compare and contrast results for one, two, and three qubits coupled to a waveguide [see Figure 1(a)], focusing in particular on the generation of photon-photon correlations.

Correlations between photons are a key signature of non-classical light. They are often characterized by the second-order correlation function (photon-photon correlation function) $g_2(t)$ where t is the observation time between the two photons (see below for

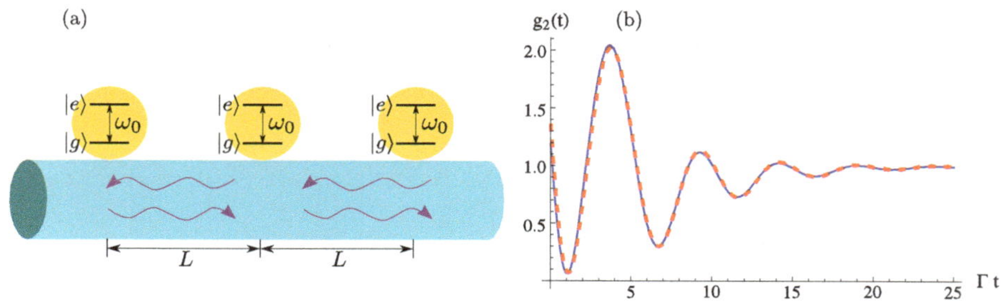

Figure 1 Quantum beats in three qubit system. (**a**) Schematic diagram of the 1D waveguide system coupled to 3 identical qubits with separation L. (**b**) Comparison between the Markovian approximation (solid blue curve) and full numerical results (red dashed line) for $g_2(t)$ of reflected photons with $N = 3$, $k_0L = \pi/2$, $T = 50\%$, and $\Gamma' = 0$.

precise definition) [50]. The uncorrelated, classical value is $g_2 = 1$ (obtained, for example, for a coherent state). Bunching of photons, $g_2 > 1$, often occurs due to the bosonic nature of photons but anti-bunching, $g_2 < 1$, also occurs [50]. In recent experiments, $g_2(t)$ of microwave photons coupled to superconducting qubits was measured, and both bunching and anti-bunching were observed [38, 51]. In a multi-qubit situation, one expects to have interference between the various scattered partial waves; interference effects in the photon-photon correlations $g_2(t)$ are known as 'quantum beats' [52].

In this paper, we first present our method of calculation, which exploits a bosonic representation of the qubits in the rotating wave approximation. We obtain a complicated yet analytic result for $g_2(t)$ in the Markovian limit and show, by comparison with the full numerical result, that it is adequate for small, experimentally accessible separations between the qubits. In presenting results, we focus on an off-resonant case in which single photons have equal probability of being transmitted or reflected, and take the separation between qubits, denoted L, to be either $\lambda_0/4$ or $\lambda_0/8$ where λ_0 is the wavelength of a photon at the qubit resonant frequency. We find several striking features in $g_2(t)$: First, for $L = \lambda_0/4$, the transmitted photons are largely bunched for all times and become more strongly bunched as the number of qubits increases, while the reflected photons oscillate between strong bunching and anti-bunching, showing particularly strong quantum beats in the three qubit case. Second, for $L = \lambda_0/8$, we find the surprising situation that both transmitted and reflected photons are bunched at $t = 0$ but then become anti-bunched for a large time interval. This suggests that the photons in this case become organized into bursts.

2 Method

The Hamiltonian describing N identical qubits coupled to a 1D waveguide [see Figure 1(a)] is, in the rotating wave approximation,

$$H_0 = \hbar(\omega_0 - i\Gamma'/2)\sum_{i=1}^{N}\sigma_i^+\sigma_i^- - i\hbar c\int dx\left[a_R^\dagger(x)\frac{d}{dx}a_R(x) - a_L^\dagger(x)\frac{d}{dx}a_L(x)\right]$$
$$+\sum_{i=1}^{N}\sum_{\alpha=L,R}\hbar V\int dx\,\delta(x-l_i)\left[a_\alpha^\dagger(x)\sigma_i^- + a_\alpha(x)\sigma_i^+\right], \tag{1}$$

where $\sigma_i^{\pm}$ are the raising/lowering operators for ith qubit, l_i is its position which is fixed by $L = l_{i+1} - l_i$, ω_0 is the transition frequency of the qubit, and Γ' is the decay rate to channels other than the waveguide. The spontaneous decay rate to the waveguide continuum is given by $\Gamma = 2V^2/c$. In the waveguide QED context, 'strong coupling' signifies that the spontaneous decay rate to the waveguide is much faster than the decay to all other modes, namely that the Purcell factor is large, $P \equiv \Gamma/\Gamma' \gg 1$.

To find $g_2(t)$, we first obtain the two-photon eigenstate of H_0. As discussed in Ref. [17], it is convenient to use a bosonic representation of the qubits that includes an on-site interaction,

$$H = H_0 + V, \quad V = \frac{U}{2}\sum_{i=1}^{N} d_i^{\dagger} d_i \left(d_i^{\dagger} d_i - 1\right). \tag{2}$$

The raising/lowering operators $\sigma_i^{\pm}$ in H_0 are replaced by the bosonic creation/annihilation operators $d_i^{\dagger}$ and d_i, respectively. One then takes $U \to \infty$ in the end to project out occupations greater than 1. In this bosonic representation, the $U = 0$ case corresponds to a non-interacting Hamiltonian and can readily be solved. In terms of the non-interacting wavefunctions and Green functions, a formal expression for the two-photon 'interacting' wavefunction in the $U \to \infty$ limit can be obtained; this then is the solution to the waveguide QED problem in which we are interested. Finally, the two-photon wavefunction together with the one-photon wavefunction yields $g_2(t)$ for a weak incident coherent state. More details of this procedure are given in the appendices.

The Markovian approximation allows a considerable simplification of the final result [17]. In the present context, the Markovian approximation consists of an approximate treatment of certain interference terms valid for small separation between the qubits. In the formal expression for the two-photon wavefunction discussed above, there is an integral over the non-interacting wavefunctions which generally must be performed numerically. The non-interacting wavefunctions naturally involve interference factors e^{i2kL} that make this integral difficult. However, if the qubits are close enough, k may be replaced by $k_0 = \omega_0/c$, allowing the integral to be performed analytically using contour integration (the analytic expression of the final result is lengthy, so we just give the steps of the derivation in the appendices as well as the $N = 2$ result as an example). All of the results in this paper are obtained in the regime where this is valid. An example of the checks we have made is shown in Figure 1(b): the full numerical result is in good agreement with that from the small separation approximation.

We compare the one, two, and three qubit cases: $N = 1, 2$, or 3. In order to make a fair comparison, the typical transmission through the system in the three cases should be the same; otherwise, the lower probability of finding a photon in one case compared to another will affect g_2. We therefore consider off-resonance cases (*i.e.* $\omega \neq \omega_0$ where ω is the incoming photon frequency) in which the single-photon transmission probability T is fixed. Because the single-photon transmission spectrum depends on the number of qubits, the frequency used is different in the three cases N = 1-3. Due to the asymmetry of the single-photon transmission spectrum in certain cases, the criterion used throughout this work is to pick up the frequency closest to ω_0 so that $g_2(0)$ is the largest.

In the following results, we consider $N = 1, 2$, or 3; $k_0 L = \pi/4$ or $\pi/2$; and $T = 50\%$. The single photon transmission curves used to choose the photon frequency ω are shown in

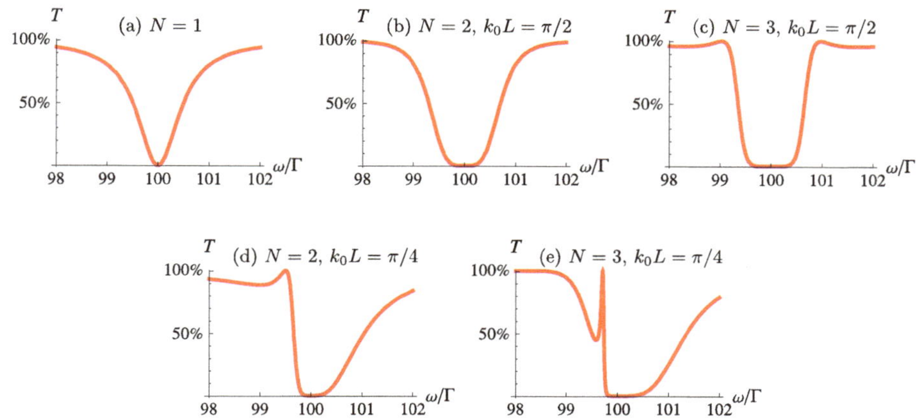

Figure 2 Transmission spectra near the qubit resonant frequency ($\omega_0 = 100\Gamma$) for an incident single-photon Fock state in the five situations studied here.

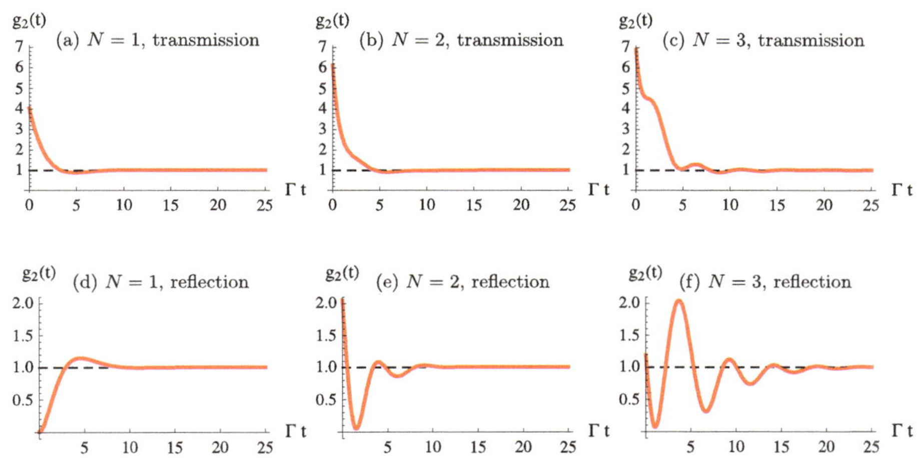

Figure 3 Second-order correlation function, $g_2(t)$, calculated with a weak incident coherent state for spacing $k_0L = \pi/2$. First row is for transmitted photons, second row for reflected photons. The columns correspond to $N = 1, 2,$ or 3 qubits coupled to the waveguide. The photon frequency is chosen so that $T = 50\%$. The result for uncorrelated photons, $g_2 = 1$, is marked (dashed line) for comparison. In the three qubit case, note the strong bunching in transmission [panel (c)] and striking quantum beats in reflection [panel (f)].

Figure 2. We use Γ as our unit of frequency, take $\omega_0 = 100\Gamma$, and consider the lossless case, $\Gamma' = 0$.

3 Results

The results for a single qubit, shown in Figure 3 panels (a) and (d), provide a point of comparison for the two and three qubit cases discussed below; throughout we consider the response to an incident weak coherent state. Non-classical light in a waveguide produced by a single qubit has been extensively investigated theoretically [1, 2, 4–6, 8, 11–13] as well as experimentally with microwave photons [38]. We see that for our chosen detuning such that $T = 50\%$, the transmitted field shows bunching while the reflected field is anti-bunched. The correlation decays to its classical value (namely, 1) quickly and with little structure. For this reason the single value $g_2(0)$ is a good indication of the nature of the

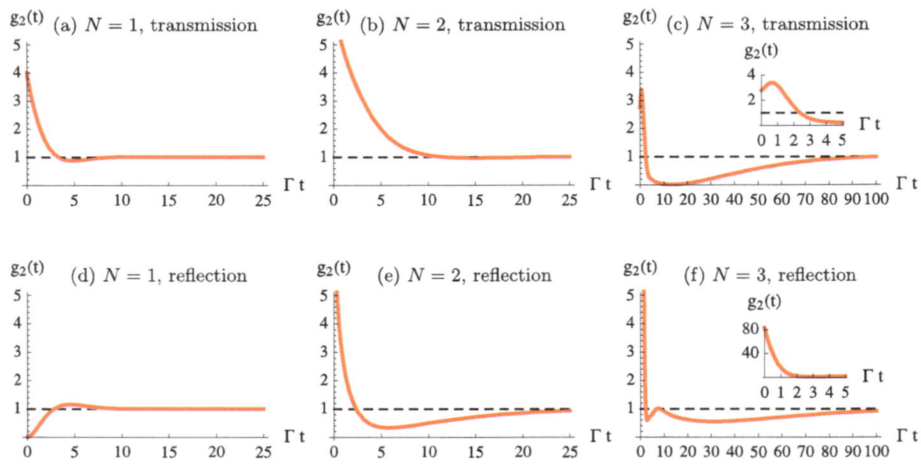

Figure 4 Second-order correlation function, $g_2(t)$, calculated with a weak incident coherent state for spacing $k_0L = \pi/4$. First row is for transmitted photons, second row for reflected photons. The columns correspond to $N = 1, 2,$ or 3 qubits coupled to the waveguide. The photon frequency is chosen so that $T = 50\%$. The result for uncorrelated photons, $g_2 = 1$, is marked (dashed line) for comparison. In the three qubit case, note the strong bunching in reflection [panel (f)] and long anti-bunching interval after the initial bunching in transmission [panel (c)].

correlations overall. Note that in panel (d), $g_2(0) = 0$ due to the inability of a single excited qubit to release two photons at the same time.

For $N = 2$ or 3, we start by considering the case $k_0L = \pi/2$, in which case the qubits are separated by $\lambda_0/4$; the results are shown in Figure 3. The presence of quantum beats coming from interference among the partial waves scattered by the qubits is clear, especially for three qubits. In the transmitted wave, photon bunching is considerably enhanced in magnitude and extends for a longer time (compared to a single qubit). In reflection, $g_2(t)$ develops a striking oscillation between strongly bunched and anti-bunched [panel (f)]. Such behavior in g_2 suggests that the photons become organized periodically in time and space.

Turning now to the case $k_0L = \pi/4$ (a separation of $\lambda_0/8$), we see in Figure 4 that the behavior is completely different. First, the quantum beats largely disappear in both transmission and reflection. Instead, for $N = 3$ we see that both the reflected and transmitted photons are initially bunched, in the reflected case quite strongly bunched. The initial bunching is followed in both cases by anti-bunching. This anti-bunching is dramatic for the transmitted photons: strong anti-bunching persists for a time interval of several tens of Γ^{-1} (the natural unit of time in our problem). Initial bunching followed by a long interval of anti-bunching suggests that the photons are organized into bursts.

The different behavior for $k_0L = \pi/4$ compared to $k_0L = \pi/2$ can be traced to a difference in the structure of the poles of the single photon Green function (see, *e.g.*, the discussion in Ref. [17]). For instance in the $N = 2$ cases, for $k_0L = \pi/2$ there are two dominant poles that have the same decay rate but different real frequencies, leading to maximum interference effects between those two processes. In contrast, for $k_0L = \pi/4$, the poles have very different decay rates; the one decaying most rapidly yields the sharp initial bunching, while the one with the slowest decay produces the long time anti-bunching.

To study how the correlations depend on the frequency of the photons, we show the initial correlation, $g_2(0)$, in Figures 5 and 6. Because of the oscillating structure in $g_2(t)$ when

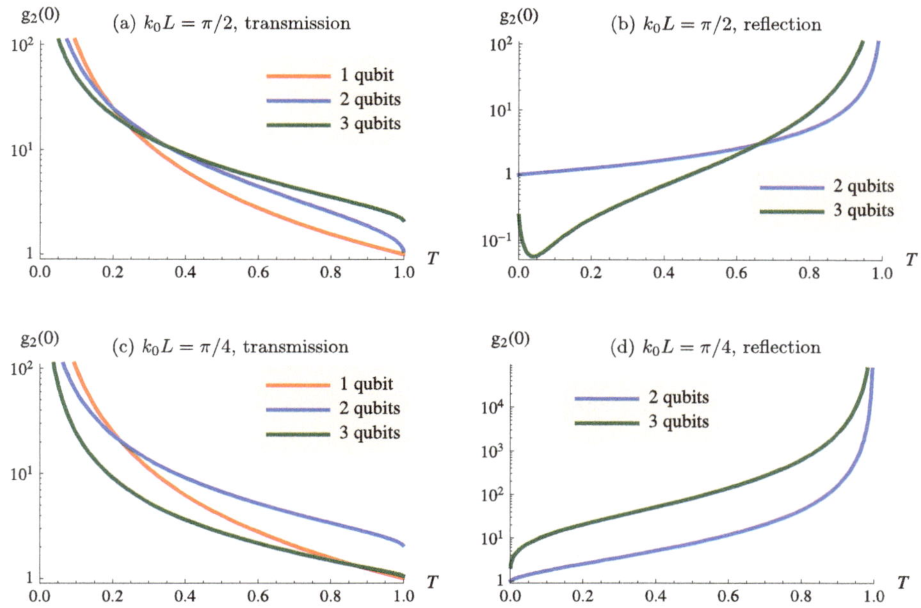

Figure 5 The initial second-order correlation, $g_2(0)$ (on a logarithmic scale), calculated with a weak incident coherent state as a function of the single-photon transmission probability, T, for different numbers of qubits. The first (second) row is for $k_0L = \pi/2$ ($\pi/4$); the first (second) column is for transmitted (reflected) photons. For reflected photons with $N = 1$, $g_2(0) = 0$ for all T and hence is not plotted. For a wide range of parameters, both transmitted and reflected photons are bunched.

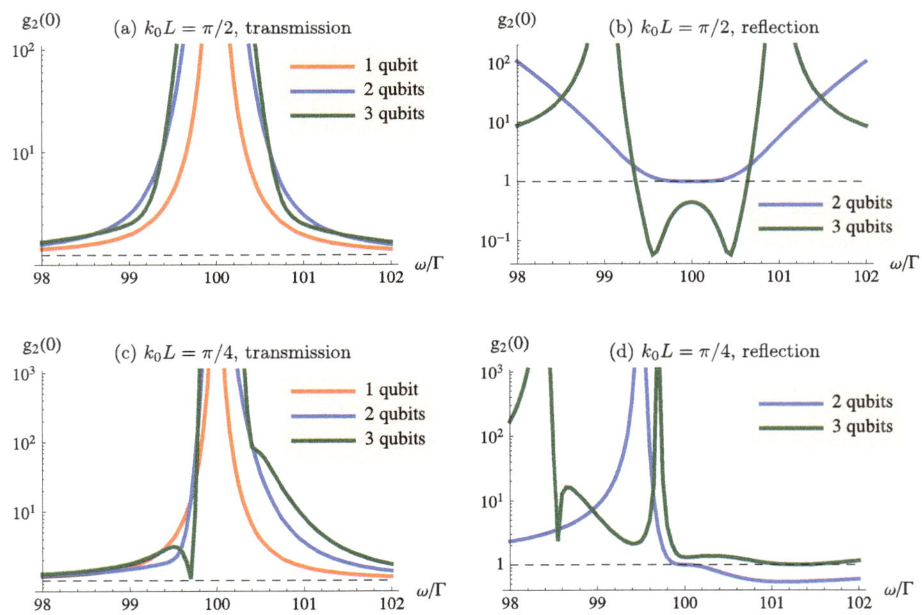

Figure 6 The initial second-order correlation, $g_2(0)$ (on a logarithmic scale), calculated with a weak incident coherent state as a function of the frequency ω near the qubit resonant frequency ($\omega_0 = 100\Gamma$) for different numbers of qubits. The first (second) row is for $k_0L = \pi/2$ ($\pi/4$); the first (second) column is for transmitted (reflected) photons. The black, dashed line indicates the classical value (*i.e.*, $g_2 = 1$). For reflected photons with $N = 1$, $g_2(0) = 0$ for all ω and hence is not plotted. Note that as for the single-photon transmission in Fig. 2, $g_2(0)$ is symmetric about ω_0 for $k_0L = \pi/2$ but asymmetric in the $k_0L = \pi/4$ case. For a wide range of parameters, both transmitted and reflected photons are bunched.

there are multiple qubits, $g_2(0)$ is not necessarily a good indication of the behavior at later times; nevertheless, the degree of initial bunching or anti-bunching is a physically important and measurable quantity. For a fair comparison between the $N = 1, 2,$ and 3 cases, we first plot $g_2(0)$ as a function of the single photon transmission, T; see Figure 5. To match the desired T with an off-resonant photon frequency we follow the following procedure: Starting near the resonant frequency ω_0 (where $T = 0.1\%$), we scan toward smaller frequencies until $T = 99.9\%$ is reached. We then use frequencies within the scanned range to calculate $g_2(0)$ for both transmission and reflection as a function of T at $k_0L = \pi/2$ and $\pi/4$. Another way of presenting the data is to simply plot $g_2(0)$ directly as a function of frequency, as in Figure 6. By comparing with Figure 2, we see that the method above for selecting the range of frequencies to use in making Figure 5 selects the range with largest $g_2(0)$ for a given value of T. Finally, note that for reflection from one qubit, $g_2(0) = 0$ in all cases, as mentioned above, and so is not plotted in panels (b) and (d) of both figures.

Several general trends are clear from Figure 5. Bunching is favored over anti-bunching for both $N = 2$ and 3. As the single photon transmission increases, $g_2(0)$ decreases for transmission but generally increases for reflection. Opposite trends for transmission and reflection are natural based on the simple argument that incoming uncorrelated photons divide between transmitted and reflected ones so that bunching in one implies anti-bunching in the other. Clearly, this simple argument does not apply here; indeed, it is striking and surprising that for a broad range of parameters *both transmitted and reflected photons are bunched.*

Trends as the number of qubits increases from 1 to 3 are also evident in Figure 5. In panels (a) and (d) the trend is monotonic: For $k_0L = \pi/4$, the reflected photons become tremendously bunched [panel (d)], whereas for $k_0L = \pi/2$ and transmitted photons [panel (a)], the curves cross at the same point indicating that the trend changes sign-increasing bunching as N increases for $T \geq 0.25$ but decreasing bunching for smaller T. In the other two cases, panels (b) and (c), the trend as N increases from 1 to 3 is not monotonic. For $k_0L = \pi/2$ and $T \leq 0.65$, the reflected photons switch from being anti-bunched to bunched to anti-bunched as N changes from 1 to 3, but show increasing bunching for larger T [panel (b)]. Finally, in panel (c) [$k_0L = \pi/4$ and transmitted photons], there is a monotonic trend toward less bunching for $T \leq 0.25$ but non-monotonic behavior for larger transmission.

From the explicit dependence on frequency shown in Figure 6, we see that bunching is generally favored even outside the frequency range chosen in Figure 5 (which in the $k_0L = \pi/4$ case is quite small ($<\Gamma$)). Comparing to the single photon transmission spectrum (see Figure 2), we point out two features: First, in the $k_0L = \pi/4$ case, $g_2(0)$ shows the asymmetry with respect to ω_0 [panel (c) and (d)] seen in $T(\omega)$; this again can be traced to the asymmetric pole structure of the Green functions mentioned above. $g_2(0)$ is larger (for reflection) and varies more rapidly on the red-detuned side ($\omega < \omega_0$), which explains why we chose the frequency range use in Figure 5. In fact, on the blue-detuned side ($\omega > \omega_0$) the structure in $g_2(t)$ is less dramatic, and it returns to 1 faster (data not shown). Second, as also shown in Figure 5, the peaks of $g_2(0)$ for transmission are located where $T = 0$, while the peaks of reflected $g_2(0)$ are located where $T = 1$. Note that the leftmost peak of $N = 3$ in Figure 6(d) is completely due to the small denominator ($R = 1 - T \approx 0$) at that point.

4 Conclusion

In this work, we have calculated the second-order correlation function, $g_2(t)$, for photons in a one-dimensional waveguide interacting with one, two or three qubits. By taking the separation between the qubits small, we are able to make a Markovian approximation which then allows an analytic solution. The small separation and small N on which we focus means that these systems are within the range of current experimental capability [19].

The interference among the partial waves scattered from the qubits leads to a variety of behavior in g_2 that is sensitive to both the separation between the qubits (L) and the frequency of the incoming photons. As examples of the rich variety accessible in these waveguide QED structures, we mention three here in conclusion: (i) For a wide range of parameters, both transmitted and reflected photons are initially bunched. (ii) For reflected photons with $N = 3$ and $k_0L = \pi/2$, $g_2(t)$ oscillates between bunching and anti-bunching [Figure 3(f)]. (iii) For transmitted photons with $N = 3$ and $k_0L = \pi/4$, initial strong bunching is followed by a long (*i.e.* $\sim 30\Gamma^{-1}$) interval of antibunching [Figure 4(c)]. These last two observations suggest that some nascent organization of the photons may be occurring, providing an interesting direction for future research.

Appendix 1: Two-photon interacting scattering eigenstate

The single photon eigenstate $|\phi_1(k)\rangle_\alpha$ with $\alpha = \mathrm{L,R}$ is by definition the eigenstate of H_0, *i.e.*, $H_0|\phi_1(k)\rangle_\alpha = \hbar ck|\phi_1(k)\rangle_\alpha$, where

$$|\phi_1(k)\rangle_\alpha = \left[\int dx\big(\phi_\mathrm{R}^\alpha(k,x)a_\mathrm{R}^\dagger(x) + \phi_\mathrm{L}^\alpha(k,x)a_\mathrm{L}^\dagger(x)\big) + \sum_{i=1}^{N} e_i^\alpha(k)\sigma_i^+\right]|0\rangle, \tag{3}$$

$$\phi_\mathrm{R}^\mathrm{R}(k,x) = \frac{e^{ikx}}{\sqrt{2\pi}}\left(\theta(l_1 - x) + \sum_{i=1}^{N-1} t_i(k)\theta(x - l_i)\theta(l_{i+1} - x) + t_N(k)\theta(x - l_N)\right), \tag{4}$$

$$\phi_\mathrm{L}^\mathrm{R}(k,x) = \frac{e^{-ikx}}{\sqrt{2\pi}}\left(r_1(k)\theta(l_1 - x) + \sum_{i=2}^{N} r_i(k)\theta(x - l_{i-1})\theta(l_i - x)\right), \tag{5}$$

$$\phi_\mathrm{R}^\mathrm{L}(k,x) = \frac{e^{ikx}}{\sqrt{2\pi}}\left(r_1(k)\theta(l_N - x) + \sum_{i=2}^{N} r_i(k)\theta(x - l_{N-i+1})\theta(l_{N-i+2} - x)\right), \tag{6}$$

$$\phi_\mathrm{L}^\mathrm{L}(k,x) = \frac{e^{-ikx}}{\sqrt{2\pi}}\left(\theta(x - l_N) + \sum_{i=1}^{N-1} t_i(k)\theta(x - l_{N-i})\theta(l_{N-i+1} - x) + t_N(k)\theta(l_1 - x)\right), \tag{7}$$

and the incoming photon travels in the α-direction with wavevector k. The single photon transmission amplitude is given by $t_N(k)$ and the reflection amplitude by $r_1(k)$. Note that the positions of the qubits are chosen to be symmetric with respect to the origin, *i.e.*, $l_{N-i+1} = -l_i$, in order to take advantage of parity symmetry. Setting $\hbar = c = 1$ from now on, we have for $N = 2$ [17]

$$t_2(k) = \frac{4(k - \omega_0)^2}{(i\Gamma + 2k - 2\omega_0)^2 + \Gamma^2 e^{2ikL}}, \tag{8}$$

$$r_1(k) = \frac{\Gamma(\Gamma - e^{2ikL}(\Gamma + 2ik - 2i\omega_0) - 2ik + 2i\omega_0)}{\Gamma^2 e^{3ikL} + e^{ikL}(i\Gamma + 2k - 2\omega_0)^2}, \tag{9}$$

$$e_1^\mathrm{R}(k) = -\frac{i\sqrt{\Gamma}e^{-\frac{1}{2}ikL}(\Gamma(-1 + e^{2ikL}) + 2ik - 2i\omega_0)}{\sqrt{\pi}((i\Gamma + 2k - 2\omega_0)^2 + \Gamma^2 e^{2ikL})}, \tag{10}$$

$$e_2^{\mathrm{R}}(k) = \frac{2\sqrt{\Gamma}e^{\frac{ikL}{2}}(k-\omega_0)}{\sqrt{\pi}((i\Gamma+2k-2\omega_0)^2+\Gamma^2 e^{2ikL})}. \tag{11}$$

The corresponding result for $N = 3$ is

$$t_3(k) = \frac{8(k-\omega_0)^3}{\eta_+^3+2\Gamma^2\eta_+ e^{2ikL}+\Gamma^2\eta_- e^{4ikL}}, \tag{12}$$

$$r_1(k) = \frac{\Gamma(\eta_+^2+2e^{2ikL}(\Gamma^2+2(k-\omega_0)^2)+\eta_-^2 e^{4ikL})}{2i\Gamma^2\eta_+ e^{4ikL}+i\Gamma^2\eta_- e^{6ikL}+i\eta_+^3 e^{2ikL}}, \tag{13}$$

$$e_1^{\mathrm{R}}(k) = \frac{\sqrt{\Gamma}e^{-ikL}(i\eta_+^2+2\Gamma e^{2ikL}(i\Gamma+k-\omega_0)+\Gamma\eta_- e^{4ikL})}{\sqrt{\pi}(i\eta_+^3+2i\Gamma^2\eta_+ e^{2ikL}+i\Gamma^2\eta_- e^{4ikL})}, \tag{14}$$

$$e_2^{\mathrm{R}}(k) = \frac{2\sqrt{\Gamma}(k-\omega_0)}{\sqrt{\pi}(\eta_+^2+i\Gamma\eta_- e^{2ikL})}, \tag{15}$$

$$e_3^{\mathrm{R}}(k) = \frac{4\sqrt{\Gamma}e^{ikL}(k-\omega_0)^2}{\sqrt{\pi}(\eta_+^3+2\Gamma^2\eta_+ e^{2ikL}+\Gamma^2\eta_- e^{4ikL})}, \tag{16}$$

with $\eta_\pm \equiv 2k - 2\omega_0 \pm i\Gamma$. Note that we do not need the other amplitudes for the rest of this section. For $N = 1$ results see, *e.g.*, Ref. [8].

We can now construct the two-photon 'non-interacting' eigenstate

$$\left|\phi_2(k_1,k_2)\right\rangle_{\alpha_1\alpha_2} = \frac{1}{\sqrt{2}}\left|\phi_1(k_1)\right\rangle_{\alpha_1} \otimes \left|\phi_1(k_2)\right\rangle_{\alpha_2}. \tag{17}$$

As described in the Supplementary Material of Ref. [17], starting from the Lippmann-Schwinger equation

$$\left|\psi_2(k_1,k_2)\right\rangle_{\alpha_1\alpha_2} = \left|\phi_2(k_1,k_2)\right\rangle_{\alpha_1\alpha_2} + G^R(E)V\left|\psi_2(k_1,k_2)\right\rangle_{\alpha_1\alpha_2}, \tag{18}$$

$$G^R(E) = \frac{1}{E-H_0+i\epsilon}, \tag{19}$$

where V is given in Eq. (2) and E is the two photon energy, one can derive the two-photon interacting eigenstate in the coordinate representation in the $U \to \infty$ limit:

$$\begin{aligned} {}_{\alpha_1',\alpha_2'}\left\langle x_1,x_2\middle|\psi_2(k_1,k_2)\right\rangle_{\alpha_1,\alpha_2} &= {}_{\alpha_1',\alpha_2'}\left\langle x_1,x_2\middle|\phi_2(k_1,k_2)\right\rangle_{\alpha_1,\alpha_2} \\ &\quad - \sum_{i,j=1}^{N} G_i^{\alpha_1'\alpha_2'}(x_1,x_2)\left(G^{-1}\right)_{ij}\left\langle d_j d_j\middle|\phi_2(k_1,k_2)\right\rangle_{\alpha_1,\alpha_2}, \end{aligned} \tag{20}$$

$$\begin{aligned} G_i^{\alpha_1\alpha_2}(x_1,x_2) &= {}_{\alpha_1,\alpha_2}\left\langle x_1,x_2\middle|G^R(E)\middle|d_i d_i\right\rangle \\ &= \sum_{\alpha_1',\alpha_2'}\int dk_1\,dk_2 \frac{{}_{\alpha_1,\alpha_2}\langle x_1,x_2|\phi_2(k_1,k_2)\rangle_{\alpha_1',\alpha_2'}\langle\phi_2(k_1,k_2)|d_i d_i\rangle}{E-(k_1+k_2)+i\epsilon}, \end{aligned} \tag{21}$$

$$G^{-1} = \begin{pmatrix} G_{11} & G_{12} & \cdots & G_{1N} \\ G_{21} & G_{22} & \cdots & G_{2N} \\ \vdots & \vdots & \ddots & \vdots \\ G_{N1} & G_{N2} & \cdots & G_{NN} \end{pmatrix}^{-1}, \tag{22}$$

$$G_{ij} = \langle d_i d_i | G^R(E) | d_j d_j \rangle = \sum_{\alpha_1,\alpha_2} \int dk_1 dk_2 \frac{\langle d_i d_i | \phi_2(k_1,k_2)\rangle_{\alpha_1,\alpha_2} \langle \phi_2(k_1,k_2) | d_j d_j \rangle}{E - (k_1 + k_2) + i\epsilon}. \tag{23}$$

Note that x_1 and x_2 here refer to the positions of the photons.

By observing the structure of these Green functions, one would realize that given the following two pieces

$$\langle d_i d_i | \phi_2(k_1,k_2)\rangle_{\alpha_1,\alpha_2} = e_i^{\alpha_1}(k_1) e_i^{\alpha_2}(k_2), \tag{24}$$

$${}_{\alpha_1',\alpha_2'}\langle x_1 x_2 | \phi_2(k_1,k_2)\rangle_{\alpha_1,\alpha_2} = \frac{1}{2}(\phi_{\alpha_1'}^{\alpha_1}(k_1,x_1)\phi_{\alpha_2'}^{\alpha_2}(k_2,x_2) + \phi_{\alpha_1'}^{\alpha_2}(k_2,x_1)\phi_{\alpha_2'}^{\alpha_1}(k_1,x_2)), \tag{25}$$

the whole prescription is complete and in principle one may numerically compute the two-photon interacting eigenstate Eq. (20) for any N.

Finally, to proceed with the Markovian approximation, we explicitly write down the integrands in Eqs. (21) and (23), replace the factors $\exp(2ikL)$ by $\exp(2ik_0L)$ therein, and do the double integral by standard contour integral techniques enclosing the poles in the upper half complex plane (for the $N = 3$ case, for example, the denominator of each transmission amplitude is a cubic polynomial in k, so there are three roots). The $N = 2$ case [17] could serve as an illustrative example owing to its relatively simple polynomial structure: For $k_0L = A\pi$ with $0 \le A \le 1/2$, we have

$$G_{11} = \frac{e^{2i\pi A}\Gamma^2 + 2\eta^2}{2(e^{2i\pi A}\Gamma^2\eta + \eta^3)}, \tag{26}$$

$$G_{12} = -\frac{e^{2i\pi A}\Gamma^2}{2\eta((-1 + e^{2i\pi A})\Gamma^2 - 4i\Gamma\omega_0 + E^2 + 2iE(\Gamma + 2i\omega_0) + 4\omega_0^2)}, \tag{27}$$

$$\begin{aligned} G_1^{\mathrm{R,R}}(x_1,x_2) = -\Big\{ & \Gamma\big[\beta_+(-i\beta_-\Gamma + E - 2\omega_0)(2E - 2\omega_0 + i\gamma) \\ & + \beta_- e^{e^{i\pi A}\Gamma t}(-i\beta_-\Gamma + 2E - 4\omega_0)(E + i\gamma)\big] \\ & \times \exp\left(iE(t + x_1) - \frac{1}{2}\gamma t\right)\Big\} \Big/ 8\left(e^{2i\pi A}\Gamma^2\eta + \eta^3\right), \end{aligned} \tag{28}$$

$$\begin{aligned} G_2^{\mathrm{R,R}}(x_1,x_2) = i\Big\{ & \Gamma\big[\beta_+(-i\beta_-\Gamma + E - 2\omega_0)\left((-2 + e^{i\pi A})\beta_+\Gamma + 2iE - 4i\omega_0\right) \\ & + \beta_- e^{e^{i\pi A}\Gamma t}\left(-(e^{i\pi A} + e^{2i\pi A} - 2)\Gamma - 2iE + 4i\omega_0\right)(E + i\gamma)\big] \\ & \times \exp\left(-\frac{1}{2}\gamma t - i\pi A + iE(t + x_1)\right)\Big\} \Big/ 8\left(e^{2i\pi A}\Gamma^2\eta + \eta^3\right), \end{aligned} \tag{29}$$

where $\eta \equiv E - 2\omega_0 + i\Gamma$, $\gamma \equiv (e^{i\pi A} + 1)\Gamma + 2i\omega_0$, and $\beta_\pm \equiv e^{i\pi A} \pm 1$. During the two contour integrations, $x_1 > l_2$ and $x_2 = x_1 + t$ (with $t > 0$) are used. Due to parity symmetry, $G_{21} = G_{12}$, $G_{22} = G_{11}$, $G_1^{\mathrm{L,L}}(-x_1,-x_2) = G_2^{\mathrm{R,R}}(x_1,x_2)$ and $G_2^{\mathrm{L,L}}(-x_1,-x_2) = G_1^{\mathrm{R,R}}(x_1,x_2)$.

Appendix 2: Two-photon correlation function $g_2(t)$

For a non-dispersive photonic field operator in the Heisenberg picture which satisfies $a^\dagger(x,t) = a^\dagger(x - ct)$, the two-photon correlation function $g_2(t)$ can be rewritten in the

Schrödinger picture as

$$g_2(t) = \frac{\langle\psi|a^\dagger_\alpha(x)a^\dagger_{\alpha'}(x+ct)a_{\alpha'}(x+ct)a_\alpha(x)|\psi\rangle}{\langle\psi|a^\dagger_\alpha(x)a_\alpha(x)|\psi\rangle\langle\psi|a^\dagger_{\alpha'}(x+ct)a_{\alpha'}(x+ct)|\psi\rangle}$$
$$\approx \frac{|_{\alpha\alpha'}\langle x, x+ct|\psi_2(k_1,k_2)\rangle_{\rm R,R}|^2}{|_\alpha\langle x|\phi_1(k_1)\rangle_{\rm R}|^2|_{\alpha'}\langle x+ct|\phi_1(k_2)\rangle_{\rm R}|^2}, \tag{30}$$

where $|\psi\rangle$ is the asymptotic output state and $\alpha = \alpha' = R$ for transmitted photons or $\alpha = \alpha' = L$ for reflected photons. The second equality holds if a weak incident coherent state (mean photon number $\bar{n} \ll 1$) with right-going photons is assumed - as is appropriate for comparison with an eventual experiment - such that we consider only two-photon states in the numerator and one-photon states in the denominator. The justification for the latter is twofold: (i) In the numerator, the 0- and 1-photon states are eliminated by the annihilation operators, leaving the 2-photon sector untouched which, then, can be described by $_{\alpha,\alpha'}\langle x, x+ct|\psi_2(k_1,k_2)\rangle_{\rm R,R}$. (ii) In the denominator, the probability of having only one photon is much larger then having two, so that the factors $|_{\rm R,R}\langle\psi_2(k_1,k_2)|a^\dagger_\alpha(x)a_\alpha(x)|\psi_2(k_1,k_2)\rangle_{\rm R,R}|^2$ can be replaced by the single photon eigenstate $|_\alpha\langle x|\phi_1(k_1)\rangle_{\rm R}|^2$ given that $k_1 = k_2 = E/2$ (*i.e.* two identical incident photons). We are thus lead to an explicit expression for the photon-photon correlations in terms of the 1- and 2-photon states found using the method outlined in Appendix 1.

Competing interests
The authors declare that they have no competing interests.

Authors' contributions
YLF carried out the calculations and made the plots. All three authors participated in the design of the research, analysis of the results, and writing of the paper.

Acknowledgements
This work was supported by U.S. NSF Grant No. PHY-10-68698. HZ was supported by a John T. Chambers Fellowship from the Fitzpatrick Institute for Photonics at Duke University.

References

1. Rupasov, VI, Yudson, VI: Rigorous theory of cooperative spontaneous emission of radiation from a lumped system of two-level atoms: Bethe ansatz method. *Zh Èksp Teor Fiz* 1984, **87**:1617-1630 . [Sov Phys JETP 1984, **60**:927-934].
2. Yudson, VI: Dynamics of integrable quantum systems. *Zh Èksp Teor Fiz* 1985, **88**:1757-1770. [Sov Phys JETP 1985, **61**:1043-1050].
3. Chang, DE, Sørensen, AS, Hemmer, PR, Lukin, MD: Quantum optics with surface plasmons. *Phys Rev Lett* 2006, **97**(5):053002. doi:10.1103/PhysRevLett.97.053002.
4. Shen, J-T, Fan, S: Strongly correlated two-photon transport in a one-dimensional waveguide coupled to a two-level system. *Phys Rev Lett* 2007, **98**(15):153003. doi:10.1103/PhysRevLett.98.153003.
5. Shen, J-T, Fan, S: Strongly correlated multiparticle transport in one dimension through a quantum impurity. *Phys Rev A* 2007, **76**(6):062709. doi:10.1103/PhysRevA.76.062709.
6. Yudson, VI, Reineker, P: Multiphoton scattering in a one-dimensional waveguide with resonant atoms. *Phys Rev A* 2008, **78**:052713. doi:10.1103/PhysRevA.78.052713.
7. Witthaut, D, Sørensen, AS: Photon scattering by a three-level emitter in a one-dimensional waveguide. *New J Phys* 2010, **12**(4):043052. doi:10.1088/1367-2630/12/4/043052.
8. Zheng, H, Gauthier, DJ, Baranger, HU: Waveguide QED: many-body bound-state effects in coherent and Fock-state scattering from a two-level system. *Phys Rev A* 2010, **82**(6):063816. doi:10.1103/PhysRevA.82.063816.
9. Ian, H, Liu, Y-X, Nori, F: Tunable electromagnetically induced transparency and absorption with dressed superconducting qubits. *Phys Rev A* 2010, **81**:063823. doi:10.1103/PhysRevA.81.063823.
10. Rephaeli, E, Kocabaş, ŞE, Fan, S: Few-photon transport in a waveguide coupled to a pair of colocated two-level atoms. *Phys Rev A* 2011, **84**:063832. doi:10.1103/PhysRevA.84.063832.
11. Roy, D: Two-photon scattering by a driven three-level emitter in a one-dimensional waveguide and electromagnetically induced transparency. *Phys Rev Lett* 2011, **106**(5):053601. doi:10.1103/PhysRevLett.106.053601.
12. Roy, D: Correlated few-photon transport in one-dimensional waveguides: linear and nonlinear dispersions. *Phys Rev A* 2011, **83**:043823. doi:10.1103/PhysRevA.83.043823.

13. Shi, T, Fan, S, Sun, CP: **Two-photon transport in a waveguide coupled to a cavity in a two-level system.** *Phys Rev A* 2011, **84**:063803. doi:10.1103/PhysRevA.84.063803.
14. Zheng, H, Gauthier, DJ, Baranger, HU: **Cavity-free photon blockade induced by many-body bound states.** *Phys Rev Lett* 2011, **107**:223601. doi:10.1103/PhysRevLett.107.223601.
15. Zheng, H, Gauthier, DJ, Baranger, HU: **Strongly correlated photons generated by coupling a three- or four-level system to a waveguide.** *Phys Rev A* 2012, **85**:043832. doi:10.1103/PhysRevA.85.043832.
16. Rephaeli, E, Fan, S: **Stimulated emission from a single excited atom in a waveguide.** *Phys Rev Lett* 2012, **108**:143602. doi:10.1103/PhysRevLett.108.143602.
17. Zheng, H, Baranger, HU: **Persistent quantum beats and long-distance entanglement from waveguide-mediated interactions.** *Phys Rev Lett* 2013, **110**(11):113601. doi:10.1103/PhysRevLett.110.113601.
18. Moeferdt, M, Schmitteckert, P, Busch, K: **Correlated photons in one-dimensional waveguides.** *Opt. Lett.* 2013, **38**:3693. doi:10.1364/OL.38.003693.
19. Lalumière, K, Sanders, BC, van Loo, AF, Fedorov, A, Wallraff, A, Blais, A: **Input-output theory for waveguide QED with an ensemble of inhomogeneous atoms.** *Phys Rev A* 2013, **88**:043806. doi:10.1103/PhysRevA.88.043806.
20. Shen, J-T, Fan, S: **Coherent single photon transport in a one-dimensional waveguide coupled with superconducting quantum bits.** *Phys Rev Lett* 2005, **95**(21):213001. doi:10.1103/PhysRevLett.95.213001.
21. Chang, DE, Sørensen, AS, Demler, EA, Lukin, MD: **A single-photon transistor using nanoscale surface plasmons.** *Nat Phys* 2007, **3**:807-812. doi:10.1038/nphys708.
22. Zhou, L, Gong, ZR, Liu, Y-X, Sun, CP, Nori, F: **Controllable scattering of a single photon inside a one-dimensional resonator waveguide.** *Phys Rev Lett* 2008, **101**(10):100501. doi:10.1103/PhysRevLett.101.100501.
23. Longo, P, Schmitteckert, P, Busch, K: **Few-photon transport in low-dimensional systems: interaction-induced radiation trapping.** *Phys Rev Lett* 2010, **104**(2):023602. doi:10.1103/PhysRevLett.104.023602.
24. Kolchin, P, Oulton, RF, Zhang, X: **Nonlinear quantum optics in a waveguide: distinct single photons strongly interacting at the single atom level.** *Phys Rev Lett* 2011, **106**(11):113601. doi:10.1103/PhysRevLett.106.113601.
25. Eichler, C, Bozyigit, D, Wallraff, A: **Characterizing quantum microwave radiation and its entanglement with superconducting qubits using linear detectors.** *Phys Rev A* 2012, **86**:032106. doi:10.1103/PhysRevA.86.032106.
26. Zheng, H, Gauthier, DJ, Baranger, HU: **Decoy-state quantum key distribution with nonclassical light generated in a one-dimensional waveguide.** *Opt Lett* 2013, **38**(5):622-624. doi:10.1364/OL.38.000622.
27. Zheng, H, Gauthier, DJ, Baranger, HU: **Waveguide-QED-based photonic quantum computation.** *Phys Rev Lett* 2013, **111**:090502. doi:10.1103/PhysRevLett.111.090502.
28. Akimov, AV, Mukherjee, A, Yu, CL, Chang, DE, Zibrov, AS, Hemmer, PR, Park, H, Lukin, MD: **Generation of single optical plasmons in metallic nanowires coupled to quantum dots.** *Nature* 2007, **450**:402. doi:10.1038/nature06230.
29. Bajcsy, M, Hofferberth, S, Balic, V, Peyronel, T, Hafezi, M, Zibrov, AS, Vuletic, V, Lukin, MD: **Efficient all-optical switching using slow light within a hollow fiber.** *Phys Rev Lett* 2009, **102**(20):203902. doi:10.1103/PhysRevLett.102.203902.
30. Babinec, TM, Hausmann, BJM, Khan, M, Zhang, Y, Maze, JR, Hemmer, PR, Lončar, M: **A diamond nanowire single-photon source.** *Nat Nanotechnol* 2010, **5**(1038):195. doi:10.1038/nnano.2010.6.
31. Claudon, J, Bleuse, J, Malik, NS, Bazin, M, Jaffrennou, P, Gregersen, N, Sauvan, C, Lalanne, P, Gérard, J-M: **A highly efficient single-photon source based on a quantum dot in a photonic nanowire.** *Nat Photonics* 2010, **4**(1038):174. doi:10.1038/nphoton.2009.287.
32. Bleuse, J, Claudon, J, Creasey, M, Malik, NS, Gérard, J-M, Maksymov, I, Hugonin, J-P, Lalanne, P: **Inhibition, enhancement, and control of spontaneous emission in photonic nanowires.** *Phys Rev Lett* 2011, **106**:103601. doi:10.1103/PhysRevLett.106.103601.
33. Laucht, A, Pütz, S, Günthner, T, Hauke, N, Saive, R, Frédérick, S, Bichler, M, Amann, M-C, Holleitner, AW, Kaniber, M, Finley, JJ: **A waveguide-coupled on-chip single-photon source.** *Phys Rev X* 2012, **2**:011014. doi:10.1103/PhysRevX.2.011014.
34. Astafiev, O, Zagoskin, AM, Abdumalikov, AA, Pashkin, YA, Yamamoto, T, Inomata, K, Nakamura, Y, Tsai, JS: **Resonance fluorescence of a single artificial atom.** *Science* 2010, **327**(5967):840-843. doi:10.1126/science.1181918.
35. Astafiev, OV, Abdumalikov, AA, Zagoskin, AM, Pashkin, YA, Nakamura, Y, Tsai, JS: **Ultimate on-chip quantum amplifier.** *Phys Rev Lett* 2010, **104**:183603. doi:10.1103/PhysRevLett.104.183603.
36. Eichler, C, Bozyigit, D, Lang, C, Steffen, L, Fink, J, Wallraff, A: **Experimental state tomography of itinerant single microwave photons.** *Phys Rev Lett* 2011, **106**:220503. doi:10.1103/PhysRevLett.106.220503.
37. Hoi, I-C, Wilson, CM, Johansson, G, Palomaki, T, Peropadre, B, Delsing, P: **Demonstration of a single-photon router in the microwave regime.** *Phys Rev Lett* 2011, **107**:073601. doi:10.1103/PhysRevLett.107.073601.
38. Hoi, I-C, Palomaki, T, Lindkvist, J, Johansson, G, Delsing, P, Wilson, CM: **Generation of nonclassical microwave states using an artificial atom in 1D open space.** *Phys Rev Lett* 2012, **108**:263601. doi:10.1103/PhysRevLett.108.263601.
39. Eichler, C, Lang, C, Fink, JM, Govenius, J, Filipp, S, Wallraff, A: **Observation of entanglement between itinerant microwave photons and a superconducting qubit.** *Phys Rev Lett* 2012, **109**:240501. doi:10.1103/PhysRevLett.109.240501.
40. Hoi, I-C, Kockum, AF, Palomaki, T, Stace, TM, Fan, B, Tornberg, L, Sathyamoorthy, SR, Johansson, G, Delsing, P, Wilson, CM: **Giant cross-Kerr effect for propagating microwaves induced by an artificial atom.** *Phys Rev Lett* 2013, **111**:053601. doi:10.1103/PhysRevLett.111.053601.
41. van Loo, AF, Fedorov, A, Lalumiére, K, Sanders, BC, Blais, A, Wallraff, A: **Photon-mediated interactions between distant artificial atoms.** *Science* 2013, **342**:1494. doi:10.1126/science.1244324.
42. Wallraff, A, Schuster, DI, Blais, A, Frunzio, L, Huang, R-S, Majer, J, Kumar, S, Girvin, SM, Schoelkopf, RJ: **Strong coupling of a single photon to a superconducting qubit using circuit quantum electrodynamics.** *Nature* 2004, **431**:162. doi:10.1038/nature02851.
43. Schoelkopf, RJ, Girvin, SM: **Wiring up quantum systems.** *Nature* 2008, **451**:664. doi:10.1038/451664a.
44. Dzsotjan, D, Sørensen, AS, Fleischhauer, M: **Quantum emitters coupled to surface plasmons of a nanowire: a Green's function approach.** *Phys Rev B* 2010, **82**:075427. doi:10.1103/PhysRevB.82.075427.
45. Gonzalez-Tudela, A, Martin-Cano, D, Moreno, E, Martin-Moreno, L, Tejedor, C, Garcia-Vidal, FJ: **Entanglement of two qubits mediated by one-dimensional plasmonic waveguides.** *Phys Rev Lett* 2011, **106**:020501. doi:10.1103/PhysRevLett.106.020501.

46. Dzsotjan, D, Kästel, J, Fleischhauer, M: **Dipole-dipole shift of quantum emitters coupled to surface plasmons of a nanowire.** *Phys Rev B* 2011, **84**:075419. doi:10.1103/PhysRevB.84.075419.
47. González-Tudela, A, Porras, D: **Mesoscopic entanglement induced by spontaneous emission in solid-state quantum optics.** *Phys Rev Lett* 2013, **110**:080502. doi:10.1103/PhysRevLett.110.080502.
48. Gonzalez-Ballestero, C, Garcia-Vidal, FJ, Moreno, E: **Non-Markovian effects in waveguide-mediated entanglement.** *New J Phys* 2013, **15**(7):073015. doi:10.1088/1367-2630/15/7/073015.
49. Roy, D: **Cascaded two-photon nonlinearity in a one-dimensional waveguide with multiple two-level emitters.** *Sci Rep* 2013, **3**:2337. doi:10.1038/srep02337.
50. Loudon, R: *The Quantum Theory of Light.* 3rd edition. New York: Oxford University Press; 2003.
51. Lang, C, Bozyigit, D, Eichler, C, Steffen, L, Fink, JM, Abdumalikov, AA, Baur, M, Filipp, S, da Silva, MP, Blais, A, Wallraff, A: **Observation of resonant photon blockade at microwave frequencies using correlation function measurements.** *Phys Rev Lett* 2011, **106**:243601. doi:10.1103/PhysRevLett.106.243601.
52. Ficek, Z, Sanders, BC: **Quantum beats in two-atom resonance fluorescence.** *Phys Rev A* 1990, **41**:359-368. doi:10.1103/PhysRevA.41.359.

17

Quantum lattice Boltzmann is a quantum walk

Sauro Succi[1*], François Fillion-Gourdeau[2] and Silvia Palpacelli[3]

*Correspondence:
succi@iac.rm.cnr.it
[1]IAC-CNR, Istituto per le Applicazioni del Calcolo 'Mauro Picone', Via dei Taurini, 19, Rome, 00185, Italy
Full list of author information is available at the end of the article

Abstract

Numerical methods for the 1-D Dirac equation based on operator splitting and on the quantum lattice Boltzmann (QLB) schemes are reviewed. It is shown that these discretizations fall within the class of quantum walks, i.e. discrete maps for complex fields, whose continuum limit delivers Dirac-like relativistic quantum wave equations. The correspondence between the quantum walk dynamics and these numerical schemes is given explicitly, allowing a connection between quantum computations, numerical analysis and lattice Boltzmann methods. The QLB method is then extended to the Dirac equation in curved spaces and it is demonstrated that the quantum walk structure is preserved. Finally, it is argued that the existence of this link between the discretized Dirac equation and quantum walks may be employed to simulate relativistic quantum dynamics on quantum computers.

Keywords: quantum walks; Dirac equation; lattice Boltzmann; operator splitting

1 Introduction

A quantum walk (QW) is defined as the quantum analogue of a classical random walk, where the 'quantum walker' is in a superposition of states instead of being described by a probability distribution. One of the earliest realization of this concept was proposed by Feynman as a discrete version of the massive Dirac equation [1]. In recent times, there has been a surge of interest for this topic, due to the conceptual and possible practical import of QW's as discrete realizations of stochastic quantum processes and because they can solve certain problems with an exponential speedup, i.e. using exponentially less operations than classical computations [2]. Moreover, QW's are amenable to a number of experimental realizations, such as ion traps [3–5], liquid-state nuclear-magnetic-resonance quantum-information processor [6], photonic devices [7] and other types of optical devices [8]. As a result, they hold promise of playing an important role in many areas of modern physics and quantum technology, such as quantum computing, foundational quantum mechanics and biophysics [9].

One of the most interesting features of discrete QW's is their continuum limit, which recovers a broad variety of relativistic quantum wave equations [10–12]. As stated earlier, this was first discussed by Feynman and is now known as Feynman's checkerboard [13]. This was originally formulated for the free Dirac equation but extensions of these ideas, which include the coupling to external fields, have been investigated [10–12]. Pioneering work have been performed in which the Dirac equation is related to cellular automata

[14, 15]. Lately, the link between QW's and the Dirac equation have been discussed extensively [10–12, 16]. In these studies, the starting point is a general QW formulation from which the continuum limit is evaluated and then related to the Dirac equation.

In this article, QW's and their relations to some known numerical schemes of the Dirac equation are reviewed from a slightly different perspective: it will be demonstrated that the most general QW's are obtained from lattice discretizations of the relativistic quantum wave equation for spin-1/2 particles. More precisely, starting from the continuum Dirac equation, it is shown that QW's can be placed in one-to-one correspondence with numerical schemes based on operator splitting and the QLB scheme. These numerical methods have been developed and employed as efficient numerical tools to solve relativistic quantum mechanics problems on classical computers [17–19]. They have a number of interesting properties: they are easy to code, they can be easily parallelized and are very versatile. Moreover, their mathematical structure and the fact that the time discretization is realized by a set of unitary transformations makes the link to QW's possible. This connection is explored below and represents one of the main purposes of this article. Much of the material discussed here, in particular the numerical simulations, is not completely new, which is in line with the main purpose of this paper, namely an attempt to bridge the techniques utilized in numerical analysis and (quantum) Lattice Boltzmann theory to the language of quantum computing.

This article is organized as follows. In Section 2, a general formulation of QW is presented, where the transfer matrix is time and space dependent. In Section 3, the split operator method for the Dirac equation is presented, along with its exact correspondence with QW. Section 4 is devoted to the QLB method and connections with QW. Section 5 is devoted to a qualitative discussion of the link between these numerical schemes and quantum computation. In Section 6, the schemes are casted in the form of a propagation-relaxation process and the notion of quantum equilibrium is introduced. Based on the analogy between QW and QLB, a new QLB scheme for the $(1+1)$ Dirac equation in curved space is proposed in Section 7. Finally, the generalization of these methods to many dimensions is briefly discussed and numerical results are presented in Section 8.

2 Quantum walks

Let us consider a $(1+1)$ quantum walk on the line for a pair of complex wave functions (bispinor ψ), obeying a discrete space-time evolution equations described by the following discrete map [10–12, 20]:

$$\begin{bmatrix} \psi_{1,j}^{n+1} \\ \psi_{2,j}^{n+1} \end{bmatrix} = B_{j,n} \begin{bmatrix} \psi_{1,j-1}^{n} \\ \psi_{2,j+1}^{n} \end{bmatrix}. \tag{1}$$

Here, the indices $j, m \in \mathbb{N} \otimes \mathbb{Z}$ label points on a discretization of space and time, respectively. The object $B_{j,n}$ is a two-by-two matrix with components [12]

$$B_{j,n} := e^{-i\xi_{j,n}} \begin{bmatrix} e^{i\alpha_{j,n}} \cos\theta_{j,n} & e^{i\beta_{j,n}} \sin\theta_{j,n} \\ -e^{-i\beta_{j,n}} \sin\theta_{j,n} & e^{-i\alpha_{j,n}} \cos\theta_{j,n} \end{bmatrix}. \tag{2}$$

This matrix is a $U(2)$ operator ($B \in U(2)$) parametrized by the three space-time dependent Euler angles $\theta_{j,n}$, $\alpha_{j,n}$, $\beta_{j,n}$ and a space-time dependent phase $\xi_{j,n}$. The latter is relevant when it depends on time and space, i.e. when it is local. If it is global ($\xi_{j,n} = \xi$), it disappears from any observables and becomes unimportant. This occurs because when the phase is space-time dependent, it does not commute with the time and space translation operators.

The matrix obeys $B \in SU(2)$ only when $\xi_{j,n} = k\pi$ for all j, n, with $k \in \mathbb{Z}$. Thus, this formulation is slightly more general than QW's considered in [10, 11, 20] where $B \in SU(2)$ is studied. As seen in the next section, the choice $B \in U(2)$ will be important to have a general connection between mass terms of the Dirac equation and QW's. Finally, the $U(2)$ QW can also be implemented on quantum computers because the matrix B is a unitary transformation: it represents the most general QW consistent with quantum computations.

In the above, the amplitudes $\psi_{1,2}$ code for the probability of the quantum walker to move up (down) along the lattice site $j \in \mathbb{Z}$ at the time step $n \in \mathbb{N}$. This is a very rich structure, which has been shown to recover a variety of important quantum wave equations, as soon as the Euler angles are allowed to acquire a space-time dependence [12]. In addition, it provides a wealth of potential algorithms for quantum computing. This was studied extensively in [10–12] by analysing the continuum limit of these QW, yielding different versions of the Dirac equation. In this work, the opposite path is taken: it is shown that specific discretizations of the Dirac equation, using either a split-operator approach or the lattice Boltzmann technique, naturally lead to a QW formulation.

3 Split-operator and quantum walks

The starting point of this discussion is the 1-D Dirac equation in Majorana representation written as (in units where $c = \hbar = 1$):

$$i\partial_t \psi(z,t) = \left[-i\sigma_z\partial_z + M(z,t)\right]\psi(z,t), \tag{3}$$

with the bi-spinor $\psi \in L_2(\mathbb{R}, \mathbb{C}^2)$. The generalized 'mass' matrix M is space and time dependent and may include contributions from the physical mass, the coupling to an electromagnetic potential or any other type of coupling. One requirement, however, is that M is a Hermitian local operator without any derivatives. Generally, it can be written as

$$\begin{aligned} M(z,t) &= \mathbb{I}_2 M_0(z,t) + \boldsymbol{\sigma}\cdot\mathbf{M} \qquad (4)\\ &= \mathbb{I}_2 M_0(z,t) + \sigma_x M_x(z,t) + \sigma_y M_y(z,t) + \sigma_z M_z(z,t), \qquad (5)\end{aligned}$$

where $\mathbb{I}_2$ is the two-by-two identity matrix, $(\sigma_i)_{i=x,y,z}$ are Pauli matrices and the coefficients $M_{0,x,y,z}$ represent the time- and space-dependent external fields which couple to the spinor.

The formal solution of Eq. (3) is given by

$$\psi(z,t) = \hat{T}\exp\left[-\Delta t\sigma_z\partial_z - i\int_{t_0}^{t} M(z,t')\,dt'\right]\psi(z,t_0), \tag{6}$$

where $\hat{T}$ is the time-ordering operator, t_0 is the initial time and $\Delta t = t - t_0$. Using an operator splitting technique, the solution in Eq. (6) can be approximated by [19]

$$\psi(z,t) = \exp\left[-i\Delta t M(z,t_0)\right]\exp[-\Delta t\sigma_z\partial_z]\psi(z,t_0) + O\left(\Delta t^2\right). \tag{7}$$

The first exponential is a translation (streaming) operator which shifts the spinor components according to:

$$\exp[-\Delta t\sigma_z\partial_z]\psi(z,t_0) = \begin{bmatrix}\psi_1(z-\Delta t,t_0)\\ \psi_2(z+\Delta t,t_0)\end{bmatrix}. \tag{8}$$

This suggests to use a spatial discretization where $\Delta z = \Delta t$ (this corresponds to a Courant-Friedrichs-Lewy (CFL) condition $C = c\Delta t/\Delta z = 1$, c being the speed of light) such that the translation is exact on the lattice. Equation (7) can then be written as:

$$\begin{bmatrix}\psi_{1,j}^{n+1}\\ \psi_{2,j}^{n+1}\end{bmatrix} = \exp[-i\Delta t M_{j,n}]\begin{bmatrix}\psi_{1,j-1}^{n}\\ \psi_{2,j+1}^{n}\end{bmatrix}, \tag{9}$$

where we defined the following quantities on the lattice: $M_{j,n} := M(n\Delta t, j\Delta z)$ and $\psi_j^n := \psi(n\Delta t, j\Delta z)$. This last equation yields a numerical scheme to solve the Dirac equation. This numerical scheme has interesting properties, as discussed extensively in [17, 19, 21]: it can be extended to higher order accuracy, it can be easily parallelized and it can be easily coded on a computer. In the following, it is also demonstrated that it is completely equivalent to the $U(2)$ QW described in the last section.

Equation (9) is in the form of Eq. (1). Moreover, the exponential is also a unitary matrix:

$$B' := \exp[-i\Delta t M_{j,n}] \in U(2), \tag{10}$$

and thus, there clearly exists a connection between B and B'. They are expressed in different representation: B uses the Euler angle parametrization while B' is expressed in the canonical representation obtained by the exponential mapping of the Lie algebra. The latter is given explicitly by

$$B' = \exp[-i\Delta t M_{0,j,n}]\exp[-i\Delta t\boldsymbol{\sigma}\cdot\mathbf{M}_{j,n}] \tag{11}$$

$$= \exp[-i\Delta t M_{0,j,n}]\left[\mathbb{I}_2\cos\left(|\mathbf{M}_{j,n}|\Delta t\right) - i\frac{\boldsymbol{\sigma}\cdot\mathbf{M}_{j,n}}{|\mathbf{M}_{i,m}|}\sin\left(|\mathbf{M}_{j,n}|\Delta t\right)\right], \tag{12}$$

where

$$|\mathbf{M}_{j,n}| = \sqrt{M_{x,j,n}^2 + M_{y,j,n}^2 + M_{z,j,n}^2}. \tag{13}$$

It is well-known that parameterizations of $U(2)$ matrices are related to each other [22] and it can be determined that the mapping between both representations is given by

$$\xi_{j,n} = M_{0,j,n}\Delta t, \tag{14}$$

$$\tan(\alpha_{j,n}) = -\frac{M_{z,j,n}}{|\mathbf{M}_{j,n}|}\tan\left(|\mathbf{M}_{j,n}|\Delta t\right), \tag{15}$$

$$\tan(\beta_{j,n}) = \frac{M_{x,j,n}}{M_{y,j,n}}, \tag{16}$$

$$\tan(\theta_{j,n}) = \tan\left(|\mathbf{M}_{j,n}|\Delta t\right)\sqrt{\frac{M_{x,j,n}^2 + M_{y,j,n}^2}{|\mathbf{M}_{j,n}|^2 + M_{z,j,n}^2\tan^2(|\mathbf{M}_{j,n}|\Delta t)}}. \tag{17}$$

When $M_0 = 0$, one recovers the $SU(2)$ QW because then, $\tilde{B} := B'|_{M_0=0} \in SU(2)$. These last equations give a one-to-one correspondence between the QW formulation, characterized by the parameters $\alpha_{j,n}$, $\beta_{j,n}$, $\theta_{j,n}$, $\xi_{j,n}$, and the discretization of the Dirac equation with a generalized mass term $M_{j,n}$. Therefore, the last results show that in the continuum limit, every space-time dependent QW on the line becomes a time-dependent 1-D Dirac equation with a specific mass matrix. A given QW can thus be fully characterized by the relativistic dynamics of an electron coupled to a space-time dependent external field. This occurs because there is an equivalence between the discretization of the Dirac equation, based on operator splitting, and the QW formulation. Moreover, this connection may be the base for the implementation of a quantum algorithm that solves the Dirac equation on quantum computers.

The operator splitting technique presented here also bears a close relationship with the QLB technique, to which we now turn.

4 Quantum lattice Boltzmann, operator splitting and quantum walks

The QLB was inspired by a direct analogy between the way the Dirac equation goes to the Schroedinger equation in the limit $v/c \to 0$, and the way that the Navier-Stokes equations of classical fluid-dynamics emerge from the Boltzmann equation in the limit of small Knudsen number, $Kn \to 0$, where $Kn = l/L$ is the ratio of the molecular mean free path to the typical macroscopic scale. In both cases, the smallness parameter controls the enslaving of the fast modes to the slow ones: *non-equilibrium* to *equilibrium* for the classical case, versus *excited* states to *ground* state in the quantum one. Of course, the quantum case shows no genuine relaxation since its dynamics is reversible. Yet, enslaving can be interpreted in the sense of fast oscillations around a local quantum equilibrium (Zitterbewegung), which average out once time is coarse-grained on a scale larger than the period of the fast oscillations. So, Zitterbewegung may be regarded as the quantum relativistic analogue of classical non-equilibrium fluctuations.

Based on this analogy, QLB was formulated as a lattice Boltzmann analogue of the Dirac equation in the Majorana representation, where the streaming matrix is *real*. To obtain the QLB scheme, it is convenient to write the Dirac equation as

$$[\partial_t + v_a \partial_z]\psi_a(z,t) = -i \sum_b M_{ab}(z,t)\psi_b(z,t), \tag{18}$$

where a, b are spinor indices and the 'microscopic velocities' are given by $v_{1,2} = \pm 1$. This is clearly in a 'Boltzmann-like' form with two discrete velocities and a collision term M. When QLB is employed in fluid mechanics to solve the continuum equations of motion, there is a family of possible numbers of discrete velocities for a given lattice [23] and each choice yields a different numerical scheme (for instance, the 9 velocities scheme in 2-D and the 27 velocities scheme in 3-D are popular choices on square lattices [24]). For the Dirac equation, this choice is dictated by the mathematical structure of the equation.

The lattice Boltzmann equation is then written as [25, 26]

$$\psi_i(z + v_a \Delta t, t + \Delta t) = \psi_a(z,t) - i\Delta t \sum_b M_{ab}(z,t)\psi_b(z,t) + O(\Delta t^2). \tag{19}$$

As for the splitting method described in the last section, this suggests to use a space discretization where $\Delta z = \Delta t$. Using a 'naive' approach in line with LB methods in fluid mechanics, the last equation would become

$$\begin{bmatrix} \psi_{1,j+1}^{n+1} \\ \psi_{2,j-1}^{n+1} \end{bmatrix} = \begin{bmatrix} \psi_{1,j}^{n} \\ \psi_{2,j}^{n} \end{bmatrix} - i\Delta t M_{j,n} \begin{bmatrix} \psi_{1,j}^{n} \\ \psi_{2,j}^{n} \end{bmatrix}. \tag{20}$$

Then, the matrix $-iM$ on the right acts as collision operator while the streaming is executed by the left part of the equation. This scheme is derived by using the formal analogy between the Dirac equation, the Boltzmann equation and the LB technique. However, the resulting numerical method is unstable and the L^2 norm is not preserved [27].

It is possible, however, to recover a stable and norm-preserving method. Instead of using the 'naive' LB equation discretization given in Eq. (20), the mass term on the right-hand side of Eq. (19) is discretized by using an implicit Crank-Nicolson average:

$$iM(z,t) \begin{bmatrix} \psi_1(z,t) \\ \psi_2(z,t) \end{bmatrix} = \frac{i}{2} M_{j,n} \left\{ \begin{bmatrix} \psi_{1,j+1}^{n+1} \\ \psi_{2,j-1}^{n+1} \end{bmatrix} + \begin{bmatrix} \psi_{1,j}^{n} \\ \psi_{2,j}^{n} \end{bmatrix} \right\}. \tag{21}$$

Reporting this into Eq. (19), one obtains the second order accurate QLB scheme:

$$\begin{bmatrix} \psi_{1,j+1}^{n+1} \\ \psi_{2,j-1}^{n+1} \end{bmatrix} = T_{j,n} \begin{bmatrix} \psi_{1,j}^{n} \\ \psi_{2,j}^{n} \end{bmatrix} + O(\Delta t^2), \tag{22}$$

where the transfer matrix is given by

$$T_{j,n} = \left[\mathbb{I}_2 + i\frac{\Delta t}{2} M_{j,n} \right]^{-1} \left[\mathbb{I}_2 - i\frac{\Delta t}{2} M_{j,n} \right]. \tag{23}$$

To link these results with the ones of the last sections, it is convenient to shift the spinor components by one lattice point (we let $\psi_{1,j+1} \to \psi_{1,j}$ and $\psi_{2,j-1} \to \psi_{2,j}$). Then, the QLB scheme becomes

$$\begin{bmatrix} \psi_{1,j}^{n+1} \\ \psi_{2,j}^{n+1} \end{bmatrix} = T_{j,n} \begin{bmatrix} \psi_{1,j-1}^{n} \\ \psi_{2,j+1}^{n} \end{bmatrix} + O(\Delta t^2), \tag{24}$$

which is in the same form as the QW and the splitting method presented previously. The transfer matrix can be evaluated explicitly. It is given by

$$T_{j,n} = \frac{1}{C_{j,n}} \begin{bmatrix} 1 - i\Delta t M_{z,j,n} + \frac{\Delta t^2}{4} M_{j,n}^2 & -\Delta t(iM_{x,j,n} + M_{y,j,n}) \\ -\Delta t(iM_{x,j,n} - M_{y,j,n}) & 1 + i\Delta t M_{z,j,n} + \frac{\Delta t^2}{4} M_{j,n}^2 \end{bmatrix}, \tag{25}$$

where $C_{j,n} = 1 + i\Delta t M_{0,j,n} - \frac{\Delta t^2}{4} M_{j,n}^2$ and $M_{j,n}^2 = M_{0,j,n}^2 - \mathbf{M}_{j,n}^2$. It can be readily verified that $T_{j,n} T_{j,n}^{\dagger} = \mathbb{I}_2$. As a consequence, the transfer matrix is unitary $T_{j,n} \in U(2)$, for any size of the time step and thus, conserves the L^2 norm. Moreover, just like the splitting method, there exists a correspondence with QW's because both have $U(2)$ collision matrices. The identification with the general QW in Eq. (1) yields

$$\tan(\xi_{j,n}) = \frac{\mathrm{Im}(C_{j,n})}{\mathrm{Re}(C_{j,n})} = \frac{\Delta t M_{0,j,n}}{1 - \frac{\Delta t^2}{4} M_{j,n}^2}, \tag{26}$$

$$\tan(\alpha_{j,n}) = -\frac{\Delta t M_{z,j,n}}{1+\frac{\Delta t^2}{4}M_{j,n}^2}, \tag{27}$$

$$\tan(\beta_{j,n}) = \frac{M_{x,j,n}}{M_{y,j,n}}, \tag{28}$$

$$\tan(\theta_{j,n}) = -\Delta t\sqrt{\frac{M_{x,j,n}^2 + M_{y,j,n}^2}{(1+\frac{\Delta t^2}{4}M_{j,n}^2)^2 + \Delta t^2 M_{z,j,n}^2}}. \tag{29}$$

These relations give a correspondence between the QLB technique and the QW. They are similar to the ones for the splitting method displayed in Eqs. (14) to (17) and actually serve the same purpose: they allow to map a numerical scheme to the QW. The differences are obviously due to the fact that QLB is based on a LB formulation combined with a Crank-Nicolson average to insure stability while the splitting scheme separates the Dirac equation into different operators which can be integrated exactly. However, both methods share the same general structure where a streaming step is followed by a collision step.

4.1 An explicit example: the free case

The Dirac equation for the massive and free case (no external field coupling) is

$$i\partial_t\psi(z,t) = [-i\sigma_z\partial_z + \sigma_y m]\psi(z,t), \tag{30}$$

where $M_y = m$ is a physical fermion mass. This representation of the Dirac equation yields real spinor components, as readily seen by writing the equation componentwise:

$$(\partial_t + \partial_z)\psi_1(z,t) = -m\psi_2(z,t), \tag{31}$$

$$(\partial_t - \partial_z)\psi_2(z,t) = +m\psi_1(z,t). \tag{32}$$

The QLB scheme, for the massive and free case, reads as follows [18]:

$$\begin{bmatrix}\psi_{1,j}^{n+1}\\ \psi_{2,j}^{n+1}\end{bmatrix} = T_{\text{free}}\begin{bmatrix}\psi_{1,j-1}^{n}\\ \psi_{2,j+1}^{n}\end{bmatrix}, \tag{33}$$

where the transfer matrix is can be obtained from Eq. (25) by setting $M_0 = M_x = M_z = 0$ and $M_y = m$:

$$T_{\text{free}} := \begin{bmatrix} a(m) & b(m)\\ -b(m) & a(m)\end{bmatrix} = \frac{1}{1+\frac{\Delta t^2 m^2}{4}}\begin{bmatrix}1-\frac{\Delta t^2 m^2}{4} & -m\Delta t\\ m\Delta t & 1-\frac{\Delta t^2 m^2}{4}\end{bmatrix}. \tag{34}$$

It is readily seen that $a(m)$ and $b(m)$ are second-order Pade'-like approximants of $\cos(m)$ and $\sin(m)$, respectively. This is the natural consequence of the implicit time-marching (Crank-Nicolson) scheme as applied to the $(1+1)$ Dirac equation in Majorana form.

For a free particle, the correspondence to QW is particularly simple and is given by

$$\xi = \alpha = \beta = 0, \tag{35}$$

$$\tan(\theta) = -\frac{m\Delta t}{1-\frac{\Delta t^2 m^2}{4}}. \tag{36}$$

This mapping of the QLB to the QW is exact for any value of m.

5 Prospects for quantum simulation

The numerical methods described in Sections 3-4 can be straightforwardly implemented on classical computers. However, the stream-collide structure of these numerical schemes makes them suitable for an efficient implementation on quantum computers as well. In particular, they can be written as:

$$\psi_j^{n+1} = B_{j,n} S_j \psi_j^n, \tag{37}$$

where S_j is the shift operator, defined as:

$$S_j \psi_j^n = \begin{bmatrix} \psi_{1,j-1}^n \\ \psi_{2,j+1}^n \end{bmatrix}. \tag{38}$$

This shift operator is a unitary operation and it can be realized experimentally by using fundamental quantum gates [6, 20]. The rotation operator $B_{j,n}$ belongs to $U(2)$ and therefore can also be realized by these quantum gates, as any other unitary transformations [28]. Therefore, it is possible to map the numerical methods to quantum walks, which can be implemented efficiently on quantum computers. This mapping is possible because each step of the scheme is a unitary transformation: this makes these schemes norm-preserving and sets the link with quantum computations.

The latter would be particularly useful for the study of relativistic quantum systems where a time-dependent solution of the Dirac equation is required, such as in very high intensity laser physics [29] or graphene physics [30].

Another subject of major interest for future research is the extension of the QLB methodology to quantum many body systems and quantum field theory, two paramount sectors of modern physics which are particularly exposed to the limitations of classical (non-quantum) electronic computing. Progress in this direction depends on the ability to replace the quantum wavefunction by the corresponding second-quantized quantum operators, and show that the dynamics of the second-quantized QLB scheme still preserves the appropriate equal-time commutation relations. Preliminary efforts along this line have been developed in [31] in $1+1$ dimensions. Extensions to strongly non-linear field theories in $d > 1$ remain to be explored. As to quantum-many body problems, LB-like methods have been recently adapted to electronic structure simulations [32]. In this work, a classical LB scheme is employed to solve the Kohn-Sham equations of density functional theory in the form of diffusion-reaction equations in imaginary time. Allied QLB schemes could prove very useful to solve the corresponding real-time quantum many-body transport problems within the framework of time-dependent density functional theory.

Finally, we wish to point out the intriguing possibility of realizing both quantum and classical LB schemes on quantum analogue simulators, as recently explored in [33].

6 Quantum equilibria

In view of quantum computing implementations, it is of interest to cast the Dirac equation in the form of a propagation-relaxation process, where the collision matrix is now interpreted as a scattering process, relaxing the spinor component around a local quantum equilibrium.

For this purpose, it is useful to reconsider the 1-D Dirac equation in the form:

$$[\partial_t + v_a \partial_z]\psi_a(z,t) = -i \sum_b M_{ab}(z,t)\psi_b(z,t). \tag{39}$$

Then, it is formally possible to define a local equilibrium as:

$$\psi_{\text{eq}}(z,t) := U\psi(z,t), \tag{40}$$

where U is a unitary matrix that depends on M. This transformation is chosen to recast the Dirac equation in relaxation form:

$$[\partial_t + \sigma_z \partial_z]\psi(z,t) = -\Omega\left[\psi(z,t) - \psi_{\text{eq}}(z,t)\right], \tag{41}$$

where $\Omega = iM[\mathbb{I} + U]^{-1}$. The explicit value of U, Ω is not unique but a convenient choice is $U = e^{iM\tau}$.

In this vest, the Dirac equation looks formally like a linear Boltzmann equation for two-component models in the single relaxation time approximation [25]. Therefore, it can be interpreted as a propagation-relaxation process in imaginary time, whereby collisions, implemented by the scattering operator Ω, drive oscillations around the equilibrium distribution ψ_{eq}. By defining a post-collision wavefunction as:

$$\psi'(z,t) := (1 - \Delta t\Omega)\psi(z,t) + \Delta t\Omega\psi_{\text{eq}}(z,t), \tag{42}$$

the Dirac equation takes the most compact form

$$\psi(z + v_a\Delta t, t + \Delta t) = \psi'(z,t). \tag{43}$$

This is particularly suitable to quantum computing implementations in the form of a classical stream-collide dynamics. The collision (relaxation) gate transforms the pre-collisional spinor ψ into the post-collisional state ψ', and the streaming gate moves the post-collisional spinor to its destination location $z \pm \Delta z$. Both operations are unitary and can be encoded in logical gates for quantum computing purposes [34, 35].

The expression (40) shows that the local equilibria are linear functions of the actual wavefunction ψ, hence itself a function of space and time.

The question is: will the actual wavefunctions ever reach this moving target, i.e., $\psi = \psi_{\text{eq}}$?

Based on its definition, this can only occur once ψ_{eq} lies in the null-space of the scattering matrix M, namely:

$$M\psi_{\text{eq}} = 0. \tag{44}$$

For the case of a free massive particle, it can be checked explicitly that the only solution is the trivial vacuum $\psi_{\text{eq}} \equiv 0$.

This means that the spinorial wavefunction is a superposition of (both slow and fast) zero-average oscillations around a local equilibrium which, consistently with the reversible nature of quantum mechanics, is actually never attained.

Although this remains to be checked in detail, we conjecture that the same holds true for the case of a massive particle in an external potential, because in this case the Dirac equation is still linear.

Based on (44), the condition for the local equilibrium to depart from a trivial vacuum is that the matrix M be singular, i.e. the local equilibrium is a zero-mode of the scattering matrix.

Non-trivial quantum zero-modes, $\psi_{\text{eq}} \neq 0$, may indeed arise for *non-linear* quantum wave equations, such as the Gross-Pitaevski or the mean-field version of the Nambu-Jona-Lasinio model, to be dicussed shortly. A non-trivial local equilibrium would then signal a spontaneously broken symmetry, which is indeed the distinctive trait of the aforementioned non-linear quantum wave equations.

Even though the notion of quantum equilibrium remains purely formal in nature, it is argued that it might nonetheless facilitate quantum computing implementations based on the compact expressions (42) and (43). This stands out a very interesting topic for future research.

7 QLB in curved space-time

Quantum walks have been shown to map into Dirac-like equations in curved space as well by evaluating the continuum limit of certain QW's [12, 36]. Here, in the same spirit as other sections, a QW structure is obtained by discretizing the Dirac equation in curved space time using a QLB-like approach. However, because the wave function propagates on a curved manifold, the structure of the resulting QW is different from Eq. (1) and should include a residency matrix that corrects the streaming step, which is strictly valid only in flat space. This is different from the result obtained in [12, 36].

The Dirac equation in a static (1 + 1) curved space writes as:

$$\gamma^a\left[e_a^\mu(\partial_\mu - iA_\mu)\right]\psi + \frac{1}{2}g^{-1/2}\partial_\mu(g^{1/2}e_a^\mu)\psi)] = -im\psi, \tag{45}$$

where $a = 0,1$ and $\mu = t, z$. In the above e_a^μ is the two-dimensional vierbein (zweibein) relating the components of the locally tangent Minkowski space basis (e_0, e_1), described by the metric η_{ab}, to the global space basis (e_t, e_z), described by the general metric $g_{\mu\nu}$. Also, $g := \det(g_{\mu\nu})$ is the determinant of the metric tensor.

The general form of the corresponding partial differential equation is

$$\partial_t\psi + \sigma_z A(z)\partial_z\psi = Q(z,t)\psi, \tag{46}$$

with $A(z) \in \mathbb{R}$ a function of space and $Q(z,t)$ the two-by-two gravitational collision matrix associated with Eq. (45). This becomes a hyperbolic system of equations where the advection speed $A(z)$ should not be confused with the vector electrodynamic potential. Since the advection term is heterogeneous, a strict QLB structure, i.e. streaming along constant lightcones $\Delta z = \mp c\Delta t$, is no longer viable. Here, the situation is very similar to classical LB schemes on non-uniform grids (unsurprisingly, since in dimension $D = 1$, gravity is basically a stretching of the metric). For this purpose, several finite-volume LB (FVLB) schemes have been formulated, whose main outcome is that the streaming operator is no longer a diagonal matrix with eigenvalues ± 1, as required by the formal QLB structure [37, 38].

In this respect, a finite-volume QLB scheme for the Dirac equation in curved space can be obtained by writing the Dirac equation as

$$\partial_t \psi + \partial_z\left[\sigma_z A(z)\psi\right] = Q'(z,t)\psi, \tag{47}$$

with $Q'(z,t) := Q(z,t) + \partial_z A(z)\sigma_z$. Then, this equation is integrated over a control volume $\mathcal{V}$ (enclosed by a surface $\mathcal{S}$), extending from $(j-\frac{1}{2})\Delta z$ to $(j+\frac{1}{2})\Delta z$. Finally, applying a Crank-Nicolson time-marching and combining with upwind finite-differences, the discretized equation is given by

$$\begin{aligned}
&\psi_{1,j}^{n+1} - \psi_{1,j}^{n} + \frac{1}{c}\oint A\psi_1\, d\mathcal{S} \\
&\quad = \frac{1}{2c}\left(Q'_{11,j-1}\psi_{1,j-1}^{n} + Q'_{11,j}\psi_{1,j}^{n+1}\right) + \frac{1}{2c}\left(Q'_{12,j+1}\psi_{2,j+1}^{n} + Q'_{12,j}\psi_{2,j}^{n+1}\right),
\end{aligned} \tag{48}$$

$$\begin{aligned}
&\psi_{2,j}^{n+1} - \psi_{2,j}^{n} - \frac{1}{c}\oint A\psi_2\, d\mathcal{S} \\
&\quad = \frac{1}{2c}\left(Q'_{22,j+1}\psi_{2,j+1}^{n} + Q'_{22,j}\psi_{2,j}^{n+1}\right) + \frac{1}{2c}\left(Q'_{21,j-1}\psi_{1,j-1}^{n} + Q'_{21,j}\psi_{1,j}^{n+1}\right),
\end{aligned} \tag{49}$$

where $c := \Delta z/\Delta t$ is the uniform lattice light speed (or CFL condition). The boundary integrals are given by

$$\oint A\psi_{1,2}\, d\mathcal{S} = [A\psi_{1,2}]_{j+\frac{1}{2}} - [A\psi_{1,2}]_{j-\frac{1}{2}}, \tag{50}$$

an thus, require an interpolation from the cell centers j, to the north and south boundaries at $j \pm 1/2$, respectively. Following a common practice in finite-volume formulations of hyperbolic problems, the flux terms are approximated by [39]:

$$[A\psi_1]_{j+\frac{1}{2}} = A_j\psi_{1,j}^{n}, \tag{51}$$

$$[A\psi_1]_{j-\frac{1}{2}} = A_{j-1}\psi_{1,j-1}^{n}, \tag{52}$$

$$[A\psi_2]_{j+\frac{1}{2}} = A_{j+1}\psi_{2,j+1}^{n}, \tag{53}$$

$$[A\psi_2]_{j-\frac{1}{2}} = A_j\psi_{2,j}^{n}. \tag{54}$$

As a result:

$$\begin{aligned}
&\left(1 - \frac{Q'_{11,j}}{2c}\right)\psi_{1,j}^{n+1} - \frac{Q'_{12,j}}{2c}\psi_{2,j}^{n+1} \\
&\quad = \left(1 - \frac{A_j}{c}\right)\psi_{1,j}^{n} + \left(\frac{Q'_{11,j-1}}{2c} + \frac{A_{j-1}}{c}\right)\psi_{1,j-1}^{n} + \frac{Q'_{12,j+1}}{2c}\psi_{2,j+1}^{n},
\end{aligned} \tag{55}$$

$$\begin{aligned}
&\left(1 - \frac{Q'_{22,j}}{2c}\right)\psi_{2,j}^{n+1} - \frac{Q'_{21,j}}{2c}\psi_{1,j}^{n+1} \\
&\quad = \left(1 - \frac{A_j}{c}\right)\psi_{2,j}^{n} + \left(\frac{Q'_{22,j-1}}{2c} + \frac{A_{j+1}}{c}\right)\psi_{2,j-1}^{n} + \frac{Q'_{21,j+1}}{2c}\psi_{1,j+1}^{n}.
\end{aligned} \tag{56}$$

It is readily seen that this reduces to a standard QLB in the limit of a uniform grid, when $A_j/c = 1$. In this case the streaming is diagonal with speed $\pm c$ and the spinors at $(j, n+1)$ are

Figure 1 Sketch of the transfer and residency matrix elements. T_{11} is the fraction of up-moving spinor jumping from $j-1$ at time $n-1$ to j at time n, while R_{11} is the fraction left in $j-1$. T_{22} and R_{22} bear the same meaning for the down moving spinor.

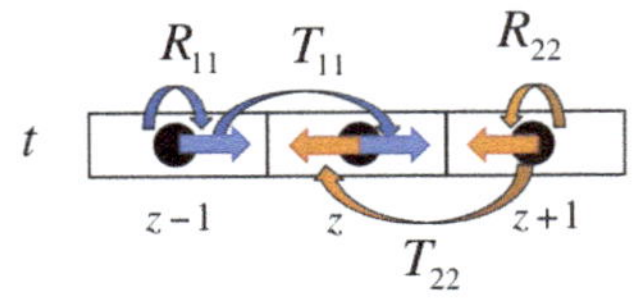

connected to the corresponding spinors at $(j \pm 1, n)$ by a local 2×2 matrix, which can be readily inverted to deliver a fully explicit map. However, when $A_j/c \neq 1$, the spinors at (j, n) also enter the map, so that local inversion delivers a slightly more elaborated structure, namely:

$$\psi_j^{n+1} = (R + TS)\psi_j^n, \tag{57}$$

which can be written more explicitly as

$$\begin{bmatrix} \psi_{1,j}^{n+1} \\ \psi_{2,j}^{n+1} \end{bmatrix} = R \begin{bmatrix} \psi_{1,j}^{n} \\ \psi_{2,j}^{n} \end{bmatrix} + T \begin{bmatrix} \psi_{1,j-1}^{n} \\ \psi_{2,j+1}^{n} \end{bmatrix}. \tag{58}$$

In the above T is the local 2×2 transfer matrix including collisions, S is the streaming operator and R the local residency matrix, expressing the fraction of spinors which are left in the cell centered about z as the quantum system advances from t to $t + \Delta t$. This is depicted in Figure 1.

Clearly, the residency matrix vanishes in the case of a uniform mesh, i.e. no gravity. The mapping (58) represents the 'gravitational' QLB. The detailed expressions of the streaming and residency matrices depend on the specific form of the metric tensor and associated vierbeins. Moreover, this analysis concentrates on the mathematical structure (streaming and collision steps) of the resulting scheme rather than on its numerical properties (convergence, stability, etc.). These topics shall make the object of a future publication.

8 Multi-dimensions

The discretization presented in this work extends to the $D+1$ dimensional case by applying the notion of operator splitting. This implies the inclusion of a new dynamic step which is entirely quantum: namely a 'rotation', designed so as to keep the spin aligned with the momentum along each of the three spatial directions. Schemes using this strategy can be found in [18, 19] and in [17] for higher order splittings.

It might be that such rotation is not needed by formulating the Dirac equation as a random walk on other lattice with more natural topologies (the diamond) lattice. The QLB-QW equivalence in multi-dimensions will be discussed in a future publication. However, to demonstrate the strength of the numerical schemes presented here and to show some possible applications for quantum computing, numerical results in 2-D are presented in the following.

8.1 Numerical results

As an example of possible applications of QLB scheme, we present two representative simulations: Klein tunnelling in the presence of random impurities and Dirac equation with Nambu-Jona-Lasinio (NJL) interactions in 2 + 1 space-time dimensions. Details on

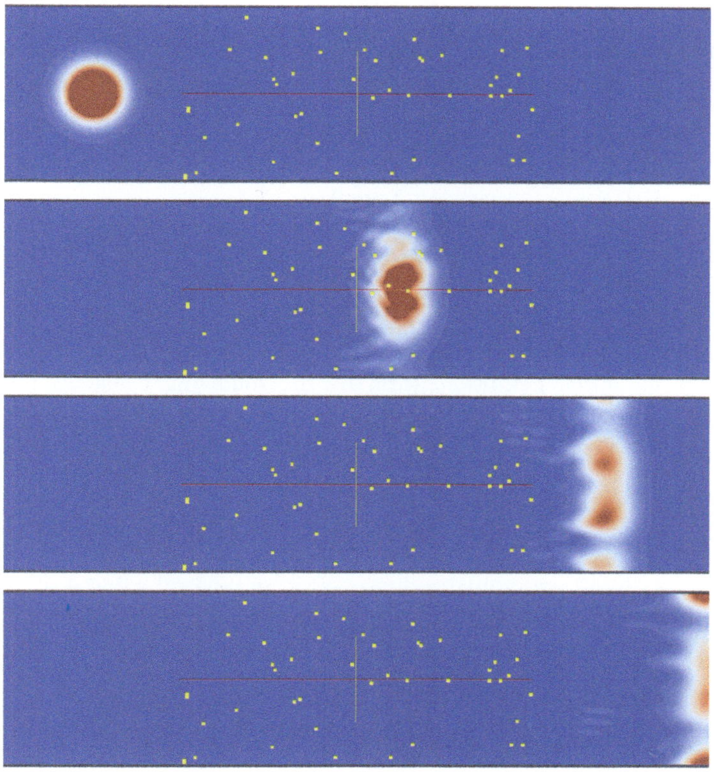

Figure 2 Wave packet density ρ at times = 0, 900, 1,500, and 1,800 (lattice units) for a Gaussian wave packet propagating through a graphene sample with randomly distributed impurities. The simulation is performed with impurity percentage $C = 0.5\%$ and impurity potential $V = 50$ MeV.

the numerical methods used to obtain these results are given in [18, 19, 40]. Also, these results are not completely new as similar systems have been studied in [19, 41].

8.1.1 Graphene with random impurities

In the first numerical test, the propagation of a Gaussian wave packet through a graphene sample with randomly distributed impurities is simulated [41]. In Ref. [41], simulations are performed for different values of the impurity concentration and the potential barrier, in order to provide an estimate of the effect of impurity concentration on the conductivity of the graphene sample. In Figure 2, we report some representative snapshots of the first 1,800 time steps of the simulation, at an impurity percentage = 0.5% and $V = 50$ MeV. A lattice of size $2{,}048 \times 512$ cells is used and the cell size is chosen to be $\Delta z = 0.96$ nm, while the spreading of the initial Gaussian wave packet is $\sigma = 48$ (in numerical units), leading to a Fermi frequency $k_F = 0.117$ (80 MeV in physical units). In this simulation, a fully relativistic particle ($m = 0$) is considered.

From Figure 2, we can see that the wave packet is scattered by the impurities, giving rise to a plane front out of the initial Gaussian configuration. As a consequence of the randomness induced in the wave function by the disordered media, there is a momentum loss and therefore the motion of the wave packet is found to experience a corresponding slow down. It is also found that the wave packet takes more time to regroup as the impurity concentration and impurity potential are increased.

8.1.2 Nambu-Jona-Lasinio interaction

As a second example, we present a 2 + 1 space-time simulation of the Dirac equation with a Nambu-Jona-Lasinio (NJL) interaction [42]. The Dirac equation with a NJL interaction term driven by the coupling parameter g, reads as follows

$$\left(i\gamma^{\mu}\partial_{\mu} - m\right)\psi + g\left((\bar{\psi}\psi)\psi + (\bar{\psi}\gamma^{5}\psi)\gamma^{5}\psi\right) = 0, \tag{59}$$

where $\psi = (\psi_1, \psi_2, \psi_3, \psi_4)^T$, $\gamma^5 \equiv i\gamma^0\gamma^1\gamma^2\gamma^3$ and $\bar{\psi} = \psi^{\dagger}\gamma^0$.

This model represents a paradigm for dynamic mass acquisition via spontaneous symmetry breaking due to the non-linear interactions.

Let us consider an initial condition given by the following Gaussian minimum-uncertainty wave packet:

$$G_0(z, y) = \left(2\pi\sigma^2\right)^{-1/2} \exp\left(-\frac{z^2 + y^2}{4\sigma^2}\right), \tag{60}$$

centered about $(z, y) = (0, 0)$, with initial width σ. Let k_z and k_y be the initial energy of the wave packet and impose the following initial condition:

$$\begin{aligned} u_1(z, y) = u_2(z, y) &= C_u G_0(z, y) \exp\left(i(k_z z + k_y y)\right), \\ d_1(z, y) = d_2(z, y) &= C_d G_0(z, y) \exp\left(-i(k_z z + k_y y)\right), \end{aligned} \tag{61}$$

where coefficients C_u and C_d obey the condition $2C_u^2 + 2C_d^2 = 1$, so that $\rho = |\psi_1|^2 + |\psi_2|^2 + |\psi_3|^2 + |\psi_4|^2 = |G_0|^2$.

A grid size of $N_z \times N_y = 1{,}024^2$ elements is used and the initial wave packet spread is set at $\sigma = 48$, a fully relativistic particle ($m = 0$) is considered.

In these simulations, we impose $g = 0$ and $g = 1{,}000$ and vary the initial energy of the wave packet $k \equiv k_z = k_y$ in order to inspect the effect of this parameter on the wave packet separation, which, in turn, informs on the effective mass acquired by the up and down propagating modes.

In Figure 3, the wave function density at time $t = 200$ for $k = 0.004$, 0.04 and 0.4 is shown for $g = 0$ and $g = 1{,}000$, respectively. The figure shows that sufficient energy, $k > 0.004$, is needed to observe the splitting of the wavepacket. The effects of non-linear interactions, fringes and distortions, are also well visible in the right column of Figure 3. A quantitative analysis in the one-dimensional case led to satisfactory agreement with asymptotic solutions for the dynamic mass as a function of the interaction strength g. A similar analysis in two spatial dimensions remains to be developed.

9 Summary and outlook

Summarizing, we have reviewed discretizations of the Dirac equation and described their mapping into QW's. These relations may allow the solution of the Dirac equation on quantum computers. In the first part, a general argument is given, using the operator splitting method. Then, the QLB scheme is studied within the same perspective and a similar relation is found. We have also shown that a similar structure remains in curved space, using a scheme based on a finite volume formulation, with the important caveat that the exact nature of the streaming operator, typical of QLB, is no longer preserved. Rather, one

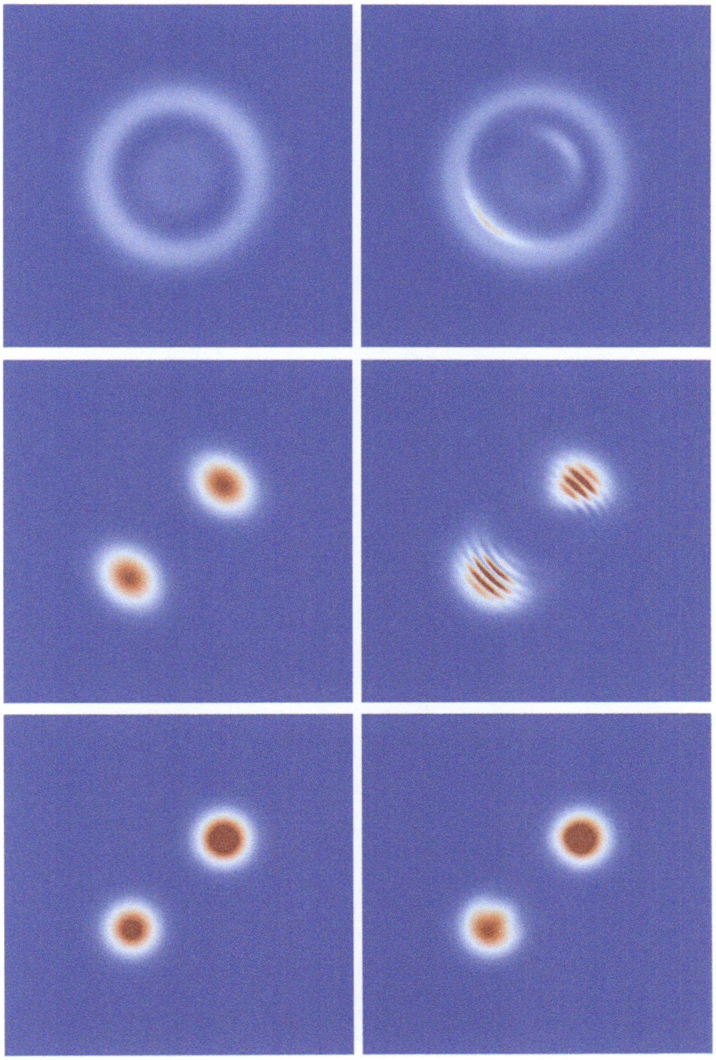

Figure 3 Wave packet density $\rho = |\psi_1|^2 + |\psi_2|^2 + |\psi_3|^2 + |\psi_4|^2$ for the scheme solving Dirac equation with NJL interaction. These snapshots are taken at time $t = 200$ for $k = 0.004$, 0.04 and 0.4 (from top to bottom). The figure shows how the initial energy affects the separation phenomenon. In particular, at low energy, $k = 0.004$, no splitting of the wavepackest is observed.

sees the appearance of the residency matrix, which characterizes the fraction of spinor which is left in the cell after one step in the time evolution. This scheme, along with its generalization to many dimensions, will be studied in future work.

Competing interests

The authors declare that they have no competing interests.

Authors' contributions

SS had the idea to link QLB to QW, and to develop a QLB-like method for the curved-space Dirac equation. He also participated in the calculation and development of the numerical methods. FFG introduced the operator splitting method and participated in the calculations and development of the numerical methods. SP developed the multi-dimensional extensions and performed the numerical calculations. All authors read and approved the final manuscript.

Author details
[1]IAC-CNR, Istituto per le Applicazioni del Calcolo 'Mauro Picone', Via dei Taurini, 19, Rome, 00185, Italy. [2]INRS-Énergie, Matériaux et Télécommunications, Université du Québec, 1650, boulevard Lionel-Boulet, Varennes, J3X 1S2, Canada. [3]Numidia s.r.l., via G. Peroni, 130, Rome, 00131, Italy.

Acknowledgements
One of the authors (SS) is very grateful to F Debbasch for introducing him to the notion of quantum walks. He also wishes to express gratitude to Marcelo Alejandro Forets for organizing a very informative workshop on Relativistic Quantum Walks, where the ideas behind this paper have been first drafted out.

References

1. Feynman RP, Hibbs AR. Quantum mechanics and path integrals. Int. series in pure and applied physics. New York: McGraw-Hill; 1965.
2. Childs AM, Cleve R, Deotto E, Farhi E, Gutmann S, Spielman DA. Exponential algorithmic speedup by a quantum walk. In: Proceedings of the thirty-fifth annual ACM symposium on theory of computing. New York: ACM; 2003. p. 59-68.
3. Schmitz H, Matjeschk R, Schneider C, Glueckert J, Enderlein M, Huber T, Schaetz T. Quantum walk of a trapped ion in phase space. Phys Rev Lett. 2009;103:090504.
4. Zähringer F, Kirchmair G, Gerritsma R, Solano E, Blatt R, Roos CF. Realization of a quantum walk with one and two trapped ions. Phys Rev Lett. 2010;104:100503.
5. Gerritsma R, Kirchmair G, Zaehringer F, Solano E, Blatt R, Roos C. Quantum simulation of the Dirac equation. Nature. 2010;463(7277):68-71.
6. Ryan CA, Laforest M, Boileau JC, Laflamme R. Experimental implementation of a discrete-time quantum random walk on an NMR quantum-information processor. Phys Rev A. 2005;72:062317.
7. Aspuru-Guzik A, Walther P. Photonic quantum simulators. Nat Phys. 2012;8(4):285-91.
8. Engel GS, Calhoun TR, Read EL, Ahn T-K, Mančal T, Cheng Y-C, Blankenship RE, Fleming GR. Evidence for wavelike energy transfer through quantum coherence in photosynthetic systems. Nature. 2007;446(7137):782-6.
9. Giulini D, Joos E, Kiefer C, Kupsch J, Stamatescu I-O, Zeh HD. Decoherence and the appearance of a classical world in quantum theory. Berlin: Springer; 2003.
10. Aharonov Y, Davidovich L, Zagury N. Quantum random walks. Phys Rev A. 1993;48:1687-90.
11. Di Molfetta G, Debbasch F. Discrete-time quantum walks: continuous limit and symmetries. J Math Phys. 2012;53(12):123302.
12. Di Molfetta G, Brachet M, Debbasch F. Quantum walks in artificial electric and gravitational fields. Phys A, Stat Mech Appl. 2014;397(0):157-68.
13. Gersch HA. Feynman's relativistic chessboard as an Ising model. Int J Theor Phys. 1981;20(7):491-501.
14. Bialynicki-Birula I. Weyl, Dirac, and Maxwell equations on a lattice as unitary cellular automata. Phys Rev D. 1994;49:6920-7.
15. Meyer D. From quantum cellular automata to quantum lattice gases. J Stat Phys. 1996;85(5-6):551-74.
16. Arrighi P, Nesme V, Forets M. The Dirac equation as a quantum walk: higher dimensions, observational convergence. J Phys A, Math Theor. 2014;47(46):465302.
17. Fillion-Gourdeau F, Lorin E, Bandrauk AD. Numerical solution of the time-dependent Dirac equation in coordinate space without fermion-doubling. Comput Phys Commun. 2012;183(7):1403-15.
18. Succi S, Benzi R. Lattice Boltzmann equation for quantum mechanics. Phys D, Nonlinear Phenom. 1993;69(3-4):327-32.
19. Fillion-Gourdeau F, Herrmann HJ, Mendoza M, Palpacelli S, Succi S. Formal analogy between the Dirac equation in its Majorana form and the discrete-velocity version of the Boltzmann kinetic equation. Phys Rev Lett. 2013;111:160602.
20. Chandrashekar CM, Srikanth R, Laflamme R. Optimizing the discrete time quantum walk using a su(2) coin. Phys Rev A. 2008;77:032326.
21. Lorin E, Bandrauk A. A simple and accurate mixed solver for the Maxwell-Dirac equations. Nonlinear Anal, Real World Appl. 2011;12(1):190-202.
22. Gilmore R. Lie groups, Lie algebras, and some of their applications. New York: Dover; 2012.
23. Qian YH, D'Humières D, Lallemand P. Lattice BGK models for Navier-Stokes equation. Europhys Lett. 1992;17(6):479.
24. He X, Luo L-S. Theory of the lattice Boltzmann method: from the Boltzmann equation to the lattice Boltzmann equation. Phys Rev E. 1997;56:6811-7.
25. Succi S. The Lattice-Boltzmann equation. Oxford: Oxford University Press; 2001.
26. Chen S, Doolen GD. Lattice Boltzmann method for fluid flows. Annu Rev Fluid Mech. 1998;30(1):329-64.
27. Succi S. Kinetic approach to lattice quantum mechanics. In: Cellular automata. Berlin: Springer; 2002. p. 114-22.
28. Kaye P, Mosca M, Laflamme R. An introduction to quantum computing. Oxford: Oxford University Press; 2006.
29. Di Piazza A, Müller C, Hatsagortsyan KZ, Keitel CH. Extremely high-intensity laser interactions with fundamental quantum systems. Rev Mod Phys. 2012;84:1177-228.
30. Geim AK, Novoselov KS. The rise of graphene. Nat Mater. 2007;6(3):183-91.
31. Succi S. Lattice Boltzmann method for quantum field theory. J Phys A, Math Theor. 2007;40(26):559.
32. Mendoza M, Succi S, Herrmann HJ. Kinetic formulation of the Kohn-Sham equations for ab initio electronic structure calculations. Phys Rev Lett. 2014;113:096402.
33. Mezzacapo A, Sanz M, Lamata L, Egusquiza I, Succi S, Solano E. Quantum simulator for transport phenomena in fluid flows. arXiv preprint, arXiv:1502.00515 (2015).
34. Meyer D. From quantum cellular automata to quantum lattice gases. J Stat Phys. 1996;85(5-6):551-74.
35. Boghosian BM, Taylor W IV. Simulating quantum mechanics on a quantum computer. Phys D, Nonlinear Phenom. 1998;120(1-2):30-42.
36. Di Molfetta G, Brachet M, Debbasch F. Quantum walks as massless Dirac fermions in curved space-time. Phys Rev A. 2013;88:042301.

37. Ubertini S, Bella G, Succi S. Lattice Boltzmann method on unstructured grids: further developments. Phys Rev E. 2003;68:016701.
38. Nannelli F, Succi S. The lattice Boltzmann equation on irregular lattices. J Stat Phys. 1992;68(3-4):401-7.
39. LeVeque RJ. Finite volume methods for hyperbolic problems. vol. 31. Cambridge: Cambridge University Press; 2002.
40. Palpacelli S, Romatschke P, Succi S. One-dimensional quantum lattice Boltzmann scheme for the nonlinear Dirac equation. Int J Mod Phys C. 2013;24(12):1340001.
41. Palpacelli S, Mendoza M, Herrmann H, Succi S. Klein tunneling in the presence of random impurities. Int J Mod Phys C. 2012;23(12):1250080.
42. Nambu Y, Jona-Lasinio G. Dynamical model of elementary particles based on an analogy with superconductivity. I. Phys Rev. 1961;122:345-58.

Permissions

All chapters in this book were first published in EPJQT, by Springer; hereby published with permission under the Creative Commons Attribution License or equivalent. Every chapter published in this book has been scrutinized by our experts. Their significance has been extensively debated. The topics covered herein carry significant findings which will fuel the growth of the discipline. They may even be implemented as practical applications or may be referred to as a beginning point for another development.

The contributors of this book come from diverse backgrounds, making this book a truly international effort. This book will bring forth new frontiers with its revolutionizing research information and detailed analysis of the nascent developments around the world.

We would like to thank all the contributing authors for lending their expertise to make the book truly unique. They have played a crucial role in the development of this book. Without their invaluable contributions this book wouldn't have been possible. They have made vital efforts to compile up to date information on the varied aspects of this subject to make this book a valuable addition to the collection of many professionals and students.

This book was conceptualized with the vision of imparting up-to-date information and advanced data in this field. To ensure the same, a matchless editorial board was set up. Every individual on the board went through rigorous rounds of assessment to prove their worth. After which they invested a large part of their time researching and compiling the most relevant data for our readers.

The editorial board has been involved in producing this book since its inception. They have spent rigorous hours researching and exploring the diverse topics which have resulted in the successful publishing of this book. They have passed on their knowledge of decades through this book. To expedite this challenging task, the publisher supported the team at every step. A small team of assistant editors was also appointed to further simplify the editing procedure and attain best results for the readers.

Apart from the editorial board, the designing team has also invested a significant amount of their time in understanding the subject and creating the most relevant covers. They scrutinized every image to scout for the most suitable representation of the subject and create an appropriate cover for the book.

The publishing team has been an ardent support to the editorial, designing and production team. Their endless efforts to recruit the best for this project, has resulted in the accomplishment of this book. They are a veteran in the field of academics and their pool of knowledge is as vast as their experience in printing. Their expertise and guidance has proved useful at every step. Their uncompromising quality standards have made this book an exceptional effort. Their encouragement from time to time has been an inspiration for everyone.

The publisher and the editorial board hope that this book will prove to be a valuable piece of knowledge for researchers, students, practitioners and scholars across the globe.

List of Contributors

Christopher Eichler
Department of Physics, ETH Zürich, Zürich, 8093, Switzerland

Andreas Wallraff
Department of Physics, ETH Zürich, Zürich, 8093, Switzerland

Erik A Tholén
Intermodulation Products AB, Solna, Sweden. 2Nanostructure Physics, Royal Institute of Technology (KTH), Stockholm, Sweden

Adem Ergül
Nanostructure Physics, Royal Institute of Technology (KTH), Stockholm, Sweden

David Schaeffer
Nanostructure Physics, Royal Institute of Technology (KTH), Stockholm, Sweden

David B Haviland
Nanostructure Physics, Royal Institute of Technology (KTH), Stockholm, Sweden

Bryce Yoshimura
Department of Physics, Georgetown University, 37th and O St. NW, Washington, DC 20007, USA

Marybeth Stork
Department of Physics and Astronomy, Washington University, Campus Box 1105, One Brookings Dr., St. Louis, Missouri 63130, USA

Danilo Dadic
Department of Physics and Astronomy, University of California Los Angeles, 475 Portola Plaza, Los Angeles, CA 90095, USA.

Wesley C Campbell
Department of Physics and Astronomy, University of California Los Angeles, 475 Portola Plaza, Los Angeles, CA 90095, USA

James K Freericks
Department of Physics, Georgetown University, 37th and O St. NW, Washington, DC 20007, USA

Robert Jirschik
Physik Department, Technische Universitaet Muenchen, James Franck Str., Garching, 85748, Germany
Department of Physics and Astronomy, University College London, Gower Street, London, WC1E 6BT, United Kingdom

Michael J Hartmann
Physik Department, Technische Universitaet Muenchen, James Franck Str., Garching, 85748, Germany

Vassilios Yannopapas
Department of Physics, National Technical University of Athens, Athens, 157 80, Greece

Jake Gulliksen
Department of Physics and Astronomy, Aarhus University, Ny Munkegade 120, Aarhus C, DK-8000, Denmark

Durga Bhaktavatsala Rao Dasari
Department of Physics and Astronomy, Aarhus University, Ny Munkegade 120, Aarhus C, DK-8000, Denmark

Klaus Mølmer
Department of Physics and Astronomy, Aarhus University, Ny Munkegade 120, Aarhus C, DK-8000, Denmark

Tupac Bravo
School of Mathematical Sciences, University of Nottingham, Nottingham, NG7 2RD, UK

Carlos Sabín
School of Mathematical Sciences, University of Nottingham, Nottingham, NG7 2RD, UK

Ivette Fuentes
School of Mathematical Sciences, University of Nottingham, Nottingham, NG7 2RD, UK

Johannes Schachenmayer
Department of Physics, JILA, NIST, University of Colorado, 440 UCB, Boulder, CO 80309, USA

Lode Pollet
Department of Physics, Arnold Sommerfeld Center for Theoretical Physics and Center for NanoScience, University of Munich, Munich, 80333, Germany

Matthias Troyer
Theoretische Physik, ETH Zurich, Zurich, 8093, Switzerland

Andrew J Daley
Department of Physics and SUPA, University of Strathclyde, Glasgow, G4 0NG, UK
Department of Physics and Astronomy, University of Pittsburgh, Pittsburgh, PA 15260, USA

Ioana Serban
Instituut Lorentz, Leiden University, Leiden, The Netherlands

IQC and Department of Physics and Astronomy, University of Waterloo, Waterloo, Ontario, Canada

Frank K Wilhelm
IQC and Department of Physics and Astronomy, University of Waterloo, Waterloo, Ontario, Canada
Theoretical Physics, Saarland University, Saarbrücken, Germany

Jiasen Jin
NEST, Scuola Normale Superiore and Istituto di Nanoscienze, CNR, Piazza dei Cavalieri, 7, Pisa, 56126, Italy

Rosario Fazio
NEST, Scuola Normale Superiore and Istituto di Nanoscienze, CNR, Piazza dei Cavalieri, 7, Pisa, 56126, Italy
Center for Quantum Technologies, National University of Singapore, Singapore, 117543, Singapore

Davide Rossini
NEST, Scuola Normale Superiore and Istituto di Nanoscienze, CNR, Piazza dei Cavalieri, 7, Pisa, 56126, Italy

Timothy J Proctor
School of Physics and Astronomy, University of Leeds, E C Stoner Building, Leeds, LS2 9JT, UK

Viv Kendon
School of Physics and Astronomy, University of Leeds, E C Stoner Building, Leeds, LS2 9JT, UK
Present address: Department of Physics, Durham University, South Road, Durham, DH1 3LE, UK

Luc Bouten
BTA VOF, Slootsekuilen 9, 5986 PE, Beringe, The Netherlands

Rolf Gohm
Dept. for Mathematics and Physics, Aberystwyth University, SY23 3BZ, Wales, United Kingdom

John Gough
Dept. for Mathematics and Physics, Aberystwyth University, SY23 3BZ, Wales, United Kingdom

Hendra Nurdin
School of Electrical Engineering and Telecommunications, UNSW Australia, Sydney, NSW 2052, Australia

Nikolas Tezak
Edward L. Ginzton Laboratory, Stanford University, Stanford, CA 94305, USA

Hideo Mabuchi
Edward L. Ginzton Laboratory, Stanford University, Stanford, CA 94305, USA

Urtzi Las Heras
Department of Physical Chemistry, University of the Basque Country UPV/EHU, Apartado 644, 48080 Bilbao, Spain

Laura García-Álvarez
Department of Physical Chemistry, University of the Basque Country UPV/EHU, Apartado 644, 48080 Bilbao, Spain

Antonio Mezzacapo
Department of Physical Chemistry, University of the Basque Country UPV/EHU, Apartado 644, 48080 Bilbao, Spain

Enrique Solano
Department of Physical Chemistry, University of the Basque Country UPV/EHU, Apartado 644, 48080 Bilbao, Spain
IKERBASQUE, Basque Foundation for Science, Maria Diaz de Haro 3, 48013 Bilbao, Spain

Lucas Lamata
Department of Physical Chemistry, University of the Basque Country UPV/EHU, Apartado 644, 48080 Bilbao, Spain

Harrison Ball
ARC Centre for Engineered Quantum Systems, School of Physics, The University of Sydney, Sydney, NSW 2006, Australia
National Measurement Institute, West Lindfield, Sydney, NSW 2070, Australia

Michael J Biercuk
ARC Centre for Engineered Quantum Systems, School of Physics, The University of Sydney, Sydney, NSW 2006, Australia
National Measurement Institute, West Lindfield, Sydney, NSW 2070, Australia

Yao-Lung L Fang
Department of Physics, Duke University, P.O. Box 90305, Durham, NC 27708, USA

Huaixiu Zheng
Department of Physics, Duke University, P.O. Box 90305, Durham, NC 27708, USA

Harold U Baranger
Department of Physics, Duke University, P.O. Box 90305, Durham, NC 27708, USA

Sauro Succi
IAC-CNR, Istituto per le Applicazioni del Calcolo 'Mauro Picone', Via dei Taurini, 19, Rome, 00185, Italy

François Fillion-Gourdeau
INRS-Énergie, Matériaux et Télécommunications, Université du Québec, 1650, boulevard Lionel-Boulet, Varennes, J3X 1S2, Canada

Silvia Palpacelli
Numidia s.r.l., via G. Peroni, 130, Rome, 00131, Italy